ALSO BY DAVID QUAMMEN

NONFICTION

The Tangled Tree

Yellowstone

The Chimp and the River

Ebola

Spillover

The Reluctant Mr. Darwin

Monster of God

The Song of the Dodo

ESSAYS

The Boilerplate Rhino

Wild Thoughts from Wild Places

The Flight of the Iguana

Natural Acts

FICTION

Blood Line

The Soul of Viktor Tronko

The Zolta Configuration

To Walk the Line

BREATHLESS

The Scientific
Race to Defeat a
Deadly Virus

DAVID QUAMMEN

SIMON & SCHUSTER
New York London Toronto Sydney New Delhi

Simon & Schuster
1230 Avenue of the Americas
New York, NY 10020

First Simon & Schuster hardcover edition October 2022

SIMON & SCHUSTER and colophon are registered
trademarks of Simon & Schuster, Inc.

For information about special discounts for bulk
purchases, please contact Simon & Schuster Special Sales
at 1-866-506-1949 or business@simonandschuster.com.

The Simon & Schuster Speakers Bureau can bring authors to
your live event. For more information or to book an event, contact
the Simon & Schuster Speakers Bureau at 1-866-248-3049
or visit our website at www.simonspeakers.com.

Interior design by Lewelin Polanco

Manufactured in the United States of America

1 3 5 7 9 10 8 6 4 2

Library of Congress Cataloging-in-Publication
Data has been applied for.

ISBN 978-1-9821-6436-2
ISBN 978-1-9821-6438-6 (ebook)

to all those who have lost loved ones in this pandemic

AUTHOR'S NOTE

About methods: Unlike other books I have written, and for reasons you will understand, this one was researched without benefit of traveling to remote places and witnessing arduous fieldwork; without walking through jungles in the footsteps of doughty biologists, visiting laboratories, climbing up cliffs and across rooftops and through caves; without watching researchers stalk gorillas with tranquilizer guns or draw blood from bats. The frisson here, if any, comes in other forms. I avoided airports for more than two years after COVID-19 exploded, and I got through the year 2020 on one tank of gas. The scientific literature has been invaluable to me. My journals from previous travels helped a bit. I am also very, very appreciative of Zoom.

About quotations: All spoken quotes demarcated by quote marks are verbatim, as selected from transcribed recordings or notes taken in the moment, without cosmetic correction for grammar or improvement for flow. Whether communicating in their first language or their fourth, people don't speak in grammatically perfect sentences and paragraphs, and my goal has been to represent real speech by real people. That I have preserved the occasional grammatical glitch should be taken as testament to my respect for the people speaking and my desire to hear them, and have you hear them, closely. I have sparingly removed tics such as "um" and "you know" and "like," but not often, and not more than that. Spoken words are data, in nonfiction, and I share scientists' respect for the sanctity of data.

About names: Chinese convention puts the surname first, the given name second, as in Yuen Kwok-Yung or Zhang Yong-Zhen. But when Chinese scientists publish in English-language journals, the Western convention is generally observed: given name first, then surname. For simplicity, because I'm writing mainly about scientists and want the authors of published work to be recognized for it, I follow here the second convention.

About honorifics: Nearly everyone quoted or cited in this book has earned the title doctor, professor, or both. I have omitted all those titles in favor of respectful informality.

CONTENTS

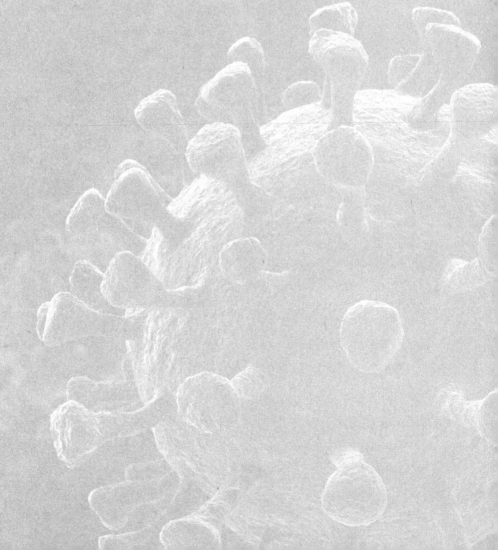

I

CITIZENS NEED NOT PANIC

1

To some people it wasn't surprising, the advent of this pandemic, merely shocking in the way a grim inevitability can shock. Those unsurprised people were infectious disease scientists. They had for decades seen such an event coming, like a small, dark dot on the horizon of western Nebraska, rumbling toward us at indeterminable speed and with indeterminable force, like a runaway chicken truck or an eighteen-wheeler loaded with rolled steel. The agent of the next catastrophe, they knew, would almost certainly be a virus. Not a bacterium as with bubonic plague, not some brain-eating fungus, not an elaborate protozoan of the sort that cause malaria. No, a virus—and, more specifically, it would be a "novel" virus, meaning not new to the world but newly recognized as infecting humans.

But if new to humans, from where would a "novel" virus emanate? Good question. Everything comes from somewhere, and new viruses in humans come from wild animals, sometimes by way of a domestic animal as intermediary. This sort of transfer, from nonhuman host to human, is known as spillover. Such viruses, including Marburg and rabies and Lassa and monkeypox, cause afflictions that are termed zoonoses—or zoonotic diseases. Most human infectious diseases are zoonotic, caused by animal-origin pathogens that reach us repeatedly (Nipah virus, spilling over from fruit bats in Bangladesh) or have reached us in the past (HIV-1 group M, the pandemic AIDS subtype, spilling over from a chimpanzee, once). Some are old to us (the plague bacterium, yellow fever virus) and hatefully familiar; some are as startlingly new and ferocious (Ebola virus) as a predatory alien in a movie.

A novel virus can be devastating if we have no vaccines to deflect it, no drugs to fight it, no history of past exposures to anything similar that might give us acquired immunity. A novel virus, if luck is good for the virus and bad for us, can go through the human population like a high-caliber bullet through marbled sirloin.

These scientists, the ones schooled in infectious diseases and savvy to zoonoses, further foresaw that it would likely be a particular kind of virus causing the next pandemic—a virus with a certain kind of genome, allowing for speedy evolution, a capacity to change and adapt fast. That genome would be written in RNA, not DNA. That is, a single-strand informational molecule, rather frangible, not DNA's double helix. Never mind for now just what RNA is, how it works, or why a single-stranded RNA genome can be especially changeable and adaptive. Suffice to say that such speedy adapters include the influenzas and the coronaviruses, two groups of viruses with histories of bringing mayhem to humans. In the years before 2019, the word "coronavirus" was unfamiliar to most people, but it already carried an ominous timbre to infectious disease scientists.

One among those scientists is Yize (Henry) Li, a China-born virologist and immunologist, now an assistant professor at Arizona State University in Tempe. Yize Li is a round-faced young man who wears stylish rectangular glasses and a splash of black bangs hanging over his forehead. He did his doctorate at the Institut Pasteur in Shanghai, under the mentorship of a French professor, and took the name Henry for convenience in the French- and English-speaking milieus he has inhabited since. He came to the United States in 2013, for a postdoctoral fellowship with Susan R. Weiss, a veteran virologist at the University of Pennsylvania's Perelman School of Medicine. Weiss is an authority on the coronaviruses, including SARS-CoV, the virus that caused the terrifying but abbreviated 2003 international outbreak of SARS (severe acute respiratory syndrome), infecting about eight thousand people and killing one in ten. Her lab also studies the MERS (Middle East respiratory syndrome) coronavirus, first recognized as a human pathogen in 2012, when a flurry of cases emerged on the Arabian Peninsula; MERS carries a fatality rate considerably higher than SARS, about 35 percent among confirmed cases. Li himself worked with Weiss both on the MERS virus and on a less dramatic coronavirus, one that causes hepatitis in mice.

He was there in Philadelphia during the latter days of December 2019 when he noticed an item on a Chinese news website, DiYiCaiJing, based in Shanghai. The item described an advisory note, supposedly confidential, that had recently circulated to staff at one Wuhan hospital and probably more than one. This advisory was said to come from the Wuhan Municipal

Health Commission. The website's reporter had somehow gotten hold of it and, contacting the commission, confirmed that it was from them. The note warned of an outbreak of an "unknown pathogen" that was bringing pneumonia cases to several hospitals in the city. Li promptly did what people do with interesting tidbits: he put the item on social media.

WeChat is an all-purpose Chinese app that combines the functions of Facebook, Instagram, WhatsApp, and Zoom. It has over a billion active users, including Henry Li and many other graduates and students of the Institut Pasteur in Shanghai. He relied on it to communicate with friends back in China. When he raised the Wuhan topic on WeChat, some of his contacts said, Yes, that's a rumor; some said, Yes, it's true. Then one of them threw down a trump card, posting an actual sequencing report that contained fragments of the genomes of multiple microbes, including bacteria and viruses, from several clinical samples. The samples—a throat swab here, a nose swab there, who knows—had been processed, RNA extracted, that RNA converted to DNA (for stability), then the DNA run through a sequencing machine in someone's lab. The samples were "dirty," as such samples commonly are, bearing smears and smudges of various genomes reflecting the microbial diversity present on human mucosal surfaces. But amid that distracting diversity, in at least one of these samples, was a patch of relevant data. This fragment was a linear sequence of roughly a thousand letters, a fraction of a genome but enough to be telling. It was contraband data. To you or to me such a sequence would have been just babble—*attaaaggtttatacc* for a thousand letters—but to scientists like Henry Li or Susan Weiss it spoke with chilling clarity. "I was amazed," Li told me later, to see that it was "very, very similar to a SARS coronavirus."

Weiss was on sabbatical in La Jolla, California, at that point, speaking with Li and other members of her lab in weekly Zoom meetings. During one of those calls in late December, to the best of her memory, Li mentioned that "something was really up" in Wuhan, China. "He probably told me," Weiss recollected, when I spoke with her more than a year later, "'Hey, there's this coronavirus circulating.'" But the term itself, "coronavirus," was not yet circulating in December 2019—not, at least, beyond such select networks of viral savvy.

Weiss returned to Philadelphia on January 2 and her crew promptly began ordering more N95 masks, the same kind they had been using in their study of the MERS virus (properly known as MERS-CoV). Other items of personal protective equipment (PPE), such as gloves and gowns, were already on backorder. Eventually they would add powered air purifying respirators (PAPRs), like space helmets without the suits. They were gearing up. She and her young colleagues had decided by then that they should work on this new coronavirus, and they knew they would need protection.

2

Marjorie Pollack is a highly sensitive alarm bell within one of the leading international alert networks on infectious disease. Stated otherwise: she is deputy editor of ProMED-mail.

ProMED (as it's commonly known) is an email service with roughly eighty thousand subscribers, devoted to detecting, gathering, and disseminating reliable information about disease events happening moment to moment anywhere in the world. It began in 1994, with a subscribership of forty, and is now run by the International Society for Infectious Diseases, a body of scientists and health care professionals. It's free. It's independent and apolitical. It's relentless, encyclopedic, and sometimes arcane. If you subscribe to ProMED, you might wake up to three or four of its emails on a given morning, one informing you of lumpy skin disease (a viral affliction) among Laotian water buffalo, another reporting shigellosis (bacterial diarrhea) among children who visited a safari park in Kansas, the third updating you on the latest Ebola outbreak in the Democratic Republic of the Congo. Pollack has been part of this operation since 1997.

She is a born-and-raised New Yorker, a graduate of NYU in that edgy time just after the 1960s ended at Altamont Speedway and Kent State. Her demeanor is mild until it is steely. Trained as a physician, now with forty-five years' experience in medical epidemiology, she does her ProMED work with the skeptical acuity of an old-school newspaper editor in

Chicago—"If your mother says she loves you, get a second source." To call Pollack an alarm bell, as I just did, is a little unfair because she channels her reports without undue noise or fanfare. She's more like a light on the dashboard that you may ignore until it glows red, strongly suggesting that you pay attention and, maybe, start to worry. But her job was to spread information, not worry.

On the evening of December 30, 2019, a Monday, after dinner with her husband at their weekend home on Long Island, Pollack went back to her computer, as she routinely does, to check email. She found a message from a colleague in Taiwan, alerting her to a statement from the Wuhan Municipal Health Commission, picked up on social media from that mainland city. The statement—probably the same advisory note about which Henry Li had read on DiYiCaiJing—mentioned some cases of unexplained pneumonia. "The email I got from this colleague," Pollack told me, "was basically, 'Do we know anything about this?'" No they didn't, not yet, but she was fervently curious, so she spent the next two and a half hours online, working her contacts and scraping the web.

"What we did was, we all searched, 'we' being the colleague in Taiwan and the colleague's colleagues," she said, "searching media for a second source." One colleague found that second source: a report from Sina Finance, a reputable Chinese-language media service, citing an "urgent notice on the treatment of pneumonia of unknown cause" from Wuhan Municipal Health. And it wasn't a single case of mysterious pneumonia; it was "patients," plural. At least one of those patients was linked to what this report called the South China Seafood Market. A reporter had phoned the health commission's hotline and confirmed that the advisory was real.

What next? "The copy editors go off at about 9:00 p.m. Eastern time and pick up again the next morning," Pollack told me. ProMED has a tiered editorial system to keep itself judicious and accurate, and Pollack herself had progressed over twenty-plus years through most of those tiers: volunteer web-searcher, moderator for a subject area, liaison editor for the regional networks, associate editor, a rotating top moderator, deputy editor. Above her was the editor, Larry Madoff, a professor at the University of Massachusetts School of Medicine, overseeing this network of critical-minded professionals from Boston. But now it was late Monday evening and Pollack was largely on her own. "We usually will not post stuff later on that hasn't been copy-edited," she said, "but we do have the occasional, *Urgent, let's get this out right away.*" She communicated with Madoff and the top moderator on duty, alerting them to the situation. She assembled a post under the headline "REQUEST FOR INFORMATION," to signal the provisionality of what she had. She took a machine translation of the Sina Finance article, with its statement about "pneumonia of unknown cause," and included the detail that some cases were linked to a market in Wuhan. At 11:59 p.m., after Pollack had submitted the report for posting, the top moderator hit SEND. That message instantly went out to eighty thousand ProMED subscribers, including me.

The next day was New Year's Eve. Pollack and her husband, as they did by tradition, were spending the holiday season at Water Mill, a little village on Mecox Bay near the east end of Long Island, where they have their getaway house. They rent out the place in summer, to avoid the Hamptons scene, which decidedly isn't their scene, and use it in winter. Their New Year's celebration is usually dinner in Water Mill at a favorite restaurant, the Plaza Café, then home to the TV and watching the ball drop in Times Square. But this night wasn't usual, not even for a New Year's Eve.

Between the appetizer and the main course, her phone rang. "I get a call, so I go outside." It was Peter Daszak, president of EcoHealth Alliance, a research and conservation organization with a mission of protecting both wildlife and humans from infectious diseases. Daszak and some of his colleagues were well connected with certain scientists in China, having worked with them in searching for the origin of the SARS virus after 2003, and on other efforts to identify dangerous wildlife viruses and warn of them, during the years since.

Pollack had spoken with Daszak earlier that day, during which call he shared an important bit of news from his sources in China, based on full sequencing of the new virus's genome, not just a fragment. "That it was SARS-like," Pollack told me. SARS-like suggested transmissible among humans and potentially quite lethal. That was ominous, and now, as Pollack stood outside in the late December night, Daszak had a discomfiting update. "I'm wearing a sweater, it's 26 degrees Fahrenheit," Pollack recollected, "and I'm pacing back and forth 'cause I didn't get my coat, and I'm talking to Peter, talking to Peter, I don't know how long I was out there." Eventually the waiter came to tell her that her main course was on the table. The conversation continued. She wanted more information, she wanted another source. Daszak couldn't help her on that, not presently. "Peter was basically telling me about how there was a total shutdown of communication with people in China at that point." After her dinner, eaten cold, she and her husband returned to the house and, in lieu of the show from Times Square, she went back to work. Finding another report in Sina Finance, and with help from another clunky machine translation, she converted it to a post in English. That one began: "Patients with unknown cause of pneumonia in Wuhan have been isolated from multiple hospitals." Then the part meant to reassure: "Whether or not it is SARS has not yet been clarified, and citizens need not panic."

3

Among those early pneumonia patients was a sixty-five-year-old deliveryman who worked at what Marjorie Pollack's machine translation, like that earlier report, called the South China Seafood Market. The market's name, 武汉华南海鲜批发市场, is also rendered in English as Huanan Seafood Wholesale Market, and the place is notorious now as an early focal point from which the virus spread. The words "Seafood Market" are misleading, in whatever order and whatever language, because the products on sale included much more than seafood: poultry, meat from livestock, and various forms of wildlife, some alive, some dead and frozen.

This deliveryman checked into Wuhan's Central Hospital on December 18, 2019. His condition worsened fast. On December 24, doctors drew fluid from his lungs and sent a sample to a private genome-sequencing company, Vision Medicals, in the city of Guangzhou. The question to Vision Medicals was basic: What nature of bug seethed in this dollop of liquid human distress? By ordinary procedure the company would have sent back results, but instead someone phoned, reaching a doctor named Su Zhao, head of respiratory medicine at the hospital. "They just called us and said it was a new coronavirus," Zhao told the Beijing-based news service Caixin.

Their concern went beyond the phone call. Several days later, executives from Vision Medicals reportedly came from Guangzhou, six hundred miles south, to discuss the genomic results with hospital people and disease-control officials in Wuhan. By one account—a social media post believed to come from an anonymous Vision Medicals employee—the hospital acknowledged having "many similar patients" and "an intensive and confidential investigation" began. Meanwhile the deliveryman was transferred to another hospital, where he later died.

Soon after the first sequencing, someone at Central Hospital took swab samples from a different patient, this time a forty-one-year-old man with no reported connection to the market. Those samples went to a different outfit, CapitalBio MedLab in Beijing. First results from this company identified the infectious agent as SARS-CoV, the original SARS coronavirus as seen in 2003, with a case fatality rate of 10 percent. That was a false positive for the SARS virus, too precise, too certain, flawed by limits of specificity in the testing tools or by careless technique. It was indeed a SARS-like coronavirus, but not a familiar one. Before the mistake could be corrected, though, that misapprehension flashed like heat lightning across private networks connecting medical professionals at the several hospitals in Wuhan. It reached, among others, Wenliang Li, a young ophthalmologist working there at Central. You've heard of him. He became the famously martyred whistleblower who alerted some people to the danger. On December 30, at 5:43 p.m. Wuhan time, Li posted on WeChat to a private group of his medical school classmates: "7 confirmed cases of SARS were reported from Huanan Seafood Market." Within an hour he had better information and corrected that to say "coronavirus infections" and that "the exact virus strain" was yet to be identified. Warn your loved ones to protect themselves, he wrote to his friends, a brave act that invited sanction by authorities, though he didn't try to warn the world at large. In fact, he wrote: "Don't circulate this information outside the group."

The following day—again, it was New Year's Eve—the Wuhan Health Commission released a statement on Weibo, another social media platform, acknowledging an outbreak of viral pneumonia that had sent twenty-seven people to Wuhan hospitals but discounting the rumor that they were cases of SARS. "Other severe pneumonia is more likely."

Further sequencing of patient samples, sent to a different private sequencing company, clarified that this was not the SARS virus, no, but about 80 percent similar in its letter-by-letter genome. Those results came back to the Wuhan Municipal Health Commission, at which point provincial authorities intervened. On January 1, according to Caixin, the health commission of Hubei province instructed the sequencing companies to "stop testing and destroy all samples." It remains unclear whether that order was meant to contain a dangerous virus or dangerous information.

4

The rumors reached Hong Kong, irrespective of any governmental order, at the speed of electricity. Hong Kong is highly attuned to any news from the mainland, but especially bad news.

As a special administrative region (SAR) of the People's Republic of China, since Britain's colonial rule ended in 1997, what we call Hong Kong encompasses not just Hong Kong Island but also Kowloon and the New Territories, both on the mainland coast. With activists fighting for democracy, and the oxymoronic ideal of "one country, two systems" slipping away as Beijing tightens its grip, there's an ambivalent relationship with the mother country. Although much of the New Territories terrain is still green and hilly, preserved as park land, Hong Kong SAR is one of the most densely populated areas on Earth, and it bristles with eminent scientists and hungry journalists as well as with political tensions, billionaires, ethnic diversity, and sheer human numbers. On December 31, the *South China Morning Post* (*SCMP*), its leading newspaper, ran a story about Hong Kong health authorities preparing emergency measures—already—against the mysterious pneumonia outbreak in Wuhan, six hundred miles away.

Hong Kong was edgy because Hong Kong remembered its outbreak of avian flu in 1997, a small but terrifying encounter with a virus fatal to one human case in three, and SARS-CoV in 2003, the first killer coronavirus known to science, which emerged in Guangdong on the mainland, got to

Hong Kong, and exploded through that city to the world. The new virus hadn't arrived yet, but medical staff were alerted, according to *SCMP*, and ready to isolate cases.

The paper also quoted Kwok-Yung Yuen, a veteran microbiologist at the University of Hong Kong. Yuen, informed by his long history of research on dangerous viruses, noted certain similarities between the Wuhan news and the 1997 and 2003 scares: links to food markets, high infection rate.

"But there's no need to panic," he told *SCMP*. Infection surveillance and control had improved since 2003, Yuen said, and so had antiviral medicines.

Information was still scarce. Up in Beijing, at that point, even the director-general of the Chinese Center for Disease Control and Prevention (CCDC), an Oxford-educated virologist named George Fu Gao, had only online reports to guide him. "I heard this on the evening 30th December," Gao told me. "China is such a big country. If any doctors—if they identified any so-called PUE, pneumonia of unknown etiology, they should report to my institute, the China CDC. But they didn't. From the very beginning they thought it's the flu."

Gao himself is an expert on the influenzas, as well as SARS-CoV, MERS-CoV, chikungunya virus, and other zoonotic viruses. His specialty is the mechanisms by which those viruses bind to and enter human cells. "This virus from the very beginning looks like flu." It did, he meant, if you were a clinician in a hospital, like the frontline doctors in Wuhan, but not if you were a molecular virologist reading its genome or an electron microscopist gazing at viral particles festooned with spikes. "There are some rumors, I heard some rumors. But I saw the news on the internet media on the 30th." So even he gave some attention to the disease chatter online. But he couldn't learn much. Those few days of delay before the CCDC was directly notified, caused by misguided caution among Wuhan city and Hubei provincial officials, were costly.

Gao alerted his bosses at the ministerial level. "And then next day we sent all our expert team to Wuhan. By then we realize, okay, that could be a problem."

As of January 1, the World Health Organization hadn't yet been notified either. Outbreak-response professionals at WHO headquarters in

Geneva had seen the ProMED posts and other online reports, and they took the initiative, contacting China's National Health Commission. *What's happening?* For two days the WHO got no response. Then came a frustratingly vague update from China: we now have forty-four cases of unspecified pneumonia, not twenty-seven.

January 1 was also the day when Wuhan authorities closed the Huanan market for "sanitation and renovation." The sanitizing was performed by technicians from a private disinfection company even while government scientists, including George Gao's team from the China CDC, gathered environmental samples from the market's runoff drains, stalls, doors, and some frozen animal carcasses left behind by hurriedly vacating merchants. That sampling began early on the morning of closure day and would continue off and on for two months. The range of sampled surfaces and creatures included trash cans, transport carts, animal cages, public toilets, and stray cats. The "renovation" of the market was left to be imagined.

Two days later, another set of samples reached another virologist, Yong-Zhen Zhang, a professor at the Shanghai Public Health Clinical Center, affiliated with Fudan University. These swabs, including one from that forty-one-year-old patient with no known links to the Huanan market, had been packed into a test tube, cradled in dry ice inside a metal box, and sent by train from Wuhan. Zhang and his group worked nonstop for most of two days and nights, extracting the RNA, converting it to DNA, sequencing that in fragments, patching the data together into a complete coronavirus genome sequence. The genome of this virus, which did not yet have a name, ran to about thirty thousand letters. "It took us less than forty hours, so very, very fast," Zhang said later, during a rare interview, to a reporter for *Time*. "Then I realized that this virus is closely related to SARS, probably 80 percent. So certainly, it was very dangerous."

Promptly he called Su Zhao, the head of respiratory medicine at Wuhan Central Hospital, the same man who had received the discomfiting preliminary call from the private sequencing company. Zhang alerted Zhao that he should be concerned and cautious, because this was a SARS-like coronavirus—not SARS-CoV itself, with its one-in-ten case fatality rate, but a new virus of the same group, and more dangerous than influenza. Implicit in that analogy, "SARS-like," and in the multiplicity of cases linked to the Huanan market, was something that wasn't yet being said publicly: the virus was likely capable of human-to-human transmission. Respiratory transmission of any virulent new virus, person-to-person, raises the possibility of a big outbreak. Soon after the call, for further emphasis, Zhang traveled to Wuhan in person and spoke with health officials there, advising them to take emergency measures toward protecting their citizens, and to begin developing antiviral treatments.

The genome sequence would be crucial to such an effort, identifying new antiviral drugs or deploying old ones, and also for preparing diagnostic tests that could tell who was infected and who wasn't. Zhang and his team had the sequence, and they had submitted it quietly to an open access international database, GenBank; but it hadn't yet been released publicly.

By one account, China's National Health Commission issued secret orders forbidding labs to publish results on the virus without official clearance. At least two other teams in China now also had the sequence, or a version of it, with only slight differences from Zhang's owing to methodological differences: a group in Wuhan, led by a scientist named Zhengli Shi, and George Gao's group at the CCDC in Beijing. "We got the materials, we did the test of the whole genome," Gao told me. "Three days later, that would be the 3rd of January, we got a whole genome sequencing and then we found it's a novel coronavirus." They also viewed it by electron microscopy, which showed the corona of protein spikes, protruding like cloves in a roast ham, that gives this viral family its name. "We saw the virus!" he said. "It looks like it's a coronavirus—you see the crown on the surface. So by 7th January, it's already confirmed." Gao spoke directly with Tedros Adhanom Ghebreyesus, director-general of the WHO, known to the world as Dr. Tedros. "And the same day, Dr. Tedros talked to our minister of health." Gao coordinated his group and others, and late on the evening of January 9 UTC (Coordinated Universal Time, what we used to call Greenwich Mean Time), Gao's deputy emailed—according to one account, which I have from a different source—full genomic sequences from three samples to the database GISAID, headquartered in Munich. The data were quickly curated and two of those sequences published on the organization's web platform, according to this source, available to anyone registered with GISAID user credentials. Late evening UTC equals early morning the next day in Beijing. So by January 10, Gao told me, "WHO, everybody, knows it's a coronavirus." Many scientists did, anyway, though there was still no publicly released sequence—depending on how you define "publicly."

The next morning, January 11, Yong-Zhen Zhang went to Shanghai's Hongqiao Airport for a flight to Beijing, where he would meet with high government officials such as Gao. Sometime during the boarding process, his phone rang.

5

It was Edward C. Holmes, calling from Sydney, Australia.

Holmes is a British evolutionary biologist at the University of Sydney, and the only non-Chinese member of Zhang's team for the sequencing, assembly, and analysis of the new virus's genome. He specializes in the molecular evolution of viruses, particularly RNA viruses and more particularly those that infect people, including the HIVs, the influenzas, measles, Ebola, hepatitis C virus, the dengue viruses, yellow fever virus, and the coronaviruses. RNA is the coding language of human pandemic, and Holmes is one of its preeminent translators.

By way of introduction to Holmes, I should say a little more about this formidable molecule, RNA, since it's so important for understanding these viruses and so central to his work and the work of Zhang and their colleagues. The initials stand for ribonucleic acid, a macromolecule that performs several functions in cells and viruses, such as coding genetic information, transmitting information that's been coded in DNA, and regulating gene expression, the process of turning such information into molecular machinery. The main structural component of RNA is a chain of four different kinds of subunit, known as nucleotide bases: adenine, cytosine, guanine, and uracil. Each nucleotide consists of a base plus two other molecules—but you can forget about those other two, as regards genetic coding. The bases are the coding elements I've been calling "letters," because they are represented by the letters as A, C, G, and U. The sequential arrangement of those bases is what constitutes genes. Three bases in an ordered triplet code for a particular amino acid (there are twenty different amino acids in biology) and amino acids linked end-to-end constitute proteins. That's how life is built. DNA is also a linear assemblage of bases, with the difference that thymine stands in place of uracil, and DNA's usual form is two strands bound together, spiraling as a helix. RNA as a genome tends to mutate more frequently than DNA in a double helix does; it lacks the stability. That's part of what makes RNA viruses so changeable and adaptable. From here on I'll refer interchangeably to the "bases" or the "letters" comprising a genomic sequence. RNA is a fascinating molecule and for someone like Holmes, who knows its vocabulary and grammar so well, it's a language of deep meanings.

Holmes is highly respected not just as a consulting wizard, a coauthor on many influential journal papers, but also for his 2009 book, *The Evolution and Emergence of RNA Viruses*, an authoritative but concise compendium. Oddly for a text that goes so deeply into the swales and the gullies of molecular evolution, the book is clear, trenchant, and readable. Two other memorable Holmes traits are his very bald and nicely rounded head, which seems almost polished as a point of pride, and the fact that everyone calls him Eddie. Speak with molecular virologists anywhere in the world, remind them "But wait, hasn't Eddie said . . ." this or that, and they may not agree with the statement but they will know who you mean. In this field there's only one Eddie.

My own first encounter with Eddie Holmes came a dozen years ago, when he held a chair at Pennsylvania State University, where he welcomed me to a small, bare office containing a desk, a computer, a couple chairs, a few books, and not much else except two wall posters, one advertising "The Virosphere," the vast dimension of Earth consisting of viruses, and the other a cartoon version of Edward Hopper's painting *Nighthawks*, with Homer Simpson in the role of a customer at the counter, gorging on donuts. Why Homer Simpson? I asked. Because he looks like me, Eddie said.

Since moving to Sydney in 2012, Holmes has collaborated on a number of projects with Chinese colleagues, teams led by Yong-Zhen Zhang and other senior figures, and those interactions have been eased slightly by his being only two time zones away from Shanghai and Beijing. Email is email, with words in cold type and the convenience of being answerable whenever one gets to it, time zones notwithstanding, but some Chinese scientists, Zhang among them, prefer the real-time immediacy and discretion of WeChat for voice. So on the morning of January 5, 2020, a Sunday, as Holmes and his family prepared for an outing at the beach, he got an email from Zhang. It said, "Call me immediately!" This was just a few hours after Zhang's lab had assembled the complete genome and seen that the dangerous new thing was a SARS-like coronavirus.

Six days earlier, Holmes had noticed what many others had noticed: Marjorie Pollack's New Year's post on ProMED, linking multiple cases of an unexplained pneumonia with the Huanan market. "Oh, shit, that's interesting," he thought. It registered because he had visited the same market himself in 2014, on a field excursion with Zhang and some colleagues from the Wuhan CDC (a regional center, distinct from, but linked to, the CCDC in Beijing). He had seen the narrow alleys crowded with people, the wildlife in cages, the butchering of meat and fish, the blood and guts flowing in open drains. "You can't think of a better place for a zoonotic event to happen," Holmes told me recently. He recalled one vendor killing a wild mammal of some sort, possibly a raccoon dog, as he stood watching. He recalled that the market sat squarely amid a city of eleven million people.

The next day, January 1, he had emailed both Zhang and George Gao. "I've read about this," he told each of them. "Are you working on it? Can I help in any way?" Gao, presumably inundated, sent a terse reply: "We are working on it. Happy New Year." Zhang replied that, no, he wasn't working on it—not yet. The week progressed, other distractions intervened, and then on the Sunday morning came Zhang's urgent message: "Call me immediately!" Holmes did, speaking with him while driving the family to the beach. It's a wonder they didn't crash.

We need to write a paper on this, Zhang said. A novel coronavirus, looking almost like the return of SARS—scientific news. Wait, no, Holmes said, there's something more urgent than a journal paper. "The first thing you've got to do, Zhang," he said, as recounted to me, "is you've got to tell public health authorities NOW. You've got to tell them exactly what it is, and you've got to release as much information as you can." Information meaning: the genome itself, the analysis that it was SARS-like, the probability of respiratory transmission. Zhang agreed, promptly alerting the National Health Commission.

"So the *same* day he got the sequence," Holmes emphasized to me, "he told the authorities what was going on." Holmes is acutely aware of accusations that Chinese scientists—not just Chinese officials—withheld facts and delayed a timely response.

In the following few days, they did write a paper, at high speed, conferring by telephone and sharing drafts by email, with Holmes editing the English text as well as contributing his views on the genome. He also contacted an editor at *Nature*, one of the world's preeminent scientific journals, to gauge interest. *Nature*'s interest was high—but they wanted the genome sequence for release along with the paper. Zhang's team sent *Nature* a draft of the paper on January 7, blistering speed for such a complex, delicate composition. But for reasons involving Zhang's situation in China and the pressures around him, the sequence remained a sticking point. Over the next two days, further reports began to emerge about what was obvious to others, as well as to Zhang's group, from sequencing efforts: that the thing was a coronavirus, somewhat like SARS. *Nature* wanted the genomic data, as well as the words, and even Holmes himself hadn't yet seen the full sequence. He was still pushing Zhang to go public with all they had. Then it was Saturday morning, January 11, and this weekend the Holmes family were *not* headed for the beach.

"I call Zhang and he's on an airplane," Holmes told me, "and I say, 'Zhang, we HAVE TO release this! We HAVE TO release the sequence, right? Everybody wants it.'"

They talked for some minutes, and by then Zhang was buckled into his seat. "I asked Eddie to give me one minute to think," Zhang recalled for *Time*. "Then I said okay." Finishing that call, he instructed one of his postdoctoral fellows to send Holmes the sequence. The plane took off, and while Zhang was airborne for two hours, 35,000 feet above northeastern China, Holmes received it.

The genome arrived by email, from the postdoc, in the form of an

attached FASTA File, a handy text format for representing genomic sequences. "No message. Just the FASTA File," Holmes told me. "Right."
No niceties, maximal speed and discretion. He opened the file and barely
glanced at the sequence, printed in six columns across, ten letters in each
column, row after row, page after page, almost thirty thousand letters,
representing almost thirty thousand bases, nothing but *a, t, c,* and *g* in gab-
bling combinations. It was written in DNA because RNA is so unstable;
genomic RNA is routinely transformed to its DNA equivalent for sequencing. "I don't even check what the hell it is. It could be bloody glowworm
DNA." This is a man who can scan a genome by eye, flick a few keys, bring
up a few comparisons, and see things others cannot see. But he didn't. "I
feel absolutely under *huge* pressure to get it out as quick as possible." The
next move was prearranged, and he made it at once.

Waiting in Edinburgh was Andrew Rambaut, another eminent evolutionary virologist and Holmes's friend of thirty years. Rambaut is the
founder and guiding elder of a website called Virological (virological
.org), which serves as a communication nexus for professional comment,
response, and thoughts that are not yet quite journal papers. "Eddie rang
me, I think, in the morning before," Rambaut recalled later. "Just to say,
you know, he was working with Zhang and hoped to have a sequence
soon." Sydney is eleven hours ahead of Edinburgh, and by Saturday morning for Holmes and Zhang it was the wee hours for Rambaut. "About one
in the morning on the 11th, I think, he finally emailed me and said, 'Okay,
let's post it. Got permission.'" Attached was the same FASTA File containing the sequence.

At Rambaut's suggestion, they composed a little introductory statement, crediting the sources in China, citing Yong-Zhen Zhang as senior
contact, and adding: "Please feel free to download, share, use, and analyze
this data." Both men know that "data" is plural but they were in a hurry.
The posting can still be found at Virological, titled "Novel 2019 coronavirus genome" and datelined "10th January 2020," though Holmes's memory as well as Rambaut's says it happened at 1:00 a.m. Edinburgh time on
the 11th. The discrepancy is unimportant except for reflecting the sense
of breathless haste in making the genome public. "I timed it," Holmes told
me. "I think I had it in my possession for fifty-two minutes from the email
arriving to when it went up online."

6

W hat was the most important decision that you made during 2020?" I asked Tony Fauci.

"Most important decision?" He thought for a moment. "There's a scientific decision and a policy decision." After decades as director of the National Institute of Allergy and Infectious Diseases (NIAID), with plenty of experience defending health care and research policies on Capitol Hill, he was now more deeply and conspicuously embedded in policy as a member of Donald Trump's White House Coronavirus Task Force. His biggest policy decision in 2020? That was "to speak up against the president, which led to a lot of other things," including death threats, harassment of his family, and the hashtag #FireFauci on social media. If you were to google the words "Fauci contradicts Trump," as I recently did, you too might get 58,400 results. An early instance of such impolitic candor, a gentle one, came during a White House press briefing on March 20, 2020, when Trump touted the drug hydroxychloroquine as a treatment for COVID-19 and Fauci noted that such reports were "anecdotal," not scientific. "I don't take any great pleasure in being in public conflict with the president of the United States," he told me. But if he hadn't done that, he added, he would have compromised his integrity and the important message that science is still the way we need to go.

And the scientific decision?

"To *immediately* say we've got to develop a vaccine and give my team all the support they needed to do it." By *immediately* he meant right after the first sequence became available from Zhang and Holmes. His "team" on that front included John Mascola, director of the Vaccine Research Center (VRC), which is part of the NIAID, and Barney Graham, a senior scientist and deputy director at the VRC, who had worked for years on the bold idea of using mRNA (messenger RNA, an information-bearing molecule within cells) in vaccines. That proof-of-principle work had reached maturity enough to be applied.

As the rumors about unexplained pneumonia leaked out of China during December, Fauci and his colleagues noted the aspects resembling

SARS-CoV. "We're all saying, 'This smells like a coronavirus,'" he told me, "but we don't know what it is. And I remember Barney Graham saying, 'Boy, just get me the sequence. We're all set to go.'" (Barney Graham remembers the moment too, but with different words. "I would not have said 'Boy,'" he told me. "I probably would have said something more like, 'If we can just get the sequences, we know what to do.'") Late on January 10 Eastern Standard Time, thanks to Zhang and Holmes, they got the sequence data.

Others were set, and knew what to do, as well. Nicole Lurie, a physician and public health professional with deep government experience in preparing for and responding to disease emergencies, had joined the Coalition for Epidemic Preparedness Innovations (CEPI), a relatively new initiative based in Oslo, as strategic advisor and lead person for preparedness planning. Her role involved, among other things, finding ways to engage developers working on other vaccines. She got CEPI started on the new virus four days before the Zhang sequence went up. "By January 7, it seemed really clear that this was something with huge pandemic potential," Lurie told me. "There were a lot of rumors circulating amongst people who were connected to China CDC, and others, that this was a novel coronavirus." CEPI contacted vaccine developers, including some at Oxford University who were working on a MERS vaccine, using a different approach—not mRNA but another—that would become the Oxford-AstraZeneca vaccine. The urgent request from CEPI, as Lurie recalled to me, was that they be ready to shift their work to the new virus as soon as its sequence was posted. They would be given a contract for such work quickly.

Emma Hodcroft was a postdoc in Switzerland, working on a project called Nextstrain, a collaboration between the University of Basel and, in Seattle, the Fred Hutchinson Cancer Research Center, toward an online platform for tracking the genomic divergence of viral and bacterial pathogens. Tracking genomic divergence allows epidemiologists to map the routes of disease transmission. Mapping transmission routes helps scientists and public health authorities to understand outbreaks and epidemics, prevent them in future, and bring them to an end. Tracking divergence also allows researchers to spot mutations that become successful, spreading throughout a population, and that sometimes aggregate with other mutations, into bouquets that we now call variants. Variants represent a virus evolving, sometimes at formidable speed, to defeat our defenses against it. But much of the work done before 2020 by Hodcroft and her colleagues, she told me, didn't rise to the level of headline news. "That is, looking at how viruses change, how they jump into humans, how they adapt to humans. It doesn't necessarily make most people's kind of radar." It didn't use to, anyway.

The new virus is different. "I remember when the sequence came out, because this was a really big deal," Hodcroft said. The post on Virological circulated quickly through her world. In a gruesome way, with dire stakes, it was exciting and impressive. "We never have had a completely unknown virus," she said, that went "from first mention to first sequence in such a short period of time." Nextstrain took note and, in the following days, as more sequences became available, began to draw a family tree. No one could predict how many branches and twigs it would grow.

7

Kwok-Yung Yuen, who had told the *South China Morning Post* on December 31 that there was "no need to panic," also felt heightened concern twelve days later, on January 11. But his new worry didn't derive from the genome sequence that Eddie Holmes had just posted. He had ominous news from a closer and more human source.

Yuen, chair of Infectious Diseases within the Department of Microbiology, which is within the medical school at the University of Hong Kong (HKU), divided his time between research and teaching, and between Hong Kong and the mainland. He served as a senior supervisor and educator at Hong Kong-Shenzhen Hospital, in the city of Shenzhen, less than twenty miles away, in Guangdong province. He was well connected there. So he got the news through personal channels, and promptly, when the hospital saw two members of a family, on January 10, and soon afterward two other members of the same family, who were suffering unexplained pneumonia after a trip to Wuhan.

None of these people had visited the Huanan market. Another family member, a grandmother who had skipped the trip and remained home in Shenzhen, also took sick and required hospitalization. All five of them, including the grandmother, tested positive for the new virus. It meant, Yuen knew immediately, that this coronavirus was not only spreading, and spreading human to human, but spreading city to city. More worrisome still, a ten-year-old grandson also tested positive, and showed lung damage on a CT scan, without having felt any symptoms. "These cryptic cases of walking pneumonia," Yuen's group declared in a paper, written quickly and published online within two weeks in a major British journal, *The Lancet*, "might serve as a possible source to propagate the outbreak."

Yuen trained as a surgeon before turning to infectious disease and virology. He was part of the team that first isolated and characterized the SARS virus in 2003. For informal communications, he goes by his first initials, K.Y. He's a forthright man, sometimes almost recklessly so, and he didn't remain silent for those two weeks. "I said to government, 'We must wear masks!'" Yuen told me. "Universal masking is very important, because there are people shedding virus without symptoms!" The ten-year-old boy suggested that likelihood, and further proof would come soon.

"You were saying that to the Hong Kong government?" I asked. There was also the Chinese national government, a much bigger dragon, to consider.

"Yes," he said, "yes."

"And they adopted that?"

"No!"

But he was an eminent scientist, drawn quickly into the national advisory huddle. On January 19 he flew to Wuhan as part of a high-level panel of experts, including George Gao, Nanshan Zhong (a revered figure, a senior pulmonologist, considered a hero for his management of the 2003 SARS event), and several others, to assess the situation at ground zero. There they learned, from officials of the Wuhan CDC, that case numbers had sharply increased, standing now at 198, with thirty-five of those people severely ill and nine in critical condition. Worse news still, fourteen health care workers at a Wuhan hospital had become infected from a single neurosurgical patient. The Wuhan CDC investigators were now calling this disease NCIP, for "novel coronavirus-infected pneumonia," and labeling the virus itself 2019-nCoV. Both of those clumsy names would soon be changed to others, more lasting though not greatly less clumsy.

One day in Wuhan was enough. Yuen went up to Beijing with the other senior advisors and there, on January 20, at a press conference in the ministry building of the National Health Commission, they announced that, yes, this virus was being transmitted person to person. It had already spread beyond Wuhan to Beijing, Shanghai, and Shenzhen, and to Thailand, South Korea, and Japan.

Yuen returned to Hong Kong and met with its chief executive, advising her that she should control the borders of Hong Kong SAR, mandate a fourteen-day quarantine for arriving travelers, and again—as he had urged earlier—get everyone into masks. On January 24, he and a group of colleagues from the University of Hong Kong and Hong Kong-Shenzhen Hospital published their important (but insufficiently noticed at the time) paper in *The Lancet*, describing the family cluster of cases in Shenzhen, with the explicit message that these cases supplied evidence of person-to-person transmission. There was also another message buried in that paper, even more ominous, and implicit in the case of the ten-year-old boy. Besides testing positive for the virus, and showing lung damage on a CT scan, he "was shedding virus without symptoms." If such asymptomatic infection was possible, asymptomatic *spread* of the virus was possible too. That would make this novel coronavirus vastly more dangerous than the original SARS virus or any other bug within recent memory.

The Chinese Lunar New Year, marked by the new moon on January 25, 2020, would be celebrated with the Spring Festival, lasting fifteen days. All over China, people would travel to be with relatives for the holiday—the migration to reunite, known as *Chunyun*—and welcome the Year of the Rat. They would gather in large, close, festive groups, sharing hot pots, dumplings, waxed duck, noodles, and germs. Perfect circumstances for a virus, if it was fortuitously adapted, with respiratory proclivities, and maybe even the capacity for asymptomatic transmission, to explore new habitat in human windpipes and lungs. K.Y. Yuen could see what was coming, and he understood that panic would be inefficient and futile.

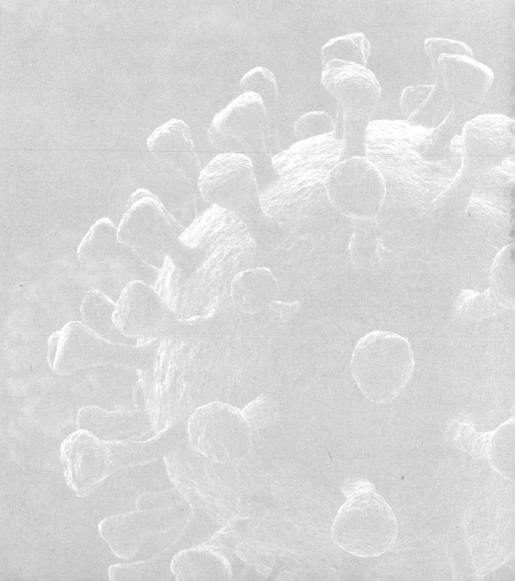

II

THE WARNINGS

8

Some people trace the warnings to SARS, hitting Singapore and Toronto, Beijing and Bangkok and Hanoi in 2003. But of course there were other warnings, still earlier: bubonic plague in the fourteenth century, caused by a bacterium carried in fleas that rode upon rats, arriving in Europe along trade routes from the East; the Great Influenza of 1918–1919, the last viral pandemic of the era before viruses could even be seen and identified, this one killing perhaps fifty million people; the first known outbreak of Ebola, mystifying and gruesome, at Yambuku Mission in Zaire, 1976, having emerged from an animal still unidentified; or HIV-1 group M, slower in effect and more subtle than Ebola, far more consequential in its death toll, becoming conspicuous in 1981, decades after its passage from one chimpanzee into one human, somewhere in or near southeastern Cameroon. These were all cautionary events, various and prominent, poorly understood in their times but offering lessons for the future.

Other warnings were quieter and more specific. One occurred in November 1997, when a man named Donald S. Burke gave an invited lecture to a branch of the American Society for Microbiology, in Atlanta. It was the Chapman Binford Memorial Lecture, named for a distinguished physician and pathologist who had specialized in leprosy. Binford and Burke had little in common beyond that they both worked on infectious diseases for the United States government—Binford in the Public Health Service, Burke in the military. Burke came down from Baltimore to give his lecture. He had lately relocated there, accepting a professorship at Johns Hopkins University as a civilian, after decades of research on HIV and other pathogens on assignment within the United States Army. We can know the gist of what he said that day because he published a version of it, the following year, as a chapter in an obscure multiauthored volume. His title was "Evolvability of Emerging Viruses."

Beware of RNA viruses, Burke wrote, because they are highly evolvable. They change and adapt fast. He explained what he meant, the basic biological mechanics, from mutation to adaptation to spillover. Viruses can only replicate within living cells (they aren't cells themselves), and all viruses mutate during replication—that is, they make small mistakes in copying their genomes to produce offspring. RNA viruses mutate faster than almost any other sort of creature on Earth. In fact, Burke wrote, they mutate about a thousand times faster than animals: roughly one mistake in every ten thousand bases of genome. Although the genomes of RNA viruses are relatively short, only a few thousand or twenty thousand or thirty thousand bases (compared to three billion bases for the human genome), that error rate is enough to put at least one mutation into every new virion (every viral particle) of the typical RNA virus. Result: each of those new virions is likely to be different, by at least one mutation, from its parent.

Mutations don't constitute adaptation, but they are raw material from which adaptation can be shaped. Natural selection does the shaping. The mutations are random changes. Most of those changes prove to be either damaging or insignificant to the virus's prospects for proliferating its offspring, and if damaging enough, the virus *has* no offspring. The mutant lineage comes to a dead end. But some mutations, by sheer chance, turn out to be helpful, virion by virion. They enhance a viral lineage, collectively. Those lineages are the fittest, and they survive.

This much, as you probably recognize, is Darwin 101.

But some RNA viruses have an additional trick that further increases their capacity to evolve. They can *recombine*, swapping sections from one viral genome to another, like railroad trains switching cars on a siding. (Coronaviruses, for instance, recombine. Influenza viruses have their own version, called *reassortment*, with the breaks occurring at regular spots along their segmented genomes.) This occurs by a sort of molecular interruption during the process of replicating their genomes within the same cell. Recombination, Burke explained, "serves both to hybridize highly fit variants and to replace defective and incompetent genes." Empty boxcars left behind, sleek Pullman cars added. In other words, recombination gives viruses major new options and clears away genetic debris. It allows them to evolve by large chunks, as well as by the tiny increments of mutation.

That may sound vaguely familiar too, from your good memory of high school biology, because recombination of a different sort occurs in animals, including humans, during the production of eggs and sperm. To put it simply (and spare you a refresher on meiosis), the chromosomes in complex creatures swap sections at a crucial moment, which reshuffles the genes received from each parent into new gene combinations for the offspring.

This process is called sex. Its value in evolutionary terms is to produce offspring that differ genetically from their parents and also (except identical twins) from their siblings. In other words, it adds variation among individuals to a population. Variation allows populations to evolve. RNA viruses are incapable of sex, so they achieve the same end by a different sort of recombination: swapping sections of RNA with other viruses while engaged in the delicate, naked act of genome replication.

Part of Burke's purpose, in the lecture and its published version, was to describe a discovery made by himself and some colleagues, also military-employed scientists, in the field of artificial intelligence and machine learning. The colleagues were computer scientists from the Navy Center for Applied Research in Artificial Intelligence. They were the wizard modelers; Burke was the concept guy. Other work of the Navy modelers included teaching torpedoes to chase other torpedoes. The common purpose for this collaboration involved teaching a virus how to evolve and survive—or, more precisely, creating a successful model of how that might happen, in order to understand how it probably does. Together with Burke, these smart Navy fellows devised a "virtual virus," a computational model to simulate viral evolution, occurring through iterations of change and challenge on a computer.

How to do that? They coded various versions of their virtual virus with differences in three key parameters: the rate of mutation, the capacity to recombine or not, and, if recombination was allowed, the matter of whether it was section-for-section swapping—that is, analogous regions of the genome, with similar functions but different details, as in switching one locomotive for another, one caboose for another—or entirely random. Random means you could have a caboose at both ends and no locomotive, not conducive to railroading success. What they found was that an idealized computer virus, one with a mutation rate like an RNA virus, and with analogous recombination also like an RNA virus, would evolve "with near-optimal efficiency." It was the perfect set of attributes for fast viral evolution in a new environment. And for "a new environment," you could understand: a new host. Even a new species of host.

The lesson Burke drew from this, which he wanted urgently to convey to his audience in Atlanta and his later readers, was that novel RNA viruses present high risk for causing pandemics. Why? Because they are so capable of adapting to new hosts. They can make the big leap, and they can thrive. It was important, he argued, to try to predict and prevent such catastrophes. So he proposed three criteria that could help identify which kinds of virus, family by family, might pose greatest risk to the global human population.

His first criterion was the most obvious: Does a family of viruses include notorious pathogens that have caused pandemics in recent human history? The family containing the influenzas, yes. The family containing the HIVs, yes.

Second criterion: Do viruses of a given family cause widespread disease among nonhuman animals? The family that includes the influenzas, again yes, considering not just the avian influenza viruses that kill many birds but also Newcastle disease (a highly contagious affliction in chickens) and canine distemper (which spreads on a sneeze and can kill noncanine mammals—ferrets, skunks, raccoons, badgers—as well as dogs). And, oh, the coronavirus family, which includes a bovine coronavirus, a feline coronavirus, a canine coronavirus, a mouse coronavirus, a rat coronavirus, a horse coronavirus, a turkey coronavirus, and a cheerful thing called porcine epidemic diarrhea virus (PEDV). That last one attacks cells in the small intestine of swine, killing most of the newborn piglets it infects. It transmits mainly by the fecal-oral route, like human polio, but it may also be capable of fecal-nasal transmission, going airborne from one pig to another, possibly even from one farm to another The name itself speaks to its contagiousness: epidemic.

Burke's third criterion, which drew on his artificial intelligence project, was intrinsic evolvability. How fast does a virus mutate? How readily and smoothly does it swap pieces of its genome? Viruses with high evolvability, he noted, have special potential to emerge from their animal hosts, get into humans, and cause pandemics. For instance? Again, a select list:

the influenza family of viruses, the HIV family, a family of viruses that cause encephalitis and meningitis, and the coronaviruses. It was 1997, remember, when he gave that talk in Atlanta.

In 2011, I spoke with Don Burke about all this—his computer modeling, his idea of evolvability, his three criteria for a dangerous virus. One major event that had occurred in the meantime, between the Atlanta lecture and my call to him, was SARS in 2003: the outbreak and speedy international spread of a lethal coronavirus, as he had warned.

"How possible is it to predict the next pandemic?" I asked. "Where it's going to come from and what it's going to look like."

"I made a lucky guess," he said.

9

Burke's career route into infectious diseases was circuitous, but not unusual within the guild of such scientists, in that it involved the United States military. He grew up in Cleveland, smart kid, good grades in high school, basketball player, class president, and began dabbling in biological research as an undergraduate at Western Reserve University, with the guidance of a mentor. In that period of the mid-1960s, with the shock of Sputnik still fresh and the U.S. government eager to improve American capacity in science and engineering, his efforts were supported by a training grant from the National Science Foundation. He studied electrical impulses in hydra (little tentacled marine creatures, distantly related to jellyfish) during a summer at the Woods Hole Marine Biological Laboratory. Then he went to Harvard Medical School, though always with a goal of doing research, not practicing medicine. As an intern in the early 1970s, with the Vietnam War blazing, he faced getting drafted—the military needed battlefield doctors—and to preempt that he volunteered for a Defense Department plan that would give him some choice. "I knew that I wanted to do infectious disease work," he told me. "I also knew that I didn't want to check for hernias at Fort Huachuca." Fort Huachuca, just his synecdoche for the regular Army, was an old garrison in the desert of southeastern Arizona, a place considered so barren that you might be AWOL for three days and the guards could still see you leaving.

Burke evaded Huachuca and the hernia practice by driving to Fort Detrick, in Maryland, and talking his way into a position at the U.S. Army Medical Research Institute of Infectious Diseases (USAMRIID). It was a famously serious post where Army researchers studied difficult tropical diseases, such as Bolivian hemorrhagic fever and Lassa fever, in maximum-containment labs. "I never even went to basic training," Burke told me. He finished his residency at Boston City Hospital one day "and showed up at Fort Detrick the next day." He began learning lab virology. Counting the time at USAMRIID, a six-year stint in Thailand, a period as head of HIV/AIDS research for the entire U.S. military, and other work on emerging disease threats, he stayed in the military twenty-three years, emerging as a colonel.

He learned a lot about viruses during that career and the second career that followed, as a professor of epidemiology and then dean of the School of Public Health at the University of Pittsburgh, a job he still held when I reached him. High on the list of his takeaways was the link between recombination, in RNA viruses, and the capacity to adapt fast and switch hosts. "The fact that the most important emerging infectious diseases are highly recombinant," he said, "is what leads to the strong hypothesis that it isn't just mutation—that it's the gene swapping—that is a critical feature of emergence." He spoke loosely, well aware that it's portions of genes, not whole genes, that generally get swapped. And the coronaviruses, he knew, are masters of recombination.

"I don't pretend to be a seer," Burke added. "Prediction is too strong a word." He preferred less dramatic terms. "'Improving the scientific basis to improve readiness,' might be a better way of thinking about it."

"Are we doing that? Is that happening?"

This was November 2011, when the George W. Bush administration had given way to the Barack Obama administration, both of which recognized a need for pandemic preparedness. The United States Agency for International Development (USAID) had launched a $200 million project called PREDICT, for the discovery and identification of animal viruses that might endanger humans. The Defense Advanced Research Projects Agency (DARPA) had mounted its own disease program called Prophecy,

devoted to predicting the rate, the direction, and the result of viral mutations. Another federal agency had recently been founded, the Biomedical
Advanced Research and Development Authority (BARDA), by provision
of a law called the Pandemic and All-Hazards Preparedness Act, to work
on developing and stockpiling vaccines, drug therapies, diagnostic tools,
and other measures for dealing with public health emergencies. Scientists
worldwide were also busy with field and laboratory studies of animal viruses that might endanger people. At that moment in political and scientific history, yes, Burke felt that readiness against the threat of pandemic
was "getting a lot better."

He made a distinction, largely lost on me at the time, that in recent
years has become a point of sharp disagreement among disease scientists.
It's a matter of strategy and funding priority toward pandemic threats:
prediction and prevention versus surveillance and response. The PRE-
DICT program, as reflected in its name, embraced the former approach.
Some eminent virologists, including Eddie Holmes, have championed the
latter, arguing that prediction is impossible or impracticable with emerging viruses, and that the money should go into surveillance and response.
(Holmes also disagrees with some of Burke's views on recombination—
it's a big tent.) And predicting which virus might get into humans from
which host is very difficult—not so easy, say, as predicting the path of
an asteroid, spotted by telescope millions of miles away and maybe, or
maybe not, headed right for Earth. Why should disease emergence be so
difficult to predict? Because ecological events such as spillover involve
something infinitely more complex and capricious than asteroid paths as
calculated from Newtonian physics: behavior of living individuals. The
surveillance-and-response approach is reactive, not predictive, but intended to be *quickly* and *forcibly* reactive. It implies networks of trained
people everywhere, in the cities and the boonies, connected at the speed
of email or WeChat with expert virologists, communications centers, public health systems, and international regulatory bodies, so that outbreaks
are spotted early, when they are small, and drastic measures are deployed
promptly to contain them and end them.

"I think that the idea is catching hold that we should move toward predict and prevent," Burke told me back then, "rather than surveillance and
response." Of course, he noted, we would always need both. The challenge

was balance and priority. The challenge was a finitude of money. And then he volunteered something that does indeed, by today's hindsight, make him seem prescient.

"If I were king, I would be investing in coronavirus diagnostics," Burke said. "I would be investing in better studies of vaccines against coronavirus."

10

Ali Khan is another expert who saw the small, dark dot on the horizon. When I first met him, in 2006, he was deputy director of the National Center for Zoonotic, Vector-borne, and Enteric Diseases (NCZVED), which is part of the United States CDC, and therefore he was tasked with dreaming pandemic nightmares in daylight. Khan is a medical doctor by training, like Don Burke, and an epidemiologist by career. He's also a man of candid, irreverent jocularity. He was wearing an epauletted uniform sweater at that first encounter in his CDC office, because he was also a rear admiral in the United States Public Health Service, which is organized into ranks like those in the Navy. But beneath the sweater was no stuffed shirt.

NCZVED (pronounced "NC Zved," like the name of a Russian basketball player) was housed in an unobtrusive gray building, behind locked gates and locked doors in the CDC's compound on Clifton Road, six miles northeast of downtown Atlanta. During a two-day visit that year, I worked my way along the corridors, interviewing scientists who knew all about ebolaviruses (yes, there are more than one) and their lethal cousin Marburg virus; about West Nile virus in the Bronx and Sin Nombre virus in Arizona; about simian foamy virus, which is carried by temple monkeys in Bali that crawl over tourists, and monkeypox, which reached Illinois in giant Gambian rats sold as pets; about Junín virus in Argentina, and Machupo virus, which causes Bolivian hemorrhagic fever; about Lassa virus in West Africa, Nipah virus in Malaysia, Hendra virus in Australia, and rabies everywhere. All these viruses are known or suspected to pass from nonhuman animals to humans. The diseases they cause are zoonoses. Most of them, once in a human body, wreak mayhem. (One exception: simian foamy virus, though vividly named, has never been implicated as an agent of disease.) Some of them also transmit well among people, bursting into local outbreaks that kill hundreds. Each of them, not so long ago, was a "novel virus."

They are still relatively new to science and to human immune systems. They emerge unexpectedly and are difficult to treat. They can be especially dangerous, as reflected in the name of the branch within NCZVED that was charged with studying them: Special Pathogens. For these reasons, some scientists and public health professionals, including Ali Khan, find them an irresistible challenge.

"It's because they keep you on your toes," he told me.

On the second day of my visit, halfway through another long schedule of intriguingly grim briefings, Khan took me to lunch at a sushi place. He surprised me with his jaunty informality. All right, Quammen, he said, you've heard all the talk from our people. Which of these diseases is your favorite?

My *favorite*? Well, Ebola is pretty damn interesting, I told him. It was an obvious answer, a beginner's answer, as though I had been asked to recommend a brilliant but underappreciated author of horror novels, and I said: Stephen King.

Aaah, Khan said dismissively, I like Ebola as much as the next person—(gallows irony: he had done frontline epidemiological work during the 1995 Ebola outbreak in Kikwit, Zaire, organizing control measures, investigating transmission, tracing the outbreak back to its Patient Zero, risking his life to help end a juggernaut of misery and death)—but for my money, he said, SARS was the one.

SARS? I knew of it only as a bad viral disease that, in 2003, came out of southern China and killed people in Toronto and a few other cities. I knew the acronym stood for "severe acute respiratory syndrome," an ugly illness that can include lethal pneumonia. I knew the numbers—about eight thousand infections, about eight hundred deaths—and that the outbreak then, for some reason, came to a halt. The virus disappeared. End of story. Not as lurid as Ebola or as consequential as a pandemic flu. Why SARS? I asked.

Because it was so contagious and so lethal, he said, and we were very lucky to stop it.

This was on our lunch break, I had set my notebook aside, and it was fifteen years ago, so I can't swear that Khan mentioned the other thing most relevant about SARS: that it was caused by a coronavirus.

11

Ali Khan is now dean of the College of Public Health at the University of Nebraska Medical Center, in Omaha. He seems an unlikely Omahan. Born in Brooklyn and raised there, by Pakistani immigrant parents, he went to Brooklyn College, followed by SUNY Downstate (also in Brooklyn) for medical school. "And then I did this crazy thing of leaving Brooklyn," he told me recently. Crazy it seemed to his family, anyway, "because I have uncles and aunts who have never left Brooklyn to go to the city." Brooklyn to Manhattan ("the city") is half an hour by subway.

His father, Gulab Deen Khan, was a self-made man of the epic sort, more adventuresome than the aunts and uncles: as a teenage peasant farmer, Gulab trekked from Kashmir to Bombay (Mumbai), lied about his age, and got work on a ship, greasing engines. His friends called him Dini, as a diminutive, because he was small. After moving to the U.S., Dini Khan stoked coal in boilers to heat apartment buildings in Brooklyn until he had saved enough to buy an apartment building himself. He made money—what seemed a fortune. Before he lost it, in another speculation, he decided that his young son, Ali, should learn about his family's culture, religion, and language. He sent Ali back to Pakistan for middle and high school. By parental miscalculation, Khan the father chose a classic British boarding school in Lahore, a better place for Khan the son to learn cricket than Urdu or Islam. Now in his mid-fifties, Ali Khan told me this story, punctuated with laughs, when I reached him for an update call. His dark hair and beard had grayed a bit, I could see on my monitor, but he looked fit and sounded jovial. He spoke of Omaha like a pitchman for the Chamber of Commerce: great city, safe, unpretentious ethos, full of billionaires such as Warren Buffett, who live in their old family homes, drive their Buicks, and write million-dollar checks to the community.

"I love being a dean," he said. "It's so much fun."

Presumably it's also a little more relaxed than his last role in Atlanta. He went to Omaha in 2014, leaving the job into which he had risen, directorship of the CDC's Office of Public Health Preparedness and Response, which included overseeing the Strategic National Stockpile of emergency medical supplies, supervising eight hundred employees, helping assemble a national biodefense strategy against pandemic threats, and much else. "The end of my career at CDC, I managed a $1.5 billion budget, so it was people and money."

Before ascending to that bureaucratic aerie, he had traveled the world on outbreak responses, from Wyoming to Bangladesh, as what is occasionally called a "disease cowboy." During a mission to southern Chile, investigating a hantavirus outbreak, he visited remote villages, sometimes on horseback, trapping rodents to determine which kind carried the virus. "We learned quickly that there were a lot of rodents," he said. After he worked on Rift Valley fever in Saudi Arabia, in 2001, the Saudi minister of health gave him a Lucite replica of a beheading sword as a token of gratitude. At one dicey moment in central Zaire, during an outbreak of monkeypox, he and his team got word that two sets of combatants in the raging civil war—Laurent Kabila's guerrillas and the opposing forces of President Mobutu—were coming. "They'll likely take your vehicles and gear," an American embassy contact advised by satellite telephone. "But they probably won't kill you." Khan's group packed fast and vamoosed on a small airplane, which rose straight into a thrashing thunderstorm. "The guy to my left was praying," Khan recounted in a book, *The Next Pandemic*, full of colorful field adventures and serious warnings, published in 2016. "I looked over and saw that the French physician sitting next to me was writing a farewell note to his family. Which got me thinking." His thought: this is a risky profession, and the work has to be worth a person's life.

For more than two decades at the CDC, it evidently was. In 1995, he did that hitch in Kikwit, Zaire, on Ebola. The following year, he went to the Sultanate of Oman to help with Crimean-Congo hemorrhagic fever. Uganda in 2001, for Ebola again. Chad was still struggling to eliminate polio in 2008, and Khan went there. Perhaps most consequentially, for his long-term perspective, there came SARS in 2003, during which he served in Singapore.

Toward the end of his tenure with the CDC, though, as a high-level bureaucrat, Khan was responsible for orchestrating, not investigating. Science was a small slice of the job. "Now it's almost all science," he told me happily from Omaha. Virology, epidemiology, ecology, and other aspects of disease science provide the substance of his mission, which is "educating the next generation of public health practitioners."

Curious about his immediate habitat, I asked that he pick up his laptop and walk me around. The eclectic décor of his current office includes electron micrographs of various pathogens hung like portraits in a rogues' gallery, two sculptures of mosquitoes as big as crows, a *Star Wars* clock, a *Big Hero 6* toy robot, cards sent from children all over the world, mementos and gifts from his travels—a Congolese incense burner, the Saudi beheading sword—and a whiteboard on which he records what he calls "my metrics." His precious metrics: measures of progress toward academic goals for his school, scientific goals, philanthropic goals to support the work. "I'm evidence-based and evidence-driven," he said.

I asked him about COVID-19. What went so disastrously wrong? Where was the public health preparedness that he had overseen at the CDC? Why were most countries—and especially the U.S.—so unready? Was it a lack of scientific information, or a lack of money?

"This is about lack of imagination," Khan said. And, of course, imagination is informed by history.

12

The history of SARS-CoV began in late 2002, when an "atypical pneumonia" of unknown origin and unknown causal agent began spreading in cities of the Pearl River Delta of southeastern China, Guangzhou among others, together constituting one of the largest urban agglomerations on the planet. In January 2003, this pneumonia reached a Guangzhou hospital in the body of a portly seafood merchant suffering a respiratory crisis. In that hospital, and then at a respiratory facility to which he was transferred, the fishmonger coughed, gasped, spewed, and sputtered, especially during his intubation, infecting dozens of health care workers. He became known among Guangzhou medical staff as the Poison King. In retrospect, disease scientists applied a different label, calling him a super-spreader.

One infected physician, a nephrologist at the hospital, experienced flu-like symptoms but then, feeling better, took a three-hour bus ride to Hong Kong for his nephew's wedding. Staying in room 911 of the Metropole Hotel, a three-star place in the city's Kowloon district, the doctor became sick again, spreading the disease along the ninth-floor corridor. In the days that followed, other guests from the ninth floor flew home to Singapore and Toronto, taking the disease with them. Cases began to flare in those cities, also in Hanoi, especially among health care workers, which was an alarming indicator that the agent, whatever it was, transmitted well from human to human. On March 12, the WHO issued a global alert about this new, severe, respiratory disease. By March 15, the WHO was reporting 150 new cases worldwide and calling the thing SARS.

Two mysteries loomed, one urgent and one haunting: What was the cause? A new virus? If so, what kind? The first mystery was soon solved by a team led in part by Malik Peiris, a Sri Lankan doctor who had done a PhD in microbiology at Oxford before moving to the University of Hong Kong. Peiris and other members of the team specialized in influenza, and they first suspected that a flu virus might be the causative agent. One worrisome possibility was H5N1, an avian flu, troublesome in birds and often lethal on those rare occasions it gets into a person, but not known as infectious human to human. It had killed a thirty-three-year-old Hong Kong man just a month earlier, after he had picked it up, evidently from direct contact with some bird, possibly a chicken or a duck, during a New Year's visit to the mainland. If H5N1 was the agent now circulating, if it had evolved into a form transmissible among humans, its case fatality rate could be terrible.

One route to identification of the SARS virus entailed culturing it—growing it within some laboratory lineage of cells and seeing it destroy them—but at first the culturing attempt got nowhere. The co-leader of this team with Peiris was K.Y. Yuen, the same fellow who, seventeen

years later, would warn Hong Kong's government of the transmissibility of the new virus, the one causing COVID-19. "We were thinking of
H5N1," Yuen said to a journalist at the time, and accordingly the team
used virus-culturing techniques specific to that virus. "So we failed to
culture the real SARS virus. It was a missed opportunity, and we have
to be honest about it." They hadn't realized that they were looking for
a novel virus, not a known one. The mistake cost weeks, crucial time in
the early phase of an epidemic, but by mid-March they had corrected
that. They found a virus in samples from two patients, sequenced a fragment of its genome from one sample, determined that the thing was a
coronavirus, and with other techniques confirmed its presence in forty-
five other patients, persuasive evidence that it was the agent of SARS.
Although earlier tradition tended toward naming new viruses by geographical association—Ebola was a river in Zaire, Marburg a city in Germany, Nipah a village in Malaysia, Hendra an Australian suburb—greater
sensitivity about stigmatization prevailed. The pathogen became known
as SARS-CoV.

This still left the second mystery: the origin of the virus. Since it was
novel and presumed animal-borne, that meant the identity of the reservoir host, the creature in which it abided before its visitation upon humans. The initial suspect was a creature called the Himalayan palm civet,
a cat-sized omnivore related to mongooses, which has the misfortune of
being highly prized and commercially traded in southern China as food.
The wildlife trade attracted attention because several of the early SARS
cases, in Shenzhen and a nearby city, Zhongshan, had occurred in restaurant workers preparing meals featuring wildlife, including civets. Swab
sampling of various animals caged at a market in Shenzhen found positives for the virus in four civets and one raccoon dog.

That evidence may have been tenuous, but in January 2004, months
after the global SARS epidemic ended, a small second outbreak occurred
in Guangzhou, this one infecting four people with a variant of the virus
distinct from all viruses sequenced during the first round. That implied
another spillover from wild animals, possibly civets or raccoon dogs
again. A new research effort, by scientists from Beijing, Guangzhou, and
the University of Hong Kong, focused on palm civets and raccoon dogs on
sale at the Xinyuan animal market in Guangzhou, a large emporium that

drew farm-raised civets from a dozen different provinces. Xinyuan was a crowded and disorderly place, with multiple animals packed into small cages, stacked atop one another, sharing their fears and their bodily fluids, while hundreds of people worked and lived and ate amid the jumble, toddlers ran back and forth amid offal from butchered animals, families slept in cramped lofts above their shops. In this study, ninety-one palm civets and fifteen raccoon dogs tested positive. The researchers also visited twenty-five source farms and tested another thousand civets, finding no trace of the virus among them. That implied exposure in transit—the civets acquiring their infections somewhere, somehow, along the chain of live-animal supply to urban markets, probably by forced proximity to creatures of some other species.

None of this information saved the civets. Guangdong's provincial government now ordered a generalized cull, for the supposed protection of consumers from the supposed source of the virus, and on the morning of January 6, 2004, animal-control officers wearing masks and smocks turned up to begin seizing civets from vendors. "All of them will be killed today," one official told a reporter for *The New York Times*. They were taken away to be executed by drowning and electrocution. Later research would confirm that palm civets were only an unlucky intermediate host for the virus, infected by some other animal, infecting humans in turn. What animal? The possibilities were broad, sampling was arduous, and to solve that second mystery would take another thirteen years.

13

In the meantime, the initial SARS outbreak went global, in a smallish but alarming way, reaching Toronto on February 23, 2003, carried by a seventy-eight-year-old woman returning from a visit to Hong Kong. She and her husband had spent the final nights of their two-week trip on the ninth floor of the Metropole Hotel, so she had presumably picked up her infection there, from the nephrologist who brought it to town. The woman sickened, then died at home on March 5, attended by family, including one of her sons, who soon showed symptoms himself. After a week of breathing difficulties, he went to an emergency room and there, without isolation, was given medicine through a nebulizer, which turns liquid into mist, pushing it down a patient's throat. "It helps open up your airways," Ali Khan told me—a useful and safe tool to prevent, say, an asthma attack. But with a highly infectious virus, unwise. "When you breathe that back out, essentially you're taking all the virus in your lungs and breathing it back out into the air—in the ER where you're being treated." Two other patients in the ER were infected, one of whom soon went to a coronary-care unit with a heart attack. There he eventually infected eight nurses, one doctor, three other patients, two clerks, his own wife, and two technicians, among others. You could call him a super-spreader. The son's ER visit led to 128 cases among people associated with the hospital. Seventeen of them died.

In Singapore, the first SARS case was also a person who recently visited Hong Kong. Two flight attendants had flown there for a shopping vacation and also stayed in a room on the ninth floor of the Metropole. Returning home, one of them developed signs of illness—fever, impaired breathing—and sought care at Tan Tock Seng Hospital, one of the city's largest facilities. She was admitted to an open ward. Antibiotics didn't work. Several days later, in distress, she was transferred to an intensive care unit. At some point, probably before that transfer, she had visitors, and when several of them came back as patients, doctors suspected something contagious, maybe related to rumors they had heard of a strange pneumonia outbreak in China. Then four nurses from the young woman's ward called in sick on one day, an abnormality noticed by Brenda Ang, a physician who oversaw infection control at the hospital. "That was the defining moment for me," said Ang, a tiny, forthright woman, when I visited her at the hospital six years later. "Everything was accelerating." The defining moment came on Wednesday, March 12, 2003, the same day that the WHO in Geneva issued its global alert about this "atypical pneumonia," henceforth to be known as SARS.

The WHO alert, coming just before Malik Peiris and his colleagues isolated the virus, warned that it didn't seem to be "bird flu" but something else—something unknown, something transmissible human to human, which therefore called for cautious isolation of patients. Singapore's

Ministry of Health formed a SARS Task Force and Tan Tock Seng Hospital set up an Ops Room as nerve center for SARS decision-making.

At about that point Ali Khan arrived in Singapore, serving as a WHO consultant (seconded from the CDC) to help organize an investigation and a response. He met daily with Dr. Suok Kai Chew, chief epidemiologist for the Ministry of Health, and along with others they developed strategy and tactics, getting governmental cooperation through the SARS Task Force. The public health strategy was isolation and quarantine. "Before this outbreak," Khan told me, "quarantine and isolation were not often invoked for infectious disease outbreaks"—at least, not in the recent past. During the medieval plagues in Europe, yes, ships arriving at ports were sometimes required to lay at anchor forty (*quaranta*) days before landing could occur, and the Mediterranean seaport Ragusa (now Dubrovnik) established a *trentino*, a thirty-day separation period for travelers arriving from plague zones. In late-nineteenth- and early-twentieth-century America, during smallpox outbreaks, victims showing pox (especially if they were poor people or people of color) could be confined in quarantine camps, surrounded by high fences of barbed wire, or in nightmarish "pesthouses," not to be treated but for the safety of the general populace. "That was a concept that had sort of gone out of vogue," Khan said dryly. He and Chew and their colleagues revived it in a more humane version.

Tan Tock Seng became the hospital for SARS and SARS alone, with other sick people diverted to Singapore General. Every suspected or probable case of SARS went into isolation at TTS, and the definition of "suspected or probable" was expanded beyond WHO guidelines to include anyone with a fever or respiratory trouble. All health care workers at all institutions suited up with PPE, including N95 masks, of which the use was strictly enforced, and they were required to check themselves for fever or other signs three times a day. Medical staff were also restricted to one institution, so they couldn't carry the virus between hospitals. During risky procedures such as intubating a patient, they wore PAPRs, those respirator helmets that pumped in purified air. Patients with other conditions, after discharge from a hospital, were placed on home quarantine for ten days.

Firm measures were also taken to limit spread in the community. As of March 27, schools closed, and the bodies of those who died of SARS were cremated within twenty-four hours. Investigators traced close contacts of each new SARS patient, also within twenty-four hours, and those contacts were consigned to mandatory self-quarantine. "Okay, you are staying home. There will be a camera we're setting up in your house, and there's a phone," Khan said, recounting the instructions. "We will call you randomly, and you're expected to turn on the camera and be there." Already, more than eight hundred people were quarantined. Flout the home quarantine, and you would be tagged with an electronic tracer, such as an ankle bracelet. But such mandatory quarantine brought logistics challenges, Khan told me. "'The moment you hold 'em, you own 'em,' is what we say." You've got to feed these people, see to their general health care, make sure they are housed and clothed. "Who takes care of them? Who pays for them?" If you're the government ministry, enforcing self-quarantine, you do.

"And Singapore is a very particular kind of place," I said. "I mean, what if you had tried that in Kinshasa?"

"Yeah, no, it wouldn't have worked."

Singapore is orderly. Singapore is rigorous and affluent. By April 24, twenty-two people had died, at which point penalties for quarantine breakers stiffened: bigger fines, the possibility of jail. Taxi drivers had their temperatures checked daily. Passengers arriving at Changi Airport were also screened, as well as people traveling in buses and private automobiles. On May 20, eleven people were fined $300 each for spitting. These measures worked. On July 13, 2003, the last SARS patient walked out of Tan Tock Seng and it was over. Some people loosely say that SARS "burned out," having killed 774 people worldwide. It didn't burn out. As Khan told me, it was stopped.

"What are you most concerned about now?" I asked Brenda Ang, the infection-control officer, during my 2009 visit.

She laughed in frustration. "Complacency," she said. "And apathy." Mundane but crucial infection-control measures—the assiduous handwashing and wiping of doorknobs with alcohol—can lapse after a crisis. "People become complacent. They think there is no new bugs around." And larger lessons, beyond the outbreak locale, beyond Singapore? Beyond this coronavirus and—I might have asked her, if I could have foreseen—applicable to the next one? "There's no point in just protecting your own turf," Ang said. "Infectious diseases are so globalized."

Ali Khan later told me the same thing. "A disease anywhere is a disease everywhere."

14

Another large lesson was not unique to SARS and not new to Ali Khan: the disproportionate importance that a single patient or a single situation can play in transmitting a virus to multiple other people. Stated otherwise: one primary case accounting for many secondary cases. This concept is now familiar, as we hear epidemiologists and public health officials talk of super-spreaders and super-spreading events. It's an old concept, a phenomenon recognized at least since the time of Typhoid Mary, the Irish-born cook named Mary Mallon, who infected fifty-one people with typhoid fever during her career in New York in the early twentieth century, despite not showing signs of illness herself. The term "super-spreader" is more recent and was probably first used, Khan told me, for highly competent transmitters of tuberculosis, such as the homeless man who apparently infected forty-one people at a neighborhood bar in Minneapolis in 1992. To Khan it has been useful ever since he did the contact tracing for the Ebola response team in Kikwit, Zaire, in 1995, as he described in a journal paper published four years later.

Two of the Ebola patients whose contacts he traced, both of whom suffered gastrointestinal hemorrhage, were named by many other patients as among their contacts. That strongly suggested that the two had played some connective role. Those two alone may have accounted for more than fifty transmissions. It could have happened by way of their bloody diarrhea; or not. "The concept of 'super-spreader' or 'high-frequency transmitter,' is novel for this hemorrhagic fever," Khan and his 1999 coauthors wrote, "and the mechanism for this high-frequency transmission is unknown." They weren't inventing the label "super-spreader" or the concept, but they were putting it into prominent use.

Other precedents existed, not just in TB or typhoid but among viral hemorrhagic fevers, and Khan's paper cited them: Lassa fever in Nigeria, in 1970, during which a single person seems to have infected fourteen other people in one ward of a hospital; Bolivian hemorrhagic fever, caused by Machupo virus, spreading in 1971 from one infected traveler to four others in Cochabamba, a city in the Andean highlands, where neither the virus nor its reservoir host (a lowland rodent) is native. There was also evidence that streptococcal bacteria, which cause strep throat and scarlet fever among other ills, transmit far better from people who carry especially high bacterial loads in their noses than from others with only moderate nosefuls, though their bodies may be full of the bug.

With SARS, the significance of super-spreaders became painfully clear during the early weeks in Guangzhou, from the case of the Poison King, and then in Hong Kong, from the visiting nephrologist who occupied room 911 at the Metropole Hotel. Khan and his colleagues saw it in Singapore too. "I was invited to come over and assist them with the investigation," he told me, "and as I got there, things became a lot more specific." He reminded me about the flight attendant who went shopping in Hong Kong. It had puzzled him, he admitted parenthetically: "Why would anybody who lives in Singapore go anywhere else to do some shopping? Because the whole country is a mall, as far as I can tell." Better prices, maybe. Anyway, she and her friend returned, both infected. "What you learn very quickly is, there are these individuals who are just *excellent* at spreading to lots and lots of other individuals." Most primary cases account for zero secondary cases. "Period. It doesn't go anywhere. But it's this small minority of people who are *so good* at transmitting to others." The first flight attendant was named Esther Mok. Her infection passed to her mother, her father, her maternal grandmother, her uncle, and the pastor of her church (who had visited her to pray), all of whom became patients at Tan Tock Seng and all of whom, except the grandmother, died. Esther Mok, unaware and blameless, also infected the four nurses whose sickouts caught the attention of Brenda Ang. Mok herself survived.

In fairness to such patients, though, Khan noted an additional factor: ecology. What he meant is that circumstances and the nature of interactions, as well as sheer biology, play a role when such broadcast transmissions occur. High on the list of dangerous situations is hospitalization of a severely infected person who is not recognized to be contagious. High among dangerous interactions, for the health care workers involved, are intubating a patient, especially one suffering a respiratory crisis, or giving medicine through a nebulizer. Therefore the Poison King in Guangzhou, and the son of the elderly woman who brought SARS home from Hong Kong to Toronto, might more generously be considered, not super-spreaders, but central figures in super-spreading events. Typhoid Mary concealed her condition through multiple jobs and name changes, but these unlucky people did not. Another dangerous circumstance, Khan added, might be sheer popularity. Esther Mok had a lot of visitors.

Khan had been urged to see this distinction, super-spreading events versus super-spreader people, by a CDC colleague named Peter Kilmarx, an infectious disease physician who was among Khan's colleagues on the Ebola response team in Kikwit in 1995. "Peter is very kind," Khan told me. Kilmarx was sensitive to the unfairness of stigmatizing anyone based on uncertain knowledge of what has happened, in the Poison King's hospital room or any other such exigent situation. "Is it a function of the person? Is it a function of the environment?" Khan asked himself and now me. "Is it a function of the virus?" Certainty is unattainable. The person could be tested, to gauge how much virus is loaded into the upper respiratory tract, ready to be spewed, versus how much is causing distress in the lower. The environment could be examined, machine by machine, surface by surface. But there would always be more to know about the virus. Any virus.

Still, 2003 was just a rehearsal. "We dodged a bullet on SARS," Don Burke told me. Super-spreading events pumped up the toll of misery, yes, elevating case numbers, increasing deaths, but the whole thing could have been far worse. If the virus had been just a little more transmissible generally, among all patients and situations, he said, "it could have been a huge problem." But that SARS-CoV virus had one feature, or the absence of a feature, standing between it and a global nightmare in 2003. "Which was, for the most part, asymptomatic people didn't transmit until they were sick. So you had time." You could identify cases, trace contacts, and quarantine. It could be stopped, for those reasons, and it was. If the virus had been a little different, "highly transmissible, with more variable disease manifestation, harder to figure out who were silent carriers, then we may never have been able to contain SARS."

He said this during our 2011 conversation, not the recent one. Was it prescience, modeling, or another lucky guess?

15

In 2012, a different coronavirus emerged on the Arabian Peninsula, serving notice that SARS had not been an anomalous event. The first recognized case was a sixty-year-old Saudi man who sought help at a private hospital in Jeddah on June 13. He was feverish, coughing and spitting, and couldn't breathe well. His chest X-rays didn't look good. Next day he was transferred to an ICU and intubated to help get him some oxygen. Samples of blood and phlegm were taken. His blood tested negative for bacteria; his phlegm tested negative for H1N1, the subtype of influenza virus that had circled the world in 2009. Other tests, starting on day three, showed his kidneys in decline. Eight days later, suffering respiratory and renal failure, the patient died. The hospital ordered no postmortem exam, which implied that his cause of death was considered to be known. But it wasn't known.

One doctor at the hospital, an Egyptian virologist named Ali Mohamed Zaki, remained interested in the case. Samples and results were turned over to the Saudi Ministry of Health, as required by law, but Zaki retained enough material to do further tests. He suspected that a virus had killed the man, maybe some sort of paramyxovirus. Paramyxoviruses are members of the family of RNA viruses that cause measles, mumps, and a list of other diseases associated with bronchitis and pneumonia; they stand high on the watch list, along with influenza viruses and coronaviruses. But no, the man's tests for paramyxoviruses came up negative. Zaki then thought about coronaviruses, because of SARS. Five coronaviruses were then known to infect humans, and four of those caused only mild, coldlike symptoms. The fifth was SARS and maybe this man had died of it—or of some other coronavirus with SARS-like lethality. In his own lab, Zaki managed to grow a virus from the sputum. He also contacted a lab in the Netherlands, arranged to send samples, and collaborated with those scientists to identify a novel coronavirus. Zaki promptly notified ProMED, even before publishing his discovery with the Dutch coauthors. In New York, Marjorie Pollack, the same ProMED deputy editor who would alert the world to COVID-19, posted Zaki's report on September 20, 2012.

Soon it became clear that the sixty-year-old man, the first case, wasn't the last. Three days later, Pollack posted another report, this one about a forty-nine-year-old Qatari citizen, in critical condition at a hospital in London, who had been brought up from Qatar by air ambulance, placed in an ICU, and who tested positive for the new virus. He had at least one thing in common with the first man: he had recently been in Saudi Arabia. Shortly after the Qatari post, Pollack heard from a ProMED subscriber working with International SOS, a health and risk management company, who recalled reading of a similar mysterious outbreak of severe respiratory illness that had occurred five months earlier, in an ICU in Jordan, affecting eleven people, with two deaths. Specimens from those two cases had reportedly tested positive for the new virus—but confirmation of the results only arrived weeks later. In the meantime, five more cases came to light, for a total of nine victims of the new virus, the new syndrome, by the end of November 2012, and neither the virus nor the syndrome yet had a name.

Marjorie Pollack herself promptly published a paper, with three coauthors, collating these case details and noting that, at present, there were more questions about the new virus than answers. What was it about these affected countries? Was there a travel history connecting the Jordan cases to Saudi Arabia? Why were the victims predominantly male?

And, after noting that five of the nine confirmed cases had a history of recent exposure to animals, Pollack and her coauthors asked: "What were the animals?"

16

Jon Epstein, a veterinarian and ecologist employed by EcoHealth Alliance, was at home in Queens for the weekend in late October 2012 when the call came. This was just after the first crackle of reports on the new virus, in the Middle East, had been heard by Marjorie Pollack and relatively few others. On the phone was Ian Lipkin, director of the Center for Infection and Immunity within the Mailman School of Public Health, at Columbia University. Lipkin is a smart and high-profile molecular biologist who describes the gist of his work as "pathogen discovery." Epstein and he had teamed frequently on the identification of new viruses, especially from bats, Epstein's specialty. To put their collaborations in simplest terms: Epstein goes to faraway places, climbs through caves and across rooftops, catches bats big and small, handles them like any gentle veterinarian would handle a kitten, and takes samples from them, mainly blood and saliva and fecal swabs; Lipkin detects and identifies viruses in the samples. "I'll never forget the day," Epstein told me. "It was kind of the last thing I ever expected to come out of his mouth."

"What are you doing tomorrow?" Lipkin asked.

"I don't know. Why?"

"We're getting on a plane to Saudi Arabia."

Epstein travels the world for his work, usually not on such short notice, but Lipkin was in a hurry. He had heard from the Saudi Ministry of Health, they wanted his help, and Lipkin wanted Epstein's, because he suspected that this latest novel coronavirus might be associated with bats.

Why bats? Because there was already a pattern of dangerous new viruses emerging from bats: Hendra virus in Australia, 1994; Nipah virus in Malaysia, 1998; Marburg virus in Uganda, traced to bats in 2009. Ebola virus, Marburg's more notorious cousin, was also thought to reside in bats, though definitive proof hadn't then (and still hasn't) been found. Rabies, a dangerous old virus, came from bats. And the SARS virus, another coronavirus, was linked fairly persuasively to bats, based on work done by a team including Epstein. Lipkin now wanted Epstein to help explore the bat-reservoir hypothesis for this new thing found in Saudi Arabia. Epstein could set up a field program to sample bats around the home of the index case, that sixty-year-old man who died at the hospital in Jeddah.

"What does it take," I asked Epstein, "to get into Saudi Arabia on twenty-four hours' notice?"

He laughed. It's doable, he said, if you have the Ministry of Health urgently inviting you and Columbia University organizing the visas. They flew into Riyadh, met with ministry officials, then connected to Jeddah, gateway city to Mecca, and proceeded by car, about six hours southeast, to a town called Bisha, where the index patient had lived. The team now included Kevin Olival, a colleague of Epstein's from EcoHealth Alliance with similar bat-handling experience, plus a technician from Lipkin's lab, who would care for the samples, and a veterinarian from the ministry named Shamsudeen Fagbo, who would help with the animal sampling, a delicate task, and serve as a cultural liaison. In Bisha, which is the hub of an agricultural area with good soil and water and famed for its fruit-producing palm trees, they located the home of the index patient, who had been a businessman there. Actually, Epstein corrected himself, they identified three different residences the man owned and variously occupied, sharing with family members in each, and the team checked those houses for evidence of bat infestation. They also met with the man's brothers. They saw camels and sheep on one of the properties.

"Camels," I said. "But at that point camels were just another animal, right?"

"Totally." He and Lipkin had discussed the possibility of livestock as an intermediate host of the virus, the direct link to humans, with SARS in mind, because of the way it had passed through civets. "We just didn't know. So everything was on the table. But we were focused on bats." They took samples from the sheep and the camels just in case. They also visited a hardware store the man had owned, and they found it fronting a garden and a date palm orchard. Epstein was well attuned to the possible linkage from date palms to fruit bats to viruses, because he had researched Nipah virus in Bangladesh, where giant fruit bats of the genus *Pteropus*, known as flying foxes, carry the virus. Date palm trees in Bangladesh don't produce edible fruit but are tapped for their sweet sap, the same way Vermonters tap maple trees, and there's a street trade in fresh sap, which people drink raw. Bats come for the sap too, and lap it up from the sap-draining cuts in the trees made by tappers. But the bats excrete virus with their feces and urine, and when those excretions fall into the little clay pots that the tappers hang beneath the cuts on the palms, the virus in fresh sap can infect humans who drink it. Detecting such linkages, and trying to interrupt them by alerting people to the dangers, is the sort of work that Epstein and EcoHealth do. But in and around Bisha he saw no evidence of any such bat-human interactions. In fact, he could find scarcely any bats at all.

It was a desert landscape, garnished only with palm groves and tilted here and there into rocky hills, not the sort of tropical forest more familiar to him. So the team began asking local people, with Fagbo as interpreter: Do you have any bats around here? The Arabic word for bats, to Epstein's recollection, sounds like "huffa-fish." So here were five men, including Epstein—who is tall and short-haired, and in khakis as he was then could pass for a major in the U.S. Marines—roaming around Bisha, showing people pictures of the bats and asking for leads. Epstein even climbed to some overlooks in the hills, scanning with his binoculars to try to spot bats flying up at dusk. Fagbo warned him: "Don't point those binoculars towards anyone's house because they will immediately assume, you know, you're looking at their family." Finally, a man in a vehicle said, Yes, come, I'll take you to them.

They jumped in the man's car, almost without thinking. Later it seemed injudicious. "If there's ever a textbook definition of what not to do when you're in another country," Epstein told me, chuckling at his lapse in caution, "never having been there before, not speaking the language—you don't get into a car with a total stranger who says, 'Yes, I'll show you where the bats are. They're just outside of town.'" Especially you don't jump into that car if you look like a Marine officer doing recon. The man drove them six miles into the desert to a derelict town, a cluster of ancient buildings, crumbling at the edges, and some of them (Epstein heard later) almost a millennium old: the Bisha ruins. In a subterranean room of one building, they found hundreds of bats, roosting sedately. "I immediately felt like I was at an archaeological site," Epstein said. "It was incredible."

Amid the desert heat, he and Olival suited up into full PPE, meaning all-body Tyvek suits, respirator helmets, boots, and gloves. Then they climbed down into the low-ceilinged room. It was sweltering, but the respirators would protect them, they hoped, in case the little known new virus was aerosolized, afloat in the acrid air of the chamber. A swirl of bats, now disturbed, flew around their heads. These weren't the big flying foxes that might feed among date palms; these were small, insectivorous bats, mostly of a group known as mouse-tailed bats, each with slender limbs, short fingers, and a long, thin tail. Mouse-tailed bats, native to dry regions of Africa, the Middle East, and southern Asia, roost in caves, crevices, along rock walls, and in tombs, including the Egyptian pyramids. Epstein and Olival laid out plastic tarps on the floor of the low room, to catch feces that fell, and at the single doorway they set up a harp trap, a standard bat-catching tool that entangles the animals gently, in its strainer of thin vertical lines, and drops them into a cloth sack. On that first night, the team caught thirty or forty bats, sampled them right there in a mobile lab setup—taking blood, fecal swabs, and throat swabs—then released them.

It was a good night's haul for Epstein's crew, and they hadn't been kidnapped, but they had no way of knowing at that stage whether they had collected anything besides bat shit and saliva and blood. The samples would be analyzed at Lipkin's laboratory back in New York, because the Saudi Ministry of Health didn't yet have the capacity to screen for this novel coronavirus. Lipkin himself, who is a laboratory leader and a scientific diplomat, not a crawler through caves and fetid basements, had made that arrangement during meetings in Riyadh, the capital, after which he flew home.

Epstein and his field team stayed three weeks, trapping at the ruins and amid some derelict buildings back in present-day Bisha. They captured and sampled, in total, ninety-six bats. Roughly a third of those were mouse-tailed bats, with another third of a kind vividly known as the Egyptian tomb bat. They also collected hundreds of samples of fecal droppings from the plastic tarps. In all, they had more than a thousand discrete samples. The samples, in labeled vials, were frozen in liquid nitrogen and, after clearing formalities in Riyadh, flown back to New York.

On arrival, the container was opened by U.S. customs inspectors and, unfortunately for the study, sat neglected at room temperature for forty-eight hours before being released to Lipkin's lab at Columbia. The samples thawed, compromising the effort to extract RNA (which can degrade quickly) and sequence it, but not ruining that possibility entirely. Among the samples, Lipkin's group detected more than two hundred fragments of various coronavirus genomes, which could be tentatively identified by matchup with already known coronaviruses, mostly from other bats, but also a few canine coronaviruses that the Bisha bats had somehow picked up, and that memorable agent of swine infection, porcine epidemic diarrhea virus. The lesson there is that viruses are restless creatures, going wherever opportunity allows—and coronaviruses, with their affinity for terrestrial mammals, maybe especially so. One sample among the thousand collected around Bisha rang up positive for the new virus the team was seeking. That sample came from the rectum of an Egyptian tomb bat caught at the Bisha ruins. It was a short fragment of RNA, only 190 bases, but it matched perfectly the equivalent sequence in an important gene, for making a crucial enzyme, within the genome of the virus taken from the Bisha man who had died.

By the time Lipkin and his group published their findings, with coauthorship by Epstein's field team and some members of the Saudi Ministry of Health, this novel virus had an official name, MERS-CoV. The disease is MERS, second of three dangerous coronavirus afflictions to emerge among humans, so far, in the twenty-first century.

17

Between the death of the Bisha man and January 2014, according to a WHO update at the time, MERS struck 178 people and killed seventy-six of them. Those cases occurred mostly on the Arabian Peninsula or in travelers coming from there. Some involved human-to-human transmission of the virus, including an outbreak among nine patients undergoing hemodialysis at a single hospital in eastern Saudi Arabia. For all other cases, the source of the virus remained uncertain, but a clue soon came from several studies that found antibodies against MERS-CoV in a kind of domesticated animal that hadn't previously been linked to zoonoses: the dromedary. That is, the one-humped, Arabian camel.

One of those studies, by a team of Dutch scientists with international colleagues, tested blood sera from various domestic livestock—cattle, sheep, goats, camels—on the Arabian Peninsula, the Canary Islands, and elsewhere, and detected antibodies specific to MERS-CoV only in camels. This team had the opportunity and wit to exploit a convenient circumstance in the Sultanate of Oman: they found a group of female dromedaries, retired racing camels belonging to different owners, that had all been put to breeding. Because of concern for their pregnancies, these cow camels were blood-tested routinely for brucellosis, an infectious bacterial disease that can cause cattle, sheep, and some other animals to abort. Camels are very susceptible, ergo the brucellosis screening. But if you tap the jugular vein of a dromedary, you get blood enough for more than one test. The Dutch-led team obtained blood sera from fifty of these retired Omani racers and bingo, all fifty tested strongly positive for the presence of antibodies against MERS-CoV.

 Camels were suddenly a thing, for virus hunters, and after that study came others, including a follow-up effort in Saudi Arabia organized by Ian Lipkin. This work, led in the field by a young Saudi scientist named Abdulaziz N. Alagaili, found serological evidence of infection by MERS-CoV or a MERS-like coronavirus, or at least exposure to such a virus, widespread among dromedaries at more than a half dozen locations across the Kingdom of Saudi Arabia. Serological evidence: that is, antibodies. Among more than two hundred camels, almost 75 percent tested positive. Alagaili's team even detected antibodies in archived serum samples dating back to 1992. This suggested that MERS-CoV may have been circulating in camels for two decades before the first recognized case in humans. Whether that implicated camels as an intermediate host from whom the virus passed to humans—or, as an alternate possibility, that humans with unrecognized infections had passed the virus to camels—was a separate question.

 Another study team included researchers from the University of Hong Kong, several of whom had worked closely on the original SARS outbreak. They found their camels in a special circumstance: awaiting slaughter for meat at abattoirs in Egypt. The researchers took nasal swab samples from 110 of these apparently healthy but doomed dromedaries, and found not just antibodies but, in four of them, fragments of MERS-CoV RNA. From one of those four, especially promising, they assembled fragments into a nearly complete genome, 99 percent identical to the virus that killed the Bisha man. This supported the hypothesis that, although the MERS coronavirus probably originated in bats, it was more likely infecting humans by way of camels.

 In 2015, a very similar strain of the virus arrived in South Korea, in the body of a sixty-eight-year-old man returning from business on the Arabian Peninsula. By that time, MERS had been nicknamed "camel flu," though it wasn't an influenza. No one knows whether the Korean businessman was sneezed on by a camel sometime during his stops in Bahrain, Qatar,

Saudi Arabia, and the United Arab Emirates, or whether he picked up his infection from a person, but that question probably mattered little to the 186 South Koreans who became infected from him, directly or indirectly, and still less to the thirty-eight who died.

Super-spreading events drove this outbreak, as they had driven SARS, but it was exacerbated by aspects of South Korea's health care system. Because citizens receive cheap medical care through a national insurance plan, with few restrictions on which hospital they can visit, people often shop for treatment. The businessman visited three different hospitals after he felt sick, and was finally admitted to a fourth, in Seoul, where he was given a diagnosis of MERS. Part of what delayed the diagnosis was that he hadn't at first reported his recent travel history in the Middle East. Eventually he infected almost forty people, of whom two became super-spreaders themselves, accounting for 106 more cases. There were sometimes four or more beds in a hospital room, and patients were allowed to receive visitors, which contributed to the spread, as did poor ventilation, poor infection control, and narrow criteria for quarantine, so that people who picked up the infection through casual contact were missed.

"They recognized at that point," Ali Khan told me, "what happens with a coronavirus that causes health-care-acquired infections within your community and hopscotches from hospital to hospital." MERS in South Korea became a textbook example of blunders that can lead to "nosocomial spread," the term for disease transmission occurring *because of*, rather than in spite of, health care circumstances. When COVID-19 arrived five years later, Khan said, "I guess maybe it was raw for them."

South Korea reacted fast to the news out of China, on January 3, 2020, with screening and quarantine measures for travelers from Wuhan. Thanks to those measures, the country's first case was detected on January 20, 2020, in a woman from Wuhan disembarking at Incheon International Airport. The government raised its crisis-management alert level from Blue to Yellow, and then a week later from Yellow to Orange. Also, on January 27, health officials summoned representatives from twenty medical companies to meet at a train station in Seoul and discuss creation of the tools for response, including privately developed diagnostic tests, with guarantees that those tests would get fast regulatory approval. Officials of the Republic of Korea took this outbreak seriously from the start. Their response was strongly in contrast to what happened and didn't happen, for instance, in the United States, which had its first detected case on January 19, one day before South Korea's.

That first U.S. case was a man who had flown home to Seattle from a family visit in Wuhan, then turned up at an urgent-care clinic in Snohomish, Washington, with symptoms. His swab samples went overnight to the CDC, tested positive there, and by January 21 his blood had been sampled too. "Every day after January 22 was a day missed—by the U.S. government," Khan told me with some frustration. January 22 was a Wednesday. Leaders of America's health agencies could have called in Becton, Dickinson, Khan added (referring to the giant multinational medical-technology company, headquartered in New Jersey), and told them: We want nationwide testing capacity ready by next week. Didn't happen. Failure of imagination. "The lack of testing defined the rest of the outbreak for us." South Korea could imagine the worst, and take prompt action, because they remembered SARS and MERS.

"South Korea is a good example for us to look at," Khan said. This conversation occurred back in March 2020, just as that country was coming out of its first wave, when its case count stood at just over six thousand and it had endured fewer than one hundred deaths, among a population of almost 52 million, for a deaths-per-population rate roughly one thousandth that of the U.S. The lockdown fatigue in South Korea, the second and third and fourth waves, and the variants were still to come. But at least they had started well, their initial pandemic response saving misery and lives. "They took a very different approach," Khan said, "and all we had to do was look at what they were doing and say, 'We're going to do the same thing.' But we did not." Scientists could describe the risks, public health officials could chart a response, but agency bureaucrats and national leadership failed to imagine how bad the outbreak, becoming a pandemic, could be. Ten days after my chat with Khan about the warnings, Donald Trump said on television, "Nobody had any idea."

18

SARS touched the United States only gently in 2003, producing twenty-seven probable cases, no super-spreading events, and no deaths, its minimal impact most likely accountable to dumb luck. Recalling that, Ali Kahn repeated the metaphor I'd heard from Don Burke. "We dodged the bullet." But he continued: "We dodged the bullet with SARS, and that may have been a bad thing, in the end, because I think we may have been better prepared if we had not dodged the bullet." Canada remembered SARS, he said, because of the deaths in Toronto, and in consequence they had taken COVID-19 more seriously, from day one, than the U.S. did.

South Korea likewise remembered MERS, and their costly, mishandled experience of 2015, another lesson with no parallel in the U.S. MERS had even less impact than SARS on American lives and awareness: two cases during 2014, both in health care workers returning from stints in Saudi Arabia, with no secondary spread among their families or other contacts. The CDC noticed, but scarcely anyone else did.

Other warnings about the potential dangers of single-stranded RNA viruses came from other scientists and other events around the world. Some people heard those warnings, some governments absorbed their lessons, and many others didn't. Influenzas are always dangerous, and the best influenza researchers live their lives with one eye open for global disaster. In 1997, as I've mentioned, a highly virulent form of avian flu struck Hong Kong. It passed from birds into people and caused eighteen cases, of which six were fatal. The government reacted strongly, a million and a half chickens were culled, and the trade in live poultry (very important in Hong Kong, where not everyone has a freezer) shut down for seven weeks. Then poultry sales reopened, but within several years the vaccination of chickens in Hong Kong became mandatory.

The flu virus that circled the globe in 2009, having evolved within pigs in central Mexico, drew special concern because it was an H1N1 virus, same subtype as the virus of the 1918–1919 pandemic. Ebola virus is ever a favorite among people eager to worry about dramatic infections. So the Ebola epidemic of 2013–2016, which caused terrible misery and eleven thousand deaths in three small West African countries, also provoked fear and xenophobia in America and elsewhere, not in disease scientists but among ordinary folk and reckless public commentators who didn't pause to absorb the fact that Ebola virus, though terrifically virulent, is not highly transmissible. It travels in liquids, such as blood and diarrhea, not in gases, floating through the air on a breath. And then there's Zika virus. Before 2015, scarcely anyone except virologists had heard of this one, while it worked its way slowly around the world in dispersing mosquitoes and the infected people they bit, until it began causing clusters of birth defects in Brazil. Zika never evoked the same degree of global anxiety as some other viruses, because those acutely at risk from its 2015 surge were a limited sector of the populace—mainly young, pregnant women who lived in or traveled to the neotropics. Throughout these minatory events, the experts largely embraced the advice that Don Burke had offered back in 1997: beware the influenzas, the coronaviruses, and any other RNA virus that evolves fast and comes out of a bird or a nonhuman mammal.

SARS had provided the most relevant predictor of the pandemic future, and one scientist who heeded that clue was a Chinese virologist named Zhengli Shi, at the Wuhan Institute of Virology (WIV). Her decades of work on coronaviruses and their reservoir hosts began haphazardly, during her university years, eventually bringing her both to international renown within the field and then into a very harsh spotlight during COVID-19.

Zhengli Shi was born in a village in Henan province in 1964, the daughter of farmers, an unpromising start with one notable advantage: her father could read. That led him to a work opportunity in building hydroelectric plants and led his family to urban life in a county town. Shi's two elder brothers received secondary educations, but the older one died young, at age twenty-one, and though the second brother aspired to university, he failed the examination. When her turn came, she passed. She went off to Wuhan University, three hundred miles south in the next province, and did a degree in genetics. Rather than returning to Henan after graduation, she stayed in Wuhan because of a boyfriend, and on short notice she decided to try to qualify for graduate study. "I was in a rush to prepare," Shi told me, "so I looked for which institute, or university, the examination probably is a little bit easier for me." She laughed—a laugh at youthful groping and the contingency of life, it seemed. "Then I decided to apply for the entrance examination at the Wuhan Institute of Virology." She got in and did her master's degree there, on a virus that infects insects of agricultural concern to tea farmers, the sort of topic that would be assigned by a pragmatic mentor. Maybe this virus—so went the hopeful logic—could be used to control the insect and save the tea trees. That degree took three years, and it was enough to earn her a job as a research assistant at the institute.

Within a few more years she was promoted to research scientist, and then worked on another virus of economic concern, the one responsible for a disease called white spot syndrome, an affliction of farmed shrimp. Known for convenience as WSSV (white spot syndrome virus), it is highly contagious and virulent, capable of killing every shrimp in a pond within ten days. Besides raising ominous white spots all over the animal's carapace, the virus attacks gills, glands, nerve tissue, bone marrow, intestinal lining, and other parts of the body, causing the cells to die and disintegrate. If you think bubonic plague and Ebola virus are scary, be glad you're not a shrimp. WSSV was a novel virus in the early 1990s as far as shrimp farmers were concerned, having recently appeared in Taiwan and then spread somehow to shrimperies on the mainland, where it nearly destroyed the industry. It was so novel, in fact, that viral taxonomists created a new family for it.

Zhengli Shi became a WSSV maven and even stayed with the topic when the Chinese government offered her, as they were doing with other promising young scientists—in an anxious effort to catch up with the West, paralleling America's anxious reaction to Sputnik—a chance to study abroad for a doctorate. She chose the University of Montpellier, in France, because a French scientist there worked on white spot syndrome. Her dissertation project involved sequencing genes of WSSV. She returned to the Wuhan Institute of Virology with higher status, as a senior scientist running her own lab, and could undoubtedly have made a long, stable career from investigations of white spot syndrome, perhaps finding a fix for the problem and gaining the quiet gratitude of shrimp farmers everywhere. But then came SARS, an alarming event for China in both epidemiological and economic terms. By one estimate, it cost China's economy $25 billion, mostly in lost tourism.

She wasn't part of the 2003 SARS response. She got involved in 2004, after a WHO team visited China and established plans with Chinese scientists to search for the origin of the virus. Civets had been identified as playing some role, at least passingly, spilling the virus into humans somewhere amid the markets or restaurants of Shenzhen and maybe other Guangdong cities. But genetic evidence suggested that the civets had been intermediates, not long-term reservoir hosts of the virus. Maybe they were even necessary intermediates—amplifier hosts, in which the virus replicated rampantly and built up to huge loads, capable of meeting the threshold for an infective dose in humans, or else perhaps transitional hosts, in which evolutionary changes prepared the virus for human infection. If so, what animal *was* the reservoir host?

Among the WHO visitors in 2003 were Linfa Wang, a Shanghai-born molecular biologist with a California PhD, a specialist in viruses causing zoonotic diseases, who now worked out of a big, high-containment, animal-health laboratory in Australia, and Hume Field, a veterinarian, environmental scientist, and epidemiologist from Australia's Department of Agriculture, Fisheries, and Forestry (DAFF). Wang was deeply versed in the genetic details of how viruses interact with the cells of their hosts. Field had played a large role in solving the mystery of the reservoir host and transmission of Hendra, a virus carried by Australian fruit bats, devastating to horses and sometimes passed from a dying horse to a trainer or veterinarian laboring to save it. The two of them had worked together for much of a decade on both Hendra virus in Australia and Nipah virus in Malaysia, Field the field man and Wang in the lab, in collaboration also with Peter Daszak of the Consortium for Conservation Medicine (his organization before its name changed to EcoHealth Alliance). On the China side of the collaboration was Shuyi Zhang, a Beijing-based zoologist and expert on Chinese bats. It was through these scientists, seeking a virologist to help them, that Zhengli Shi entered the realm of bats and the viruses they carry.

In March 2004 an international team assembled to begin field research. The group included Jon Epstein and his boss, Peter Daszak, as well as Craig Smith, another Australian from DAFF, standing in for Hume Field, who was recovering from surgery. They set off to catch bats and take samples, in search of a virus resembling SARS-CoV, but with a concomitant purpose: for the visitors to train their Chinese colleagues, neophytes at such work, in the capture and sampling methods. They started with fruit bats that roost in the caves of two southern provinces, Guangdong and Guangxi, because fruit bats are known to carry Nipah virus as well as Hendra virus, and so perhaps, it was thought, coronaviruses too.

Zhengli Shi recalled it distantly, the way you might recall learning to ride a bicycle. "My first attempt was in Guangxi," she told me. "The first cave I visited." Was it strange? Was it exciting? "I think I felt a little bit scared." She had never touched a bat. She and the others wore masks and gloves, not full PPE, which was standard precautionary procedure for bat sampling at the time, and considered adequate protection. Another reason was that this was a busy tourist cave, not far from the big city, and someone had noted that scientists in full protective regalia might scare the tourists. (It's an absurdity that recurs in the annals of bat research, in Africa and elsewhere as well as in China: researchers in protective gear capturing animals amid the same caves through which tourists stroll in T-shirts and flipflops, or guano miners in sweaty work clothes haul their loads. To approximate the sense of absurdity, you might picture yourself wearing an N95 mask into a restaurant crowded with blithe vaccine-refusers and Covid-is-a-hoaxers during the early months of 2021.) The team caught bats inside the cave with fixed nets and also outside, as the animals flew out for a night's foraging, then swab-sampled them, took blood, and released them. Handling a fruit bat can be difficult because some of them are large and strong, they have sharp teeth and big claws, and in their understandable fervor to escape, they will climb up your arm to your face if you hold them the wrong way. Now imagine extracting a few drops of blood from a very small vein along the animal's arm, or in the membrane that attaches to the leg. It's a two-person job with a steepish learning curve, perilous for both the bats and the people. If you get bit, you'll want a rabies booster pronto.

"But after the second time, and then later on," Shi said, "I didn't feel any scared when I handled bats. I even found some bats beautiful." She amended that: "I think *most* of bats are beautiful." It left me wondering what singular form of bat visage might be required for this woman, now after almost two decades of chiropteran virology, to find the creature *not* beautiful.

But they had no luck, not from the first cave or the second and in fact not after repeated field sampling and laboratory screening spread over nine months. Plenty of things were afloat in the blood and saliva of these animals, plenty embedded in their feces, but not any detectable evidence of the SARS virus. "We found we had a wrong direction," she said. They were sampling from the wrong kind of bats—an understandable mistake, given that such fruit bats were sometimes sold in wet markets, including markets where SARS-CoV had been found—and they were using the wrong method to screen for virus. Wrong method: they were looking for fragments of RNA that were very specific to SARS-CoV, with a method known as RT-PCR (reverse transcription polymerase chain reaction, but let's skip that explanation and move on), which might have enabled them to convert the viral RNA into its DNA equivalent (for stability), then amplify those small traces of DNA (using PCR) into workable amounts and sequence what was there. That approach may have failed because RNA fragments can disappear quickly in a host, as the viral presence rises and falls; or it may have failed because their molecular probes were *too* specific to the SARS virus known in humans, as distinct from its progenitor in bats; or maybe both. "We found nothing positive," Shi told me. It was a low point, after eight months of work. "We needed to make a decision if we will continue, or we stop there." Her decision was to labor on, with a change in method. They would try a different sort of test, known as ELISA, which detects antibodies that tend to linger in a host, rather than RNA fragments that tend to disappear. If the antibody test still returned negatives, Shi said, "probably we would give up the study."

Their other mistake: they had been sampling almost entirely from fruit bats. With the new method, they targeted smaller insectivorous bats, including a diverse group known as horseshoe bats. Like other insectivorous bats, the horseshoe bats use echolocation to home in on prey, and their common name comes from the large, fleshy, horseshoe-shaped structures around their nostrils, which seem to help them focus their echolocating squeaks. These adjustments brought success, and by the end of 2004 Zhengli Shi and her colleagues had found antibodies to viruses very much like the SARS virus in horseshoe bats of three different species. They cross-tested their positive samples using the PCR method and got positives from that test too. Although they didn't manage to grow any live virus from their samples, the antibody and PCR results gave strong confidence that SARS-like coronaviruses reside in horseshoe bats. From the fecal sample of one bat, which contained a special richness of coronavirus RNA, they assembled a complete genome. They labeled that dab of guano Rp3, signaling that it was their third sample from a Pearson's horseshoe bat (*Rhinolophus pearsonii*). The genome was 92 percent identical to SARS virus as sampled from a patient in Toronto.

It was a notable discovery, notable enough for publication in a major journal, but Shi and her colleagues didn't make it alone. "In fact, there were two teams," she said. The other was from Hong Kong and included again K.Y. Yuen, the outspoken microbiologist who reappears at many interesting points of this story, again in a senior role.

Earlier I had asked Yuen too about the convergence on this important discovery: Were you working together, the two teams, or competing, or just entirely independent? "Independent. Independent," he insisted. "I don't see her as a competitor at all." His team operated unaware of what Shi and her colleagues were doing, and Shi's group were likewise unaware about the Hong Kongers. Shi and Epstein and their colleagues sampled horseshoe bats in Guangxi and Hubei provinces, while Yuen's team collected their samples—also from horseshoe bats—in the New Territories of greater Hong Kong. Yuen's team published their paper in *Proceedings of the National Academy of Sciences*, a highly respected U.S.-based journal, in September 2005, and Shi's group published theirs in *Science* a month later, with Epstein, Daszak, Hume Field, Craig Smith, and Shuyi Zhang among the coauthors, Linfa Wang listed last, the position of senior author. The title of the *Science* paper, "Bats Are Natural Reservoirs of SARS-Like Coronaviruses," could as easily have been the title of the other.

This near-simultaneous publication of such similar work in two major journals suggested three things: the results were important; the solution to the reservoir host mystery was eagerly sought by multiple scientists; and SARS-like coronaviruses lurked within horseshoe bats across a large swath of southeastern China.

19

Z hengli Shi's work on coronaviruses had only begun. Within a year she had assembled two more complete genomes, based on samples from members of two other horseshoe bat species, both genomes again about 90 percent identical to the SARS virus. That degree of similarity was enough to suggest a shared ancestor in the relatively recent past, but 10 percent difference still implied decades of evolutionary divergence. The new paper Shi coauthored, again with Linfa Wang and Shuyi Zhang as collaborators, noted what seemed to be a robust diversity of SARS-like viruses in horseshoe bats, all sharing common ancestry with the SARS virus itself. That paper, published in the *Journal of General Virology*, went some way into analyzing, gene by gene, how these SARS-like genomes may have evolved. And this time Shi was the senior author.

Shi became a specialist in the coronaviruses of bats. She gave lectures at international meetings, received grant support from both Chinese and (through collaborations, including one with EcoHealth Alliance) American governmental bodies, and appeared as a coauthor on more than forty coronavirus papers published over the next dozen years. In 2008, for instance, she led a comparative study of how the human SARS virus and some of those SARS-like coronaviruses from bats manage to latch on to and enter the cells of their respective hosts. The key question it addressed was whether these SARS-like bat viruses, discovered by her group and others, were capable of infecting human cells and therefore spilling over from bats into humans, as the SARS virus itself had done, and causing human disease outbreaks. To explore that question, Shi's group performed laboratory manipulations with parts of the genomes of those viruses, creating what's called a pseudovirus system, a population of particles that can enter a cell like viruses, but not replicate themselves and burst out. One advantage of that approach is that a pseudovirus is tame and hobbled; it can't proliferate and cause a chain of infections. So it can be safely used as a proxy in the lab to investigate certain properties.

The segment of interest was the gene responsible for producing the spike proteins, the complex knobs on the surface of each spherical virion, which form a corona-like fuzz and give the family its name. The spikes protrude from the envelope (exterior wrapper) of the virion. Each spike consists of three copies of an identical protein bundled together like an inverted tripod. It's an elaborate, three-dimensional molecular feature that allows a virion to catch hold of a receptor molecule on the exterior of a cell, fuse its envelope with the cell membrane, and gain entry for its RNA genome. You often hear the simile—it's a go-to cliché—that the spike fits to the receptor like a key to a lock, opening the cell. That's too simple, because the spike is so intricate and dynamic, but it does capture the fact that precise matching between a spike and a cell receptor is what determines which coronaviruses can infect which cells within which particular host. Matching of spike and receptor also, therefore, affects the capacity for host switching. In a word, spillover.

Previous work, since 2003, had established that SARS-CoV uses a receptor called (never mind why) ACE2, which dangles from the exterior of certain human cells, including some lining the blood vessels, some in the small intestine, some in the heart and kidneys and other organs, and some (most fatefully) along the upper airway. ACE2 is there because it's an enzyme that serves functions in human metabolism, one of which is to help regulate blood pressure. But incidentally it makes cells vulnerable, providing an opportunity for certain viruses. The spike protein of SARS-CoV is known simply as S, and the small portion of the spike most keenly interesting to Shi's group was the receptor-binding domain (RBD), a short stretch of amino acids (the units of a protein) critical for the spike in latching on to its ACE2 cell receptor. The apt metaphor here could be, not key and lock, but Velcro—a Velcro in which the hooks side requires a specific gauge of fuzz. Might that little section, the RBD, be crucial in determining which SARS-like coronaviruses can infect which hosts? The researchers tested SARS-like bat coronavirus S proteins, mounted in a pseudovirus, against human ACE2 receptors and found them unable to use those receptors for cell entry. They tested the spike of SARS-CoV against bat ACE2 receptors and, likewise, found no functional match. Then they clipped out the RBD from one of those bat virus spikes and replaced it with the RBD from SARS-CoV. Would that make a critical difference? (This RBD and ACE2 stuff may seem a little weedy but it will be useful, some pages along, in understanding the controversy that arose around the origins of SARS-CoV-2.) The answer was yes. They saw evidence, in lab cell cultures, that a bat virus thus modified would likely be able to infect humans.

Had they created a dangerous new virus? No. They were working with pseudoviruses. Had they learned something significant? Yes. It was conceivable, Shi's group concluded, that those other coronaviruses residing in bats "may become infectious to humans" if they somehow swap in a little patch of genome that alters their receptor-binding domain. How might that happen? By recombination, of which coronaviruses are notoriously capable.

Five years later, after many more field trips for catching and sampling bats, many more months of laboratory experimentation and analysis, they announced their discovery of a bat virus that *could* attach to human ACE2. They found it in a fecal sample from a horseshoe bat in a cave near the city of Kunming, capital of Yunnan province, a thousand miles southwest of Wuhan. The site, known as Shitou Cave, supported a resident colony of Chinese rufous horseshoe bats, and Shi's team had sampled from them repeatedly, season by season, over more than a year. They recovered enough fragmentary RNA to assemble two more unique coronavirus genome sequences, but one sample was especially productive—a bounteous dab of bat shit, from which they managed not just to recover RNA but to isolate (that is, grow) a live virus.

It's hard to grow viruses from bat feces, and this was the first SARS-like coronavirus ever cultured. They named it WIV1, for the Wuhan Institute of Virology. Its genome was a 95 percent match to human SARS-CoV, making it the closest known relative to that original SARS virus. In the genome blip for the receptor-binding domain, the essential little patch of Velcro within the spike protein, the match to human SARS was even higher, about 96 percent. This much alone constituted a hefty discovery, but the paper appeared in *Nature*, arguably the world's most respected scientific journal, probably because there was more. WIV1, the novel bat virus crackling in cell cultures in Shi's lab, showed itself quite capable of grabbing and penetrating cells by way of human ACE2. That meant it might be road-ready to infect humans. It might not need to pass through civets or any other intermediate host.

"Our results provide the strongest evidence to date," Shi's team wrote, "that Chinese horseshoe bats are natural reservoirs of SARS-CoV," and that intermediate hosts might not be necessary for SARS-like coronaviruses to pass from bats to humans. "They also highlight the importance of pathogen-discovery programs targeting high-risk wildlife groups in emerging disease hotspots as a strategy for pandemic preparedness." Six years before the pandemic, Zhengli Shi was saying: People, get ready.

20

In 2012, around the same time Shi's team worked at Shitou Cave near Kunming, they also took samples from bats in an abandoned mine about three hours farther south. This site, in a town called Tongguan, in Mojiang County, Yunnan, had a peculiar history that drew Shi's attention in stages. The details are important because "the Mojiang mine" would later figure in some dark narratives about the origin of the pandemic.

Shi heard about the Mojiang mine sometime that summer, roughly July, to her best recollection. The first news was just a puzzling bit of scuttlebutt from other researchers, who heard it from someone at a hospital in Kunming: six workers from the Mojiang site had gotten sick, five of them suffering badly, and been admitted for treatment of severe respiratory disease. These were laborers hired to clear bat guano from the cave so it could be reactivated for production of copper ore. They had worked for days underground, shoveling guano, breathing its dust, breathing also whatever else hung in the air of the shaft. At least one of them had already died by the time Shi heard about the situation. Two others would, after forty-eight days' hospitalization, in one case, and 109 days in the other. The doctors had groped to treat them because the cause of their illness was unknown. Maybe fungal infection, maybe a virus? The descriptive diagnosis at time of death, for one of them, was "severe pulmonary infection; sepsis; septic shock, abdominal infection; respiratory cardiac arrest." The hospital had taken serum samples from four of the patients, and Zhengli Shi, because of her reputation, was asked to test those for bat viruses.

"They sent us the sero sample," she told me, using shorthand for a serological specimen. Thirteen serum samples from the four patients, in fact, but no fecal samples, no nasal swabs. "We only had the sero." Her lab crew tested the samples for Nipah virus, Ebola virus, SARS virus, and found nothing. But meanwhile she had gotten curious.

So she took her team to Mojiang and began capturing and sampling bats. It became a secondary research site, to which she returned intermittently over the next four years. "We sampled—totally, we sampled seven times in this cave," she said. Bats of six different species roosted and mingled in the mine, two horseshoe bat species and four others, and Shi's team found some evidence of coronaviruses in all six. Among their 1,322 samples, they detected fragments of RNA representing a huge diversity of coronaviruses, 293 different viruses, 284 of which belonged to the alphacoronaviruses, a genus that contains no known threats to humans. The other nine viruses, each recognized by its sequence for one crucial gene, fell within the betacoronaviruses, the same genus that harbors SARS-CoV and MERS-CoV. Those nine, because they more closely resembled the SARS virus, held the greatest interest for Zhengli Shi.

The sample in which Shi's team found one of those nine, especially notable in light of its later history, was given the number 4991. The sample number is distinct from the label assigned to any genomic sequence that may come from that sample, just as gazpacho is distinct from a cucumber; and the genomic sequence of a virus is distinct from any live virus that is grown, just as the genomic sequence of a lion is distinct from a live lion that might walk into your laboratory—but those distinctions got lost amid later criticisms of Zhengli Shi's work. Sample 4991, which came from an intermediate horseshoe bat (*Rhinolophus affinis*), contained just enough material for Shi's group, two years later, when they had better equipment, to pull out and sequence a near-complete genome. They named that genome RaTG13, from *Ra* as in *Rhinolophus affinis*, TG as in Tongguan the town, and 13 as in 2013, the year the sample was collected. Why should you care? Because these relatively simple facts of variant naming would later bring accusations of sinister obfuscation against Zhengli Shi, and RaTG13 would become the single most important, and most poorly understood, piece of data amid the controversy over the origin of the virus.

When Shi's group first published these results, in 2016, the salient point was not sample 4991 alone, or the genomic material it contained. The salient point was that so many different coronaviruses coexisted, circulating among half a dozen different kinds of bats, in one mine. The authors described that rich mix of viral diversity and bat diversity as "a phenomenon that fosters recombination and promotes the emergence of novel virus strains. Our findings highlight the importance of bats as natural reservoirs of coronaviruses and the potentially zoonotic source of viral pathogens."

One final paper in this series, published by Shi and her colleagues in 2017, drew on five years of their work on virus discovery and infectivity experiments in the laboratory. Although they had continued sampling multiple kinds of bats at multiple sites across China, this study summarized their findings from one place, Shitou Cave, on the outskirts of Kunming, which mainly harbored horseshoe bats. From RNA fragments in the samples, they assembled the full genome sequences of eleven novel coronaviruses, all fairly similar to SARS-CoV. What made those eleven most illuminating was that, among them, they included in almost precise form all the genomic elements—a gene region here, a gene region there, a receptor-binding domain—of SARS-CoV itself. Analysis of those genomes showed that recombination had been mixing and matching the parts from one genome to another. Scientists around the world recognized this as almost rock-solid evidence for the source of the SARS virus of 2003: from horseshoe bats, by recombination, if not in that very cave, then in another containing similar ingredients. A commentator in *Nature* called it the "smoking gun." After a mere fourteen years, the mystery of origins had been solved, probably. But that still left at least one question unanswered, as another Chinese virologist noted in a comment to *Nature*: If the SARS virus emerged from a bat near Kunming, how did it get from there to Guangzhou, more than six hundred miles away in the next province, without leaving a trail of sick people in its path?

And if that question sounds familiar, it's because doubters have asked nearly the same thing about SARS-CoV-2. There's a range of possible answers, enough to satisfy and dissatisfy everybody.

"This work provides new insights into the origin and evolution of SARS-CoV," wrote Shi and her coauthors of the 2017 study, "and highlights the necessity of preparedness for future emergence of SARS-like diseases." The alarms had been ringing loudly and long by then, into a void of uninterest and deafness.

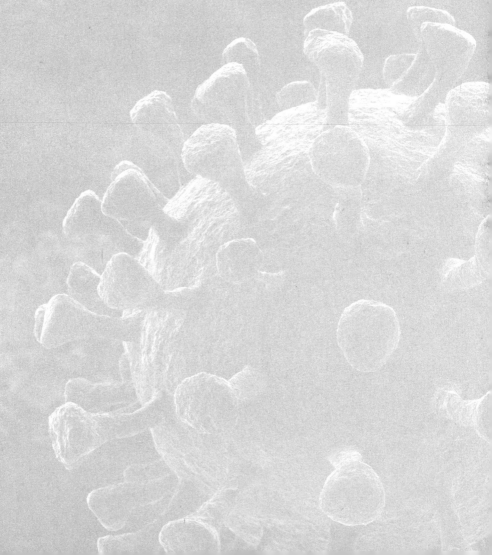

III

MESSAGE IN A BOTTLE

21

Those were the strategic warnings: urgent but somewhat generalized, and spread across more than two decades. Then began the tactical warnings, such as those from Marjorie Pollack on ProMED, in late December 2019. With them came the responses, and lack of responses, to the novel coronavirus that we now call SARS-CoV-2.

On the evening of December 30, Zhengli Shi was in Shanghai attending a conference when her boss in Wuhan, director of the Wuhan Institute of Virology, reached her by cell phone. "Around ten o'clock in the evening," as Shi told me later. "I have never heard about an unknown pneumonia before that." Now she heard. There was an atypical form of pneumonia, manifest in a smattering of cases across the city, its cause so far unknown, with preliminary lab results indicating that a coronavirus could be involved. Some samples from patients had just arrived at the WIV and the director wanted Shi's lab to work on them.

"I was asked to take action, yeah," she said. "To do the detection"— to identify the virus more conclusively. Shanghai time is thirteen hours ahead of New York's, so Pollack's first posting on ProMED hadn't yet gone up; most of the world was still oblivious, except those, like Henry Li at the Susan Weiss lab in Philadelphia, connected to Wuhan by WeChat or other social media. In Wuhan itself, word was out, but only to a select few. Shi immediately called her own lab, found that three night-owl students were still there, and asked them not to go home but to wait, despite the hour, and receive an extract of viral RNA, drawn from hospital samples, when that arrived as promised from another laboratory, at any moment. She instructed them to begin work, using two methods, to identify what sort of virus it was. The first method was a broad PCR test that would detect any form of coronavirus. The second PCR method was more specific for detecting SARS-related coronaviruses.

Shi herself had a meeting next morning in Shanghai, but as soon as that ended, on December 31, she grabbed a train home to Wuhan. She went straight to her lab and saw the PCR results, which the students had gotten that morning. "The machines read the data," she said, and from that "we know that it's a SARS-related coronavirus." They hadn't yet sequenced the full genome themselves, Shi's group, but they possessed partial sequence data from another lab. "My first reaction is we need to compare the sequence," she said—compare the new virus's genome, that is, with the genomes of bat coronaviruses detected from samples in her own lab, to see whether, by some horrible mischance, there was a match. "It's normal!" she said to me with some vehemence, reacting against criticisms that have come at her since. If she had "frantically" gone to her own data, as reported, didn't that imply a guilty awareness that the new virus had likely leaked from her lab? No, she said, it did not. What it implied was normal diligence on an important point. And the new virus did *not* match anything from her sequence records, if she can be believed (and I think she can, though I can't prove it). "So, the afternoon of December 31, I already know it's nothing related to what we have done in our laboratory." She felt great relief. That evening she met with local officials, from the Wuhan Municipal Health Commission, and reported her lab results.

Then her team plunged back to work. Within two days, they had a provisional draft of almost the full genome sequence. They may not have been the first, but they were *among* the earliest, to sequence a near-complete genome. Why didn't they publish it at once? Because of concern for accuracy over speed. The first SARS genome as published, back in 2003, had contained mistakes; sequencing technology was less precise and reliable then, and haste had preempted confirmation. This time should be different. The Health Commission asked two other institutions besides Shi's lab to produce sequences, all working independently, and then they compared versions and resolved technical disparities. By January 6, she had a complete genome, correct and confirmed. But still she didn't release it due to continuing caution. And so the Zhang version, released by Eddie Holmes through the Virological website on early January 11 (UTC), and the sequences submitted by George Gao's team to GISAID on late January 9 (UTC), were the first SARS-CoV-2 genomes made widely available.

The loss of priority on that point does not seem to have bothered her.

Neither did the early questioning about whether this novel virus might have leaked from her laboratory. "I think it's normal," she said of such questioning. People would speculate, would make accusations, but that was because they didn't know the intricacies of coronaviruses. She could see differences, a complex of features, distinguishing this one from any bat virus she had sequenced, let alone anything she had grown. "But at the beginning, I think, Okay, it's not necessary to explain too much." Maybe. If there was any delay in demands for explaining, though, it wouldn't last long.

22

Through the next couple weeks, Shi and her group were busy in the lab. Besides assembling a full genome sequence of the virus from the fragments they had detected by PCR, they compared that sequence to the sequence of SARS-CoV and found them to be 79.5 percent identical. So this was another SARS-like virus, but it wasn't the original SARS virus, because 20 percent difference implies many decades of divergent evolution. They assembled four more complete genomes, from four other patients, and each of those was an almost exact match to their first. That helped confirm what they were seeing. They gave their virus the tentative label nCoV-2019, a slight variant of what the WHO had begun calling it; but the time was still early, and naming was in flux. By way of colleagues at Wuhan Jinyintan Hospital, which treated many of the first few dozen cases, they obtained a sample from the deep airways of one patient, and from that they grew live virus. They tested that virus against cells in culture, finding that it could use the same receptor for cell entry as SARS-CoV, the ACE2 receptor. Furthermore, it could use the ACE2 of a horseshoe bat, of a civet, and of a pig, as well as the ACE2 of a human—so this virus was already broadly adapted, it seemed, for infecting a variety of hosts.

One further bit of Shi's lab work, from these early weeks of 2020, would draw continuing attention. That's putting it mildly. In fact, this discovery would become like a Rorschach inkblot, susceptible to drastically different and subjective and, in some cases, impassioned interpretations. (Is it coincidence that Hermann Rorschach's Card 5 looks so much like a bat—or is that just me?) Having noticed a strong similarity between one region of the new genome and something naggingly familiar, they gave the similarity a closer look: they retrieved their full genome sequence of a bat virus from the Mojiang mine, the one they had labeled RaTG13, and compared that with their genome from Jinyintan Hospital. The similarity was 96.2 percent. That made RaTG13, at least for the moment, the closest known relative of the pandemic virus.

On January 23, 2020, Zhengli Shi and her colleagues announced these findings to the world. They did it in the form of a preprint (a draft paper, made available on a website, not yet peer-reviewed and published in a journal) posted through the preprint repository bioRxiv (pronounced "bio-archive"), which is hosted by the Cold Spring Harbor Laboratory, an august institution on Long Island. They also submitted the paper to *Nature*, where it was peer-reviewed quickly and published on February 3.

In the meantime, case numbers rose quickly in China—from forty-one lab-confirmed cases at Jinyintan Hospital on January 2, then outbreaks in other parts of the country by January 19, exploding to 11,791 Chinese cases by January 31—and the virus escaped, by way of travelers, beyond national boundaries. Thailand reported a confirmed case on January 13, in a woman from Wuhan who had come to Bangkok on a visit. Japan confirmed a case two days later and then, on January 20, both South Korea and the United States reported their first recognized cases. Early reporting from the Wuhan Municipal Health Commission linked many of the cases to the Huanan Seafood Wholesale Market, as mentioned earlier. That connection, to a market including wildlife, drove the provisional narrative about how and from where this novel virus might have gotten into humans. But because the market had been closed by Wuhan authorities, and the place cleaned, on January 1, its potential role was never thoroughly investigated. Concern grew around the world, as people gradually realized that this tiny microbe could become a global problem. Fragmentary data and anecdotal scraps fed speculation, venturesome hypothesizing, hasty conclusions, and confusion, especially regarding the origin of the virus. Where had it come from, how had it taken shape, and how had it gotten into people? January was a feverish month.

Two early research papers exemplify the dizzy eagerness among some scientists to tell an arresting story. The first, from a Chinese team affiliated with Peking University, Guangxi University of Chinese Medicine, and other institutions, noted certain parallels between the genome of the new virus and the genomes of snakes. Those parallels involved something called codon usage, referring to the various ways by which the letters of a genome, in three-letter clusters (called codons), can specify a given amino acid be inserted as the next element in a protein being built. In other words, codon usage is spelling. All you need to know about it, for purposes here, is that there are alternate possible ways to spell the coding for each amino acid—just as there are alternate possible ways to spell the English word "color." If you see it spelled "colour," that delivers a hint: British. Similarly, these Chinese researchers claimed to see a hint in the codon usage of the new coronavirus: snake. Since the codon usage in the virus seemed to resemble the codon usage in some snakes, could that mean the virus had been a longtime infection of snakes? It was tenuous.

The scientists looked at two kinds of snake, both native to Hubei province roundabout Wuhan: the many-banded krait and the Chinese cobra. Both snakes spelled some of their amino acids with codon usage resembling the usage in the new coronavirus—more similar than that seen in birds, hedgehogs, marmots, humans, or bats. "Snakes were also sold at the Huanan Seafood Wholesale Market," the authors noted, although they don't seem to have known which kinds. Many-banded krait and Chinese cobra might well have been market-available, because they are favored for that ancient Cantonese delicacy, snake soup. But the researchers were cautious, merely suggesting that their codon-usage analysis "provides some insights to the question of wildlife animal reservoir" of the virus, "although it requires further validation by experimental studies in animals." For instance, experiments to test whether the new virus could even survive in snakes—experiments that these researchers didn't do. This study appeared in a peer-review monthly, the *Journal of Medical Virology*, but it wasn't warmly embraced by the scientific community, to say the least. It made headlines in tabloids, it got its moment on CNN, it appealed to nonscientists with a certain taste for the lurid, but the hypothesis came and went quickly. Other scientists looked at the evidence, such as it was, and essentially said: phooey.

The second incendiary paper that month came from a group of scientists in New Delhi, who posted it as a preprint on bioRxiv on January 31. These authors purported to have discovered four "unique" stretches of amino acids in the spike protein of the new coronavirus, each stretch six to twelve aminos long, that bore an "uncanny similarity" to amino acid placements in corresponding proteins in HIV-1, which includes the pandemic subtype of the AIDS virus. Such similarity, they claimed, was "unlikely to be fortuitous in nature." They called these stretches "insertions" into the coronavirus, implying that it had been assembled in a lab, possibly using portions of the HIV-1 genome to make it more infectious to human cells. But as expert critics soon pointed out, the "uncanny" coincidence was not uncanny at all. The "insertions" were not insertions; they were commonplace, resembling stretches seen in many other creatures (including the bat virus RaTG13). The whole paper was a bollox, trumpeting a coincidence about as improbable and suspicious as finding (as you can do) the words "mischievous," "players," "overcharged," and "countrymen" within the complete works of Shakespeare. Was the clever playwright from Stratford-upon-Avon slyly bragging that his mischievous players had overcharged their countrymen? Doubtful. Equally doubtful that the new virus had grabbed bits of its genome, by some form of implausible recombination, from HIV-1.

Very quickly, this paper was taken down, and if you find it online now, you'll see a large gray stamp across each page: WITHDRAWN. The authors issued a statement saying, "To avoid further misinterpretation and confusions world-over, we have decided to withdraw the current version of the preprint and will get back with a revised version after reanalysis, addressing the comments and concerns." But it seems they never have.

"I was very angry," Zhengli Shi told me. The snake story and the "uncanny similarity" paper are just a sampling of the distractions, false leads, and misapprehensions that crackled across the internet in early 2020. Some of them targeted her lab, implicitly or explicitly. "So I tried to isolate this information, misinformation," she said—to block it out, to quiet herself, to focus. "And continue to work."

23

Kristian Andersen watched all this with intense interest from La Jolla, California. "I would say it probably took me, maybe, the first week of January before I got significantly more concerned," he told me. "That has sort of only accelerated since then."

Andersen is a computational geneticist, meaning that he uses deep-dive mathematical modeling and computer analysis and simulations to investigate the secrets of genomes. He's a lean and athletic man with a small, controlled smile. Born in Denmark, trained at Cambridge and Harvard, he is now a professor at the Scripps Research Institute. His chosen mission is to understand how viruses evolve, emerge, evolve further, and cause trouble in humans. He has worked on Lassa and Ebola in West Africa, including Sierra Leone during the 2013–2016 Ebola epidemic, helping develop diagnostic tests and trace the spread of infections by genomic sequencing, a discipline known as genomic epidemiology.

No, Andersen assured me, he certainly did not coin the term "genomic epidemiology." That discipline got its start in 2003 (he was an undergraduate in Aarhus, Denmark), when it enabled scientists to trace, in retrospect, the movement of the SARS virus out of Guangdong province and through Hong Kong to the world. Sequencing had been laborious and slow in the late twentieth century, then became automated but expensive—the first human genome cost $2.7 billion—but with ever-better technology the cost went down and the speed went up. The 2013–2016 Ebola nightmare was the first epidemic event for which genomic epidemiology could be a vital tool in the immediate crisis response. Andersen was then a postdoc in the lab of Pardis Sabeti, a brilliant Iranian American with a joint appointment at Harvard and the Broad Institute. Andersen spearheaded a large team led by Sabeti and several other senior researchers. One of those other leaders was Andrew Rambaut, from the University of Edinburgh. Sequencing strains of Ebola, as sampled from human patients *in extremis*, amid a chaotic public health catastrophe, is a task for brainy people with big hearts and strong nerves. You make friends you can trust. No doubt that's partly why Andersen reconnected with Rambaut, during the early weeks of 2020, as this new catastrophe began, to coauthor a paper on what genomic analysis of the novel virus might suggest about its probable origin.

"As soon as we had, like, maybe ten or fifteen or so genomes from Wuhan," Andersen said, "and they were basically all identical," he recognized something was amiss. That access to genomes from early cases occurred by mid-January, and he was concerned to see that those genomes differed by few if any mutations. Why? "Because it tells you that this is probably spreading human to human." If a virus rides to a city market within reservoir animals, it is liable to diversify at least a little bit by mutation among the different animals, and then, if it spills over many times to different humans, it will show that modicum of diversity in the human samples. Lack of diversity suggests few spillovers, animal to human, from which nearly identical virus strains have passed quickly from human to human. And a notable lack of diversity is what Andersen saw in the early genomes. It accorded with other evidence for human-to-human transmission, unknown to him then, such as the family cluster of five cases in Shenzhen that raised concern in K.Y. Yuen.

"And that's what started myself and Eddie and Andrew and Bob on the proximal origin," Andersen said.

"Bob" was Robert F. Garry, a virologist at Tulane University in New Orleans, whom Andersen knew from the Lassa and Ebola work in Sierra Leone, where Garry had a long-term association with a government hospital to do virus research and training. Garry is an expert in the structural biology of viral proteins—how they fold, how they function—and he can deduce things about such protein mechanics by looking at a genome sequence. He's a seventyish fellow from Terre Haute, Indiana, with tawny hair except for the sideburns and thick mustache, which have faded toward white; an old-school virologist, who worked on HIV when it was new to the world, who remembers the old computers and the models they ran, and who continues to patrol the forefront of structural biology using what are now better machines and fancier techniques. "There aren't too many of us left that can take a look at a protein sequence," Garry told me, "and start to discern what that protein might be doing. You know, where the dangerous parts are."

When he saw the sequence of the new coronavirus on the website where Holmes had posted it, Garry at once discerned what might be a dangerous part. "What popped up there was a furin cleavage site," he told me. "And that meant I didn't really sleep very well that night."

A furin cleavage site is a sort of trigger within the spike protein of a coronavirus (or an equivalent protein in other viruses) that enhances the capacity of the virus to latch on to and enter cells. First the spike catches hold of a receptor protein on the exterior of the cell, such as ACE2. Then the cleavage site comes into play. It's located at the junction between the two major lobes of the spike, and when it's triggered, the spike changes shape—like a Transformer robot metamorphosing suddenly into a truck—so that the spike is "cleaved," split open, in a way that allows it to fuse with the membrane of the cell. This enables the viral genome to squirt into the cell and begin replicating itself. What triggers the furin cleavage site is the touch of a furin molecule. Furin is an enzyme with important functions, never mind what, and our bodies are full of it. So the furin cleavage site is well suited for triggering, by bodily furin, and helping the virus to penetrate cells.

Some viruses possess furin cleavage sites as part of their cell-entry machinery, and those sites seem to help make them highly virulent in humans. SARS-CoV had no such site, and that virus, fortunately, was not very efficiently contagious. So the inclusion of a furin cleavage site in the new coronavirus caused Bob Garry some concern, on the grounds that it might make the thing better able to spread.

"I started talking with some of my virologist colleagues," Garry told me. One of them was Andersen, from whom he heard that Andersen had already conferred with Rambaut and Holmes. Garry foresaw that any notable features or apparent anomalies in the virus's genome would provoke curiosity—legitimate curiosity—and suspicions and theories, whether legitimate or not. For instance, did the virus look engineered? Had the furin cleavage site perhaps been put in there by humans, rather than by mutation, recombination (natural swapping of sections with other viruses), evolution? "It seems that it is hard to separate virology from politics sometimes," Garry said, "but even there, back in early January, I mean, there were questions about, 'Who are we going to blame for this?'"

Eddie Holmes, by then in Switzerland at a virology conference, heard similar concern from an old friend, Jeremy Farrar, a medical researcher who now ran the Wellcome Trust, a London-based foundation devoted to health research. "Jeremy Farrar emailed me and said, 'There's discussion about it might have come out of a lab. Can you take a look at the sequence?'" Although Holmes had midwifed that sequence to publication, he hadn't yet studied it. Now he pulled up Zhengli Shi's preprint on the bat virus RaTG13, the one that was a 96.2 percent match to the novel virus, and scanned a figure comparing sections of the genomes. He saw nothing that struck him as weird. He caught his flight back to Australia. "And then, almost the next day, Kristian Andersen emailed me and said, 'I've seen something very strange in this sequence, can you take a look?'" Andersen by now was alert to the furin cleavage site, and he had also noticed another possible anomaly: the receptor-binding domain, that Velcro patch on the spike, so crucial for the initial attachment to a cell. It's a two-piece arrangement, the RBD catching hold of the cell, the furin cleavage site facilitating entry. The RBD of the new virus, as coded in the genome, looked distinctly well suited for grabbing ACE2 receptors of the sort found in humans, ferrets, and some other animals. It bore little resemblance to the RBDs in SARS-CoV and in Zhengli Shi's bat virus, RaTG13. So where the hell had it come from? "And I thought, 'Oh, shit,'" Holmes told me. "We alerted the authorities."

One of those authorities was Jeremy Farrar again, who urged him to gather his thoughts on the genome, what it might indicate about the origin of the virus, and share them. "Just write a report." Two other authorities in the loop by this point were Tony Fauci, director of the NIAID, and Francis Collins, director of the National Institutes of Health (NIH), within which the NIAID is embedded. Farrar arranged a secure conference call, scheduled for February 1, on which Andersen, Holmes, Fauci, Collins, and a select few other scientists scattered internationally could be brought together to confer on this question—what the genome might say about where and how the virus originated.

In the meantime, Andersen and Holmes were discussing that also with Andrew Rambaut. "Kristian and Eddie got in contact and said, 'We're looking at the virus,'" Rambaut told me, "'and trying to look at what we can see about the origins, and there are a number of interesting, quite unusual characteristics.'" They wanted Rambaut's help because of his wisdom on viral evolution, and Bob Garry's too, for his expertise in structural modeling and his knowledge of what can and can't be engineered.

On January 31, 2020, as these efforts began, an article appeared in *Science*, from the trusted staff writer Jon Cohen, about the same general effort at which they were engaged. It was titled "Mining Coronavirus Genomes for Clues to the Outbreak's Origins." It described Zhengli Shi's paper about the bat virus RaTG13, it cited a phylogenetic analysis of different strains of the new virus (depicted as family trees) led by a computational biologist named Trevor Bedford in Seattle, it noted and dismissed the snake hypothesis, and it quoted a molecular biologist named Richard Ebright, a longtime critic of what he considered unacceptably dangerous viral research carrying risks of laboratory escape, to the effect that the new virus could have reached humans either by a natural spillover, direct from a nonhuman animal, or by a laboratory accident. Ebright's own research specialty is the transcription process (genes making messenger RNA, the blueprints for proteins) in bacteria. Cohen also quoted Peter Daszak of EcoHealth Alliance, responding to Ebright. "It seems humans can't resist controversy and these myths, yet it's staring us right in the face," Daszak said. "There's this incredible diversity of viruses in wildlife and we've just scratched the surface. Within that diversity, there will be some that can infect people and within that group will be some that cause illness." Cohen was right to place Ebright and Daszak as antithetical voices on the question of SARS-CoV-2 origins, and they have retained those positions since.

January 31, 2020, was a Friday. Early that evening, Tony Fauci emailed Kristian Andersen and Jeremy Farrar, calling the Cohen article to their attention: "This just came out today. You may have seen it. If not, it is of interest to the current discussion"—the discussion that would continue next day, on the conference call. Andersen wrote back, saying yes, he had read it and noting politely that both he and Holmes were, in fact, quoted by Cohen. (Fauci, a busy man, may have noticed that or not, but if so, wanted to ensure that they had seen the piece as published.) Andersen then mentioned the difficulty of assessing small, unexpected features in a coronavirus of provenance unknown. Recombination could insert this or that. It looked like a bat virus, but there were relatively few bat coronavirus genomes to which it could be compared. He alluded to the receptor-binding domain and the furin cleavage site, those two unexpected features. Were they "unusual" or not? "On a phylogenetic tree the virus looks totally normal." The close branching with bat viruses, he added, suggested that bats might well be the reservoir.

But there was a "but," a qualification. "The unusual features of the virus make up a really small part of the genome"—less than a tenth of a percent, Andersen said—"so one has to look really closely at all the sequences to see that some of the features (potentially) look engineered."

The next day, Saturday, February 1, they held their call, beginning at 7:00 p.m. UTC (for Farrar, in London), which was 2:00 p.m. in Washington (Fauci and Collins), 11:00 a.m. in California (Andersen) and 6:00 a.m. Sunday for Holmes in Sydney. Also on the call were Rambaut and Garry, as well as Marion Koopmans (a distinguished Dutch virologist in Rotterdam), Christian Drosten (director of a virology institute in Berlin), Patrick Vallance (chief scientific advisor to the United Kingdom government), and several other scientists. Andersen and Holmes made a short presentation on what they had seen in the genome. The group exchanged thoughts, considering all possible scenarios for how the virus might have reached humans—natural spillover from a wild animal, lab leak, engineered virus. Koopmans and Drosten argued that natural spillover was the most probable explanation, according to Farrar's account of this conversation in his subsequent book, *Spike.* That's correct, Koopmans told me, and her judgment was based on the presence of furin cleavage sites in other coronaviruses in the wild, as well as additional evidence. Farrar himself was not so sure. Soon after the call he described himself as halfway between the natural and laboratory options. (Later he too would conclude, based on further data, that a natural origin was likeliest.) The call lasted an hour. In days following, Andersen and Holmes and Garry and Rambaut intensified their deliberations toward drafting a paper.

They began shooting ideas back and forth. They conferred by phone and in video conferences, and started putting down sentences, fragments of text, provisional paragraphs, shared through Google Docs. Their resolve was to consider the various candidate scenarios objectively, to remain open-minded and let the evidence guide them wherever. "Try to be agnostic about the whole thing," in Garry's phrase. By now they knew they were writing a scientific paper, not just a "report" for some indefinite audience, and that they would try first to publish it in *Nature*. They had already contacted the editors, who were interested in seeing it.

As the four men worked through a first draft, the receptor-binding domain seemed inexplicable. They knew from structural modeling of the sort Garry did that it must be a very well-optimized RBD for binding to ACE2 in human cells. It wasn't perfect, but it looked almost too good. And it was unique. "None of the other viruses we had seen previously had that exact receptor-binding domain," Andersen told me. "To be honest, that was quite concerning." They carefully considered, he said, "the possibility of this having been through a lab." They deliberated together and shared thoughts with some colleagues.

That email exchange between Andersen and Fauci, on the day just preceding the conference call, is worth attention within this context, because later it would be brought into public view, through the Freedom of Information Act, as one of "the Fauci emails," and bruited as revelatory, absent that context. Proponents of dark theories would argue that it showed Andersen and Fauci conspiring to conceal their belief that the virus was engineered in a lab. Those accusations were further fueled by Andersen's added comment to Fauci that, after discussions among the analysis team, at least some of them "find the genome inconsistent with expectations from evolutionary theory."

What did he mean? Well, expectations from evolutionary theory did not include an RBD and a furin cleavage site bearing little resemblance to anything seen in SARS-CoV, or in any SARS-like coronaviruses thus far known from bats. "But we have to look at this much more closely," Andersen ended the email, "and there are still further analyses to be done, so those opinions could still change." New phenomena with familiar aspects—as the giant Galápagos tortoises were new to Charles Darwin, though their form was familiar—required closer investigation and fresh thinking.

"What the email shows," Andersen said later, on Twitter, in response to criticisms, "is a clear example of the scientific process." They were considering possibilities, lab leak included, engineered virus included, natural spillover included, and they hungered for more data that might support or negate one or another. "It's just science. Boring, I know,"

Andersen wrote, "but it's quite a helpful thing to have in times of uncertainty." Three weeks passed, without resolution, carrying the effort into February. Then came the pangolins.

24

"This to us is huge, huge evidence," Andersen told me, that the virus was natural. What they suddenly had, thanks to a team of three Chinese scientists in Kunming, and to others who saw and reanalyzed the Kunming data and also found more, was a close match to the new virus's receptor-binding domain from a wild coronavirus. This other virus had been found in some Malayan pangolins, which were trafficked through Guangdong and confiscated by wildlife officials earlier in 2019. A report of the RBD match, detected from the accessible data by scientists working in Houston, was posted to the Virological site on January 30, 2020, and noticed a few days later by Andersen and his colleagues, just in time to influence their paper. Once they saw that, Andersen told me, they realized that "this thing we think is really unusual is already out there in nature."

Pangolins are strange and charming animals. Most people in the Western Hemisphere have never seen one, not even in a zoo. They are loosely known as scaly anteaters because of their armored skin and their diet, their elongated heads and their toothless mouths, though they aren't closely related to true anteaters. There are eight living kinds, four native to Africa and four to Asia. The Malayan pangolin (*Manis javanica*, also known as the Sunda pangolin) has a natural distribution from Java and Borneo up through Southeast Asia, and barely across the Chinese border into Yunnan. The eight species constitute a very distinct group, one of the oddest of mammalian orders, the Pholidota. They are similar to carnivores by descent and to armadillos by convergent evolution. They eat termites as well as ants, but they are virtually incapable of harming any other form of living creature, except in their own defense.

Their gentle passivity makes them disastrously susceptible to capture by humans. When attacked or challenged, a pangolin's default mode of defense is to roll into a ball, like a pill bug, scales on the outside, tender parts within. The name pangolin comes from *peng-goling*, which in Malay means "roller" or "that which rolls up." This defense works well against such predators as leopards but not against a two-legged enemy with a larger brain and a pair of hands, capable of battering a pangolin open or carrying it back to a village. Why would a person batter one open? Because the flesh is wanted as food. Why would a person carry one back to a village? Perhaps to sell it, because the scales as well as the flesh are prized in some cultures, and flow at catastrophic volumes through the illicit international trade in wildlife products.

Pangolins are solitary animals, each one foraging on its lonesome, the adults coming together briefly to breed. The female carries her single offspring piggyback for some months, and sleeps with it curled tenderly within her armor. Although pangolins are hard to find, they must have once seemed endlessly abundant. Between 1975 and 2000, according to a German biologist named Sarah Heinrich and her colleagues, drawing on the database of the Convention on International Trade in Endangered Species of Wild Fauna and Flora (a multinational compact known as CITES), roughly 776,000 pangolins became merchandise that was traded legally on the international market. That flow of products included almost 613,000 pangolin skins, exported from countries including Indonesia, Thailand, and Malaysia.

Pangolin scales are a separate commodity, highly valued for their supposed efficacy in traditional medicines. Between 1994 and 2000, almost nineteen tons of pangolin scales (accounting for roughly 47,000 pangolins) were exported from Malaysia for use in traditional Chinese medicine (TCM) in China and Hong Kong. Chinese tradition, as inscribed in old texts, holds that pangolin scales, ground to powder or burned to ash, can be useful against ant bites, midnight hysterias, evil spirits, malaria, hemorrhoids, and pinworm, and for stimulating lactation in women. Science doesn't support these claims—the scales consist merely of keratin, the same material as your hair and your nails.

"There's a lot of finger-pointing at other cultures," Sarah Heinrich told me, from her home near Potsdam. The finger could point in many directions. Most of the pangolin skins exported between 1975 and 2000 went to North America, where they were turned into handbags, belts, wallets, and fancy cowboy boots. Pangolin leather was especially coveted because the animal's skin bears an eye-catching, almost reptilian, diamond-grid pattern. The Lucchese Boot Company, bootmaker to Lyndon Johnson among others, produced pangolin-leather boots before 2000, when CITES set the export quota for wild-caught Asian pangolins to zero, essentially making the international commerce illegal.

By then, the pangolin populations in China and parts of Southeast Asia had been drastically depleted, not just to make cowboy boots for affectatious Americans but also by regional consumption. At one point, some 150,000 pangolins in China went to the knife monthly, their meat eaten and their scales used in TCM. "Such was the magnitude of this exploitation," Oxford-based pangolin expert Daniel Challender and three coauthors wrote, "that it apparently led to the commercial extinction of pangolins in China by the mid-nineteen-nineties." Importing pangolins was more practical than hunting down the few indigenous ones that remained.

Challender did some of his doctoral fieldwork in Vietnam, conducting market surveys, gathering price data on pangolin scales, visiting restaurants where the meat was served. "If you go into a restaurant in Ho Chi Minh City," he told me, "you're going to be paying $350 a kilo for a pangolin." It might be grilled, or boiled in a hot pot with ginger and spring onions. He recalled sitting in a restaurant in 2012 watching three diners enjoy a $700 pangolin meal. A server carried the animal, alive, into the restaurant in an old sack. It was balled up in its defense posture, showing only scales and claws. "They took out a large rolling pin and clubbed it unconscious," Challender said. Then "they took some scissors and used the scissor blades to cut the throat." The blood was drained out and mixed with alcohol for the diners, and the flesh was cooked.

As the Asian populations declined, African pangolins began flowing east in large quantities. Since early times, many peoples of sub-Saharan Africa have "harvested" pangolins, trapping the animals with snares, tracking them with dogs, or coming across them in the forest. The hunters traditionally consumed their catch or sold it into local bushmeat markets. Eventually the meat became popular in African cities too, such as Libreville, in Gabon, and Yaoundé, in Cameroon, and that led to rising prices around the start of the twenty-first century. The only live pangolin

I've ever seen was in Yokadouma, a remote town in southeast Cameroon, many hours down a long, unpaved road that led toward the Republic of the Congo.

This doomed creature was in the possession of a young man from the kitchen staff of the Hotel Elephant, where I was staying. He had just brought it back from the town market. He carried it by its tail as it dangled, groggy and helpless. It was reddish brown, like the roadside trees along the highway, and for the same reason—it was caked with the laterite clay dust that suffused the air around Yokadouma, pounded out and sent up by the logging trucks rumbling north from Congo forests. The scales covering its head, body, and tail looked like rusty metal feathers. The kitchen worker dipped it into a storm sewer to revive it, then let it walk a few steps. Its snout was pointy, essentially an aiming device for its long, noodle-like tongue. Its eyes were dark little beads, shiny but uncomprehending. Its belly, unprotected by scales, was a pale-cream color. This was a white-bellied pangolin, one of the four African kinds, three of which are native to southern Cameroon. It tried to hide, pushing its head into a small hole in the ground near the wall of the hotel. But even with its sizable front claws, and the strength and instincts of a burrower, it had no chance of digging its way to safety. What will you do with it? I asked the young man. It would be eaten, he said.

That was in May 2010. In the years following, the suck of international commerce might well have carried a pangolin such as that on a longer but still dooming journey, to Yaoundé and beyond. Or the animal's flesh might have been consumed locally, only its scales, more conveniently transportable, trafficked onward. African pangolin scales move in quantity through the ports and airports of Cameroon and Nigeria to Asia, especially Vietnam and China.

"I know we're serving as a transit point," Olajumoke Morenikeji told me. She's a zoologist, and founder of the Pangolin Conservation Guild of Nigeria. To judge from the thousands of kilograms of scales seized, Morenikeji said, "you can't have all that just coming from Nigeria."

Luc Evouna Embolo, an officer for TRAFFIC, which is an international network that monitors the wildlife trade, gave me a similar account from Yaoundé. Increasingly, middlemen pay local people to collect pangolins from the field. The middlemen sell to urban businessmen who illegally export the animals. A villager might get 3,000 CFA francs (roughly $5) for a pangolin that will be worth $30 in Douala, Cameroon's economic capital, and much more in China. In 2017, police made one seizure amounting to more than five tons of scales, for which two Chinese traffickers were arrested.

In late 2016, CITES decided to make all international trade of wild-caught pangolins and their parts illegal, but the traffic continued. Its scope could now be gauged only from the fraction seized by customs officials and other national enforcement authorities or detected by nongovernmental investigators. By one estimate, almost 900,000 pangolins have been smuggled during the past two decades. Some were alive. Some were dead, peeled of scales and frozen gray. The scales were concealed in sacks or boxes within shipping containers, sometimes labeled as cashews, oyster shells, or scrap plastic. Those who track this commerce, such as Challender and Heinrich, say that pangolins seem to be the most heavily trafficked wild animals in the world.

There is a vogue in urban China for *ye wei*, or "wild tastes"—wildlife meat, supposedly imbued with healthful, invigorating properties. Some consumers cherish the notion that eating pangolin is a revered national tradition. But that notion has lately been challenged. Early in 2020, a Chinese journalist named Wufei Yu published an op-ed in *The New York Times* highlighting old texts that advise against consuming the flesh of certain wild animals, notably snakes, badgers, and pangolins. Yu found that, in the year 652 (by the Western calendar), during the Tang dynasty, an alchemist named Sun Simiao warned about "lurking ailments in our stomachs. Don't eat the meat of pangolins, because it may trigger them and harm us." A millennium later, in a compendium of medical and herbal lore now considered foundational to TCM, the physician Li Shizhen cautioned that eating pangolin could lead to diarrhea, fever, and convulsions. Pangolin scales could be useful for medicines, Li Shizhen allowed, but beware the meat.

Zhou Jinfeng, a noted conservationist who heads the China Biodiversity Conservation and Green Development Foundation, in Beijing, added a caustic dismissal. "It's not a matter of tradition," he told me. "It's a matter of money."

And now, along with the traffic of pangolins into China, this new concern has arisen: the traffic of certain viruses. There was an unheeded signal in 2019. On March 24 of that year, the Guangdong Wildlife Rescue Center, in Guangzhou, took custody of twenty-one live Malayan pangolins that had been seized by Customs Bureau agents. Most of the animals were in bad health, with skin eruptions and in respiratory distress; sixteen died. Necropsies showed a pattern of swollen lungs containing frothy fluid, and in some cases a swollen liver and spleen. A trio of scientists based at a Guangzhou governmental laboratory and at the Guangzhou Zoo, led by Jinping Chen, took tissue samples from eleven of the dead animals and searched for genomic evidence of viruses. They found signs of Sendai virus, harmless to people but known for causing illness in rodents. They also found fragments of RNA from coronaviruses. Still, this was not big news when the Chen group published its report on October 24, 2019—before the pandemic. The scientists noted that either Sendai or a coronavirus might have killed these pangolins, that further study could help with pangolin conservation, and that such viruses might be capable of crossing into other mammals. They did not express anything like the urgent warnings about potential coronavirus emergence seen in Zhengli Shi's pre-2019 papers.

Three months later, the word "coronavirus" carried a much different ring. The novel virus had been isolated and sequenced, the world was on alert, China had 1,287 cases of COVID-19, nine other countries (now including France and Vietnam and Singapore) had their first confirmed cases, and Shi's group had just posted their paper reporting the bat virus RaTG13, with its 96.2 percent similarity to the coronavirus. This was strong evidence that the new virus probably originated from bats, but a 4 percent difference between the genomes was far from a perfect match. It implied decades of evolutionary divergence—maybe twenty years, maybe sixty, depending on the method of calculating and the assumptions about mutation rate. Where had the new virus spent that time—in what population of bats or other animals—and how had it evolved during the interlude, and how had it then spilled into its first human host?

Where, how, how? With those questions pending, another link in the chain of suppositions emerged. On February 7, the president of South China Agricultural University, Yahong Liu, declared at a press conference in Guangzhou that a team from her institution, in work not yet published, had found what may be the intermediate host of the virus, bridging the gap between bats and humans: pangolins.

According to a report from Xinhua, the official Chinese news agency, the pangolin virus that these researchers had investigated was a 99 percent match with the novel coronavirus showing up in people. The announcement was an overstatement of what the researchers had found, but it caused a flurry of headlines. Even the CITES secretariat, in Geneva, echoed the claim, tweeting the next day that "#Pangolins may have spread #coronavirus to humans," and sugaring that sour tweet with video footage of cute pangolins—one of them a female with a juvenile on her back—climbing tree branches and snooping for ants. The implication was: these adorable animals carry lethal viruses, so best to leave them alone. When the study from South China Ag went online, with its tables and graphs and carefully chosen language, the big result was not quite so big as President Liu had advertised, though still dramatic. The coronavirus genome that these researchers had assembled, from pangolin lung tissue samples taken from some of the same woebegone animals that died in the Guangdong Wildlife Rescue Center, contained some gene regions that were 99 percent similar, yes, to equivalent parts of the SARS-CoV-2 genome, but the overall match wasn't that close. Maybe two coronaviruses had converged in a single animal, the researchers wrote, and swapped sections of their genomes, a recombination event. Such an event may even have proved fateful, by patching one genomic section of a pangolin coronavirus into a bat coronavirus. That section was the receptor-binding domain.

Other teams in other parts of the world were meanwhile heatedly following the same leads. This was possible, even without direct access to pangolin samples such as the South China Ag group had, because in the present era of globalized genomics, researchers have recognized the value—to themselves, and to science at large—of sharing genome data quickly and freely, even before they publish. They do this by uploading genomic sequences, partial or complete, to open access databases, supported by national governments as a scientific service and known to molecular biologists and geneticists by their shorthand labels, such as GenBank, GISAID, SRA, and RefSeq. Several of those databases are maintained, wholly or in partnership, by the U.S. National Center for

Biotechnology Information (NCBI), in Bethesda, Maryland. Other scientists are at liberty to download the full sequences or partial segments and scrutinize them, using their own computational tools, algorithms, and hypotheses.

One team, led by Joseph F. Petrosino at the Baylor College of Medicine, in Houston, did that with the viral genome sequences gathered originally by Jinping Chen's group from the dead pangolins in Guangzhou. This team's involvement began when a lively young bioinformatician (a data handler and analyzer) named Matt Wong, employed in Petrosino's lab but with broad and roving interests beyond it, became intrigued by the question of what the new coronavirus out of Wuhan might resemble. That question *What might it resemble?* is, of course, an on-ramp to the questions *Where did it come from?* and *How was it made?*

25

Joe Petrosino is chair of the Department of Molecular Virology and Microbiology at Baylor. He began his career as a biodefense researcher, using genomics to work toward vaccines against potential biowarfare agents such as anthrax and the tularemia bacterium. After a few years, Petrosino shifted his focus to the genomics of the human microbiome, the aggregation of all microbial creatures that inhabit the human body, some of them potentially harmful, some benign, and some performing valuable services. In 2007, the U.S. National Institutes of Health (NIH), the country's largest funder of medical research, launched a Human Microbiome Project, in recognition, as Petrosino told me, of the fact that these microbes "are hugely important for all sorts of areas of human health and prevention of disease." And not just infectious disease, he said, but things ranging from cancer to autism to diabetes. That community of passengers—the ecosystem that is us—includes a great diversity of bacteria, but also viruses, fungi, archaea (simple-celled creatures like bacteria, but distinct, undiscovered until 1977), and protozoans. It's a complex jumble. To discern what is present, and to investigate how certain microbiota might trigger this disease or prevent that one, Petrosino's laboratory employs powerful software tools to pull relevant genomic bits out of the seething variousness of a blood sample or a smudge of stool. Using computers to grasp and interpret biological data, including genomic sequences—that's what bioinformatics is, and Petrosino had a whole team of bioinformaticians and genome analysts in his lab. Matt Wong joined that team, charged with developing a software tool to sort relevant genomic data from the welter of fragments.

Wong wasn't a graduate student. He had no interest in doing a PhD, at least partly because that would require him to obsess on a single project rather than moving from one to another. He was a gun for hire, and this was his third lab job since graduating from the University of California at Davis with a BS in biochemistry. During the Davis years, he took "a lot of programming courses on the side" and then "sort of like tripped into bioinformatics" as a professional. He joined the Petrosino lab in 2012 and worked there, unnoticed by the wider world, for eight years.

"How would you describe your role?" I asked Wong when I caught him by Zoom. He wasn't in Houston at the time; he was in Las Vegas for a few days, he explained, because his pool team back home had triumphed locally and won a free trip to a tournament at the Rio Hotel.

"I wrote pipelines to analyze the FASTQs that came off the Illumina machine," he said. Then he explained what the hell that meant. A pipeline is a series of procedural steps in validating and analyzing data. An Illumina machine is a sequencer. FASTQs are . . . never mind.

Petrosino had explained Wong's job too, somewhat more simply, as creating "computational tools to be able to mine viral genomic data from complex mixtures of data." Think of a needle in a haystack, said Petrosino. The haystack is a biological sample. The hay is all sorts of DNA, from bacteria, from other microbiota, from the human host, plus perhaps RNA from various viruses. The needle is a virus of interest. The tools that Matt Wong helped develop represent a powerful magnet, which can pull the needle out of the hay.

Petrosino then switched metaphors. At the start of 2020, he explained, his lab was working, as always, on the microbiome. They were funded, with grant money, to work on the microbiome. But along came this novel coronavirus out of Wuhan. As is typical for Matt (and a lot of other bioinformaticians he had worked with, said Petrosino), such a novelty can lead down a sidetrack. "It's like 'Squirrel!' and he went off." A squirrel can be an irresistible distraction. Each morning I walk a smart young Russian wolfhound who is also fascinated by arboreal rodents, so I understood.

At that time, January 2020, Petrosino himself happened to be gone on a professional trip, meeting with other chairs of microbiology departments in Belize. Wong emailed him, asking, "Hey, can you get your hands on the Wuhan strain genome, so that I could use it?" He was unaware, and so was Petrosino, that Eddie Holmes had already posted that genome on Virological. No matter, Wong decided. Like a hound after a squirrel, he plunged cheerfully into the bushes. "I just wanted to see how well my pipeline would generate the genome without actually having the genome in the database." He wanted to do it the hard way. "I mean, somebody had already built the full genome. I just wanted to see whether I could do it without having known what the genome was." What he had, what he could access from his computer in Houston, was a haystack of raw data, in the form of short segments of genomic code—human, viral, microbiome, what have you—drawn from liquidy samples flushed up from the lower respiratory tract of one of the first patients in Wuhan. Those segments, known as *reads*, were on file at the database called SRA, the Sequence Read Archive, a collaborative effort of the National Center for Biotechnology Information and its European and Japanese equivalents. Using the SRA reads, plus methods and tools of his own devising ("my pipeline"), Wong assembled a version of the SARS-CoV-2 genome that closely matched the genome he found, in a database, labeled "Wuhan seafood market pneumonia virus"—the one put there by Yong-Zhen Zhang, Eddie Holmes, and their Chinese colleagues. This turned out to be, for Matt Wong, the warm-up exercise. And then, on January 26, a helicopter crashed on a hillside near Calabasas, California.

Nine people were killed, the pilot and all passengers, including the retired basketball player Kobe Bryant and his thirteen-year-old daughter. The news strongly registered on Matt Wong, as he prepared to take his turn as a presenter at one of the weekly lab meetings. "Kobe Bryant just died and I was thinking, like, what am I doing with my life?" The squirrel became larger and more significant, more beckoning, like a rodent transmogrified into a snow leopard. "Up until that point," he told me, "I actually had no idea about coronaviruses." He had never studied them. He had never explored how the genome of a coronavirus might translate into functional aspects of infection. He had never wondered about their origins. Now he did.

Wong knew little more, at that point, than what most of us had heard in January 2020: that the novel virus probably came to humans from a bat, possibly after passing through some other animal. He thought, "Maybe I can try and help find an intermediate host." He began downloading datasets that might contain something relevant—genomic fragments of viruses and other creatures sampled from various animals in southern China, including fruit bats and horseshoe bats and pangolins. One of those was RaTG13, the sequence of a coronavirus that Zhengli Shi had found in a Yunnan bat. Included amid this vast jumble of fragments—like the pieces of two hundred jigsaw puzzles mixed and tumbled in a laundromat dryer—were some sequences of coronavirus from the dead pangolins at the Guangdong Wildlife Rescue Center, which Jinping Chen's group had sampled the previous year. "Just grabbing the datasets is not enough to truly figure out what's going on," Wong told me. You would have to combine the raw information of the datasets with data on known viruses, take account of viral mutation that changes some fragments slightly, and weigh the varied reliability of the sequencing that had been done. "Luckily for me," he added, "I had already written a pipeline that was designed for exactly that task." With his quicksilver software tools and his brilliantly nerdy skills, he cleaned up the sequences, assembled a genome, and aligned it with "Wuhan seafood market pneumonia virus," to see what matched closely and what didn't.

The closest overall match was RaTG13, Shi's sequence from a horseshoe bat. But there was another match, an anomaly, that caught Wong's attention. One small region of the pangolin virus genome much more closely resembled the corresponding region in the new human virus. If you translated the RNA letters into the amino acids they signified, it amounted to about two hundred amino acids, from a total of roughly ten thousand. He described this much in his talk at the lab meeting, without drawing any strong conclusions. "I was like, 'Yeah, this is pretty cool. I found a random coronavirus in a pangolin dataset that's kinda close to the outbreak strain. But these other Chinese scientists found one that's even closer. So, you know, probably did come straight from a bat.'" No memorable reaction from the group. He had noticed something odd, insignificant, and "cool." But immediately afterward, Wong found himself wondering, "What if that region actually means something important?"

Kobe Bryant had died, life was short, and he wanted larger meaning. He dove into the literature, downloading "a whole bunch of papers" on coronaviruses and reading about their genomic structure and functions. Quickly he saw that his anomalous region fell within something called the spike protein. "The spike protein—that sounds pretty important," is how he recollected his thinking to me. "So then you read about the structure of the spike protein, and you're like—wait a second. Receptor-binding domain! That sounds *really* important!" It had implications for human immune response, he realized, and potentially for vaccine development. Without a functional receptor-binding domain, this thing couldn't infect human cells. "That's how you would neutralize the virus."

There were also implications regarding the origin of the virus. If it came from a bat, why did it have a receptor-binding domain that looked so much like the RBD from a virus in pangolins?

Next morning, Wong wrote a brief statement, just one paragraph describing what he had found, and attached two figures showing the comparisons among RaTG13, and the new virus, and his pangolin virus sequence in that region of the receptor-binding domain. RaTG13 was 90 percent similar to the Wuhan virus's RBD. His pangolin segment was 97 percent similar. "This result indicates a potential recombination event" that shaped the new virus, he concluded. Wong posted this on Virological, signing only with a nonsense username he had favored since his boyhood, "torptube." Virological existed for such postings, according to Andrew Rambaut, its founder. Rambaut never intended it as an outlet for fully drafted papers; there were other sites, such as bioRxiv, offering that. Rambaut saw Virological as a forum for speculative thoughts, partial data, interesting figures, and provisional ideas, exchanged freely among scientists, as they exchange brainstorms in the hotel bar at a conference. "I'm quite resistant to it becoming a preprint server," Rambaut told me, and then, jocosely, "Maybe it's a *preprint* preprint server."

Having posted his own little pre-preprint, Wong headed for another lab meeting and ran into Joe Petrosino, his boss, in a discussion with other members of the department. By Wong's recollection, one of them asked Petrosino whether his lab was doing anything on the novel coronavirus. It was a natural question—virology labs all over the world were shifting priorities, setting aside other work, and starting efforts on the menacing new virus. Petrosino said, No, nothing major in his lab, but Matt here has a little side project. At which Wong piped up and said, Yes, he'd found something interesting and just posted it on Virological that morning.

"I'm pretty sure Joe like literally face-palmed at that moment," Wong recalled. A bioinformatician in his lab, a data wrangler with no PhD, had just spoken on the world's smartest virology website about the world's scariest new virus? What if the post was completely wrong? What if it looked bad? "Oh, don't worry," Wong told Petrosino, "I didn't use, like, my name." It wasn't an announcement from the Joseph Petrosino lab. It was just a little FYI from torptube. Petrosino's recollection is a little less colorful, while giving Wong full credit.

In any case, Kristian Andersen saw the torptube post on Virological, and it helped persuade him, at a crucial moment, that SARS-CoV-2 was naturally evolved.

Joe Petrosino also recognized the merit in what Wong had found, and over the next few days he and Wong, with two other coauthors, produced a manuscript paper describing the RBD match and its possible implications. They promptly posted that as a preprint and submitted it to a major American journal. But at the journal to which it was sent, where the editors were deluged with COVID-19 papers, it languished behind submissions with clinical relevance, and in the meantime other scientists also noticed the pangolin connection and published papers in other leading journals. By the time the Wong-Petrosino manuscript received attention, the moment of freshness and priority had passed. Their paper has never been published. But torptube's brief message on Virological was a key influence on the paper that Andersen and his coauthors were drafting on the origin of the virus.

"I remember having a conversation with Bob and Eddie and Andrew," Andersen told me. "It was like, 'Oh my God, there it is.'" So the receptor-binding domain of SARS-CoV-2 had not been designed in some laboratory, they concluded. And there were no signs that it had been inserted by some malevolent or reckless scientist. It had been designed by natural selection and inserted, perhaps, by recombination. It was out there, in nature, facilitating the coronavirus infection of pangolins, and maybe of other animals too. Torptube had found it.

They continued drafting their paper. By this time, they had added one other coauthor, Ian Lipkin at Columbia. This paper went up promptly as a preprint on Virological (notwithstanding Rambaut's preference for *pre*-preprint ruminations) on February 16, while it made its way through the editorial process more slowly at an important journal. It was titled "The Proximal Origin of SARS-CoV-2." To say that it became a focus of contention would be understatement.

26

They were candid: the genome of SARS-CoV-2 as it had emerged in Wuhan carried coding for two notable features that required explaining. These two features were unexpected, insofar as they hadn't been seen before in other known SARS-like coronaviruses (although they were not unique among coronaviruses generally, and they had parallels in other viruses too). Both features lay on the spike protein, the cell grabber. The first was the receptor-binding domain. The second was the furin cleavage site, which allowed the spike to cleave, and thereby fuse to the cell membrane, in response to a tickle from the host's furin. The effect of both features was to make the virus more capable of infecting humans. For that reason, the five authors noted, there had already been "considerable discussion" in the scientific community, and beyond, on the origin question. (But scarcely a murmur, compared to what would come later.) So they had looked closely at the two features, and they had discussed four conceivable scenarios for how those features might have arisen.

The first scenario was genetic manipulation in a laboratory. Andersen and his coauthors dismissed that one, on grounds that the main body of the SARS-CoV-2 genome bore no resemblance to any virus backbone ever known to have been used in viral engineering, and that the RBD was an anomalous concoction of which the effectiveness couldn't have been foreseen. Only one form of trial and error could have produced it, they judged, a form that is tireless, blind, and infinitely persistent: evolution by natural selection.

That was the second scenario: natural selection acting on the virus, within its animal host, before spillover into humans. The pangolin coronavirus was important evidence here, demonstrating that an RBD almost identical to the RBD in SARS-CoV-2 *could* evolve—because it *had* evolved—in a wild animal. And of course the cross-border trade in live pangolins for meat and medicine, with thousands of the animals passing from trader to buyer to butcher to consumer, provided plenty of opportunities for a pangolin virus to spill into some person. The furin cleavage site was another matter, and nothing like that had turned up in a pangolin virus.

In the third scenario, a progenitor form of the virus had passed by spillover from an animal host into one or more persons, and *then*, during some period of slow, inefficient, unrecognized transmission among humans, acquired its nifty cleavage site by evolutionary steps that vastly improved its mediocre transmissibility. This period of sputtering, unnoticed human infection might have happened in October and November of 2019, possibly stretching back even earlier. There was no known evidence for such a silent prelude to the pandemic, but some might be found. Blood samples drawn for other purposes, during that earlier period, should be checked now for antibodies against the virus, the authors suggested. That suggestion would soon be followed by other scientists, with ambiguous results. (I'll circle back to those ambiguous studies, below.) The work done so far has only begun to explore whatever clues might exist in archived samples. One value of frozen blood serum is that it can retain evidence of viral presence much longer than, say, a doorknob or an elevator button or a cutting board in a wet market.

Scenario four was the most intricate. It imagined that some team of scientists had performed "passaging" experiments with a SARS-like coronavirus—that is, intentionally infecting a series of laboratory animals, each animal given the virus in the form that emerged from a previous animal, and thereby inviting the virus to adapt better and better to those animals as it went. Or the virus might have been similarly passaged not in live animals but in cultured cells, dish by dish, using laboratory-captive strains of once human (or at least once primate) cells. That might have allowed such an RBD to evolve (but the pangolin RBD rendered this notion unnecessary). And then maybe the virus got into a lab worker by accidental infection, and then maybe that worker coughed on another. But it was a string of fractional unlikelihoods, and when you multiply them together, as with any fractions, the fractions get smaller; the unlikelihood gets larger. Finding similar RBD adaptations in a wild virus infecting pangolins, the authors wrote, "provides a much stronger and more parsimonious explanation of how SARS-CoV-2 acquired these via recombination or mutation."

The furin cleavage site was even more difficult to explain by this passaging scenario. Experimental work (some published, and more would come) seemed to show that a structure so complex and improbable as a furin cleavage site just wouldn't arise from running a virus through cultured cells, no matter how much those cells might resemble human cells. One problem, only one, was that certain features of the cleavage site (the way it seems to shield itself from the antibodies of a host) suggest "the involvement of an immune system." In other words, the pressure of natural selection to shape such a defense. Gazelles would not run so fast if they had evolved in the absence of lions and cheetahs. Tortoises would not possess shells if they didn't need protection from foxes and coyotes. But a petri dish of cells contains no immune system. Therefore, passaging a virus in cell culture offers no selection pressure to sculpt defenses against antibodies, such as the defenses that SARS-CoV-2 seems to have.

"The Proximal Origin of SARS-CoV-2" was published on March 17, 2020. That was the final version of the paper, but it wasn't the final word on the subject, as Andersen and his coauthors acknowledged. It was provisional, as scientific explanations, however well supported, should always be. They found the natural origin hypothesis more "parsimonious" than the alternatives—meaning simpler, less cluttered with improbabilities and tenuous suppositions—and parsimony is another cardinal value in science. This was early, they knew; time and further research might add clarity. "More scientific data could swing the balance of evidence to favor one hypothesis over another," they wrote. More argument might swing the balance of opinion too, but that's another matter.

27

More scientific data came soon, not on the origin of the virus but on its behavior, notably aboard the cruise ship *Diamond Princess* in the waters off China and Japan. This boat became almost a floating laboratory for the study of SARS-CoV-2 infection and transmission.

The *Diamond Princess*, an American vessel owned and operated by Princess Cruises, departed from Yokohama, Japan, on January 20, 2020, for a voyage along the coasts of China, Vietnam, and Taiwan. It's a big luxury liner, among the biggest in the world, specializing in vacations in the seas around Japan and Southeast Asia, and catering (like most luxury cruise ships) to a senior clientele, primarily sixty and older. That day it carried 3,711 people, including 2,666 passengers and 1,045 crew. One of the passengers was an eighty-year-old man from Hong Kong who had flown into Japan several days earlier to catch the ship at its start. While waiting for embarkation, he developed a cough. He boarded anyway. Five days later the octogenarian disembarked back in Hong Kong when the ship docked there, cutting his cruise short, evidently because he was feeling sick. Seven days after that, he was hospitalized with a fever and tested positive for SARS-CoV-2. The backstory on this man, according to one report, is that he had visited Shenzhen in mainland China a week before traveling to Yokohama. The report doesn't mention what he did in Shenzhen, but the implication is that he picked up the virus there, somehow.

The *Diamond Princess* proceeded south to Vietnam for a few days, then back up to Keelung, a port on the north coast of Taiwan, on January 31, one day before the man in Hong Kong would test positive. Keelung is a historic city, just half an hour from the capital, Taipei, and most of the passengers went ashore for a day tour. Taiwan had recorded its own first case of COVID-19 on January 21, and an epidemic command center had been activated, with some responsibility for border control and quarantine. A week later came entry restrictions on foreigners from areas where the virus was rife. But those apparently weren't restrictive enough to prevent shore day for the *Diamond Princess* passengers. On

February 1, the Hong Kong Department of Health announced through a government website that the feverish eighty-year-old was a COVID-19 case. Next day, Hong Kong notified Japan, to which the ship by then was returning, and when that news echoed back to Taiwan, it created "a temporary public panic about community spread," according to one group of Taiwanese scientists. Passengers aboard the *Diamond Princess* were still oblivious or unconcerned, enjoying their crowded dance parties, casinos, and buffets.

The ship returned to Yokohama a day earlier than scheduled, on the evening of February 3, but it wasn't allowed to dock, so it anchored offshore in the harbor. A quarantine team from the Japanese Ministry of Health, Labor and Welfare went aboard and, working that night, identified 273 people (mostly passengers, a few crew members) who showed Covid-like symptoms or reported close contact with the eighty-year-old man. The health officers began testing the passengers and crew, with throat swabs and PCR, but results came slowly because of laboratory limitations. Of the first batch, comprising thirty-one tested individuals, ten were positive. So the ship was presumably aflame with Covid. The authorities declared that this cruise was over, and everyone aboard would be in quarantine for fourteen days. It's safe to assume that no one was dancing now.

Full results for the 273 tested people arrived on February 7, showing sixty-one positives, or a rate of 22 percent. That was alarming enough that the ministry team proceeded, over the next ten days, to test everyone. It had been obvious for roughly a month that this new virus could transmit from one human to another, but questions remained about how readily, under which circumstances (indoors, outdoors, casual contact or close contact), by what modes (respiratory droplets, airborne virions, doorknobs, handshakes), and in what patterns (uniform spread or super-spreading events). And there was another important question almost too gloomy to be asked: Could the virus transmit from asymptomatic cases? Were there "silent" spreaders—people walking around, feeling fine, shedding virus everywhere they went?

On February 17, the Japanese government emended its shipboard quarantine order to the extent of allowing other countries to take citizens off and air-evacuate them to be quarantined in their own countries. Two planes, chartered by the United States government and carrying more than three hundred people, promptly departed for the U.S. Some of those Americans were flown to Omaha and entered a quarantine facility on a National Guard base along the Platte River, where members of Ali Khan's faculty from the University of Nebraska Medical Center managed their further isolation, screening, and care. Two other flights transferred Hong Kong residents back to that city, where they reentered quarantine in a recently built and otherwise unoccupied public housing complex called Chun Yeung, a quintet of high-rises amid the forested hills of the New Territories. That's where K.Y. Yuen and a team of colleagues pounced on them for a study.

Yuen recognized that the *Diamond Princess* event was a scientific opportunity, a sort of natural experiment in controlled contact, contagion, and silent spreading. "At that time," Yuen told me, "we already have some other data that help us to think about asymptomatic or presymptomatic infection." Those other data, sparse but important, included the family cluster detected early on by Yuen's associates at the Shenzhen hospital. One of those family members, a ten-year-old child who had refused to wear a mask when the family visited Wuhan, had been confirmed as a positive case despite showing no clinical signs. Reporting on that cluster, back on January 24, Yuen's team had warned that it would be crucial "to isolate patients and trace and quarantine contacts as early as possible because asymptomatic infection appears possible." But that observation was haphazardly captured and anecdotal compared to the people aboard the *Diamond Princess*, every one of whom was Covid-tested by Japanese health workers before stepping off the ship. "All these cases," Yuen said. "They're very controlled."

Almost four hundred Hong Kong residents had been aboard the cruise, a tenth of the passenger total. Seventy-six of those tested positive during the shipboard screening and went into hospital isolation in Japan. Two of the hospitalized patients died. Almost three hundred other Hong Kongers tested negative and were allowed to disembark. Some stayed in Japan, the rest flew back to be quarantined at the Chun Yeung housing complex. The scientists obtained voluntary participation from 215 adults, then began testing them all again, and retesting them every four days for the length of the two-week quarantine. Nine of those people tested positive, by multiple measures, including both PCR and antibodies. Translated: their bodies contained the virus, and their immune systems knew it and didn't like it. Of the nine, six remained asymptomatic throughout the whole fourteen-day quarantine period.

"If the cruise ship epidemic is a microcosm of the community outbreak scenario," Yuen and his colleagues concluded in their published report, "then individuals with or without pneumonia could carry the virus for a long period but remain asymptomatic." And they could do more than carry it; they could spread it. To me he said, "That is very important. That is why you cannot control a pandemic"—could not control *this* pandemic, anyway—"because there's so much asymptomatic cases spreading infection around." Think of it, he said: we found nine positives from the *Diamond Princess*, six of them asymptomatic. Taking that as a very rough guide, he said, for each case you identify from symptoms, "there's at least another two cases." Spreading the virus invisibly through your ship, your city, your country. By the time symptoms appear, in fact, the load of virus in a person's nose and throat has already risen to its peak.

"At what point did you know this?" I asked.

End of March, he said, but he couldn't recall the exact date.

Who did you tell?

"Well, of course, the government immediately." So the chief executive of Hong Kong knew the danger, as well as anyone who saw the preprint that Yuen and his colleagues posted online. Among political leaders and public health officials elsewhere, the guiding wisdom remained: check people for fever, test those who are coughing, and the rest of your populace should be okay.

Was this the first warning to the world about asymptomatic spread?

No, Yuen said. "The first warning is in January." That asymptomatic child in Shenzhen, the one who refused to be masked during the visit to Wuhan—that child was the first warning, as described in the paper his team had published back on January 24. The *Diamond Princess* was the second warning on asymptomatic spread, for those who needed two.

28

When you're a virologist or somebody involved in microbes," Tony Fauci told me, "you try to . . ."—he hesitated, putting the next word in careful tonal italics—". . . anthropomologize, I guess, a virus." It was his neologism for "anthropomorphize," but I knew what he meant. You've probably done that yourself, he said, in your former writings about viruses.

"Yeah."

"You turn it into a metaphor."

"Yeah."

"So if this virus were a really nefarious person, it would say, 'What do I want to do, and how can I do the most damage?' Well, first of all, 'I have to be *extremely* efficient in replicating. But I don't want to kill everybody.'" He left unspoken the reason not to kill everybody: because he cared only about evolutionary success. America's most trusted disease scientist, with his calming candor and his Brooklyn accent, thinking strategy like a nefarious virus—it was persuasive. Tony Fauci has so much steel in his spine and antifreeze in his veins, this compact man from Bensonhurst, a pharmacist's son, director of a huge federal research agency (the NIAID) since 1984, veteran of many rough days testifying to Congress, that it seems he could just as well have bossed the Gambino crime family, if he weren't so moral, or maybe served as superior general of the Jesuits. The force of his will and ambition echo forward from the fact that, in high school, at five-foot-seven, he captained the basketball team. "'I want to be a really unique bad guy,'" he continued now, in the voice of the virus. "'I want to have a situation where 40 percent of the people I infect, I don't even want them to know they're infected. I want them to be completely asymptomatic.'"

Forget the quote marks within the quote marks, he was now channeling this virus. "I want the infections to be transmitted," Fauci said, so that "50 percent of the infections are transmitted by people who are without symptoms. So. All these young, asymptomatic people—I don't really care about them." Killing is not the point, he meant. Killing is irrelevant, so long as there remain many susceptible individuals to accommodate him, the virus. "I'm not going to get rid of the population. I'm just going to be doing a lot of damage."

"Mm hm." *Keep talking,* I thought. I didn't want to break his stride.

"The elderly. And people with underlying conditions." Collateral casualties, irrelevant to the viral mission: proliferate, spread, survive.

"Mm hm."

"So at the same time," he said, in the patient tone of a high school biology teacher addressing a class of medium-bright juniors—then back into the viral persona: "I'm a hybrid. I'm a virus that causes very, very little harm, doesn't give many people symptoms." Highly capable of transmission, mild in most infected persons, invisible in many, therefore lulling the host population and its more doltish, obdurate leaders toward complacency. "At the same time that I can be absolutely deadly to a large number of people, who happen to be vulnerable."

"Yeah," I said admiringly.

"And that's the nefarious, insidious nature of this virus."

29

N efarious" and "insidious" are relative terms as well as anthropomorphic ones, of course. SARS-CoV-2 is a horrific human pathogen, inimical to our health and welfare, hateful in our eyes; nefarious and insidious *to us.* But it's still "just" a virus, doing what viruses do: obeying what I call the three Darwinian imperatives, which govern all creatures that replicate by way of variable genomes, whether they are viruses or fennel plants or rats or dandelions or kangaroos. Those imperatives are 1) copy yourself as abundantly as possible, 2) expand yourself in geographical space, and 3) extend yourself in time. Most people tend to consider *all* viruses hateful, as though these little self-replicating packets of DNA and RNA have no role on the planet except to make humans sneeze, cough, bleed, asphyxiate, suffer, and die. Some viruses do that, true. Most of the famous viruses do, and that's why they are famous. But to understand just what SARS-CoV-2 is, where it comes from, how it functions, and even *why* it functions as it does, a broader perspective is useful—even broader than Tony Fauci's "I, Virus" perspective. For starters, let's imagine Earth without any viruses at all.

We wave a wand, and they all disappear. Rabies virus is suddenly gone. The polio virus is gone. Ebola is gone. Make that: all six ebolaviruses, including Sudan virus, Taï Forest virus, Bundibugyo virus, and Reston virus—gone. Measles virus, mumps virus, and the various influenzas are gone. Immediately this brings vast reductions of human misery and death. HIV-1 is gone, so the AIDS catastrophe never happened, and HIV-2 is gone also. Nipah and Hendra and Machupo and Sin Nombre viruses, with their records of ugly mayhem, are gone. The dengue viruses, gone. All the rotaviruses, gone, a great mercy to children in developing countries who die by the hundreds of thousands each year from diarrhea and dehydration. Zika virus, gone. Yellow fever virus, gone. Herpes B virus, carried by some monkeys, often fatal when passed to humans, gone. Nobody suffers anymore from chicken pox, hepatitis, shingles, or even the common cold. Variola, the agent of smallpox? That virus was eradicated in the wild by 1977, but now it vanishes from the high-security freezers where the last spooky samples are stored. SARS-CoV is gone. MERS-CoV is gone. Five other coronaviruses also known to infect humans but causing only mild symptoms, such as OC43, are gone. All the bat-borne coronaviruses and all the pangolin-infecting coronaviruses are gone. And of course SARS-CoV-2, nefarious and insidious and catastrophic for us, is gone. Do you feel better?

Don't.

This scenario is more equivocal than you think. The fact is, we live in a world of viruses—viruses that are unfathomably diverse, immeasurably abundant, and ambivalent in their effects, even upon human health and welfare. The oceans alone may contain more virions than there are stars in the observable universe. Mammals may carry at least 320,000 different viruses. When you add the viruses infecting nonmammalian animals, plants, terrestrial bacteria, and every other possible host, the total comes to . . . lots. And beyond the big numbers are big consequences of a sort we wouldn't expect: many of those viruses bring adaptive benefits, not harms, to life on Earth, including human life.

We couldn't continue without them. We wouldn't have arisen from the primordial muck without them. There are two lengths of DNA that originated from viruses and now reside in the genomes of humans and other primates, for instance, without which—an astonishing fact—successful pregnancy would be impossible. There's viral DNA, nestled among the genes of terrestrial animals, that helps package and store memories in tiny protein bubbles. Still other genes co-opted from viruses contribute to the growth of embryos, regulate immune systems, resist cancer—important effects only now beginning to be understood. Viruses, it turns out, have played crucial roles in launching major evolutionary transitions. Eliminate all viruses, as in my thought experiment, and the immense biological diversity gracing our planet would collapse like a beautiful wooden house with every nail abruptly removed.

A virus is a parasite, yes—a genetic parasite, to be more precise, using the resources of other organisms to replicate its own genome—but in some instances that parasitism is more like symbiosis, mutual dependence that profits both visitor and host. Notwithstanding the horrors, miseries, and sorrows inflicted on humans by SARS-CoV-2, it behooves us to recognize and remember that viruses, like fire, are a phenomenon that's neither in all cases bad nor in all cases good; they can deliver advantage or destruction. They are the dark angels of evolution, terrific and terrible. That's what makes them worth understanding, rather than just fearing and deploring.

To appreciate the multifariousness of viruses, you need to start with the basics of what they are and what they are not. It's easier to say what they are not. As I noted earlier, they are not living cells. A cell, of the sort assembled in great number to make up your body or mine or the body of an octopus or a primrose, contains elaborate machinery for building proteins, packaging energy, and performing other specialized functions, the details depending on whether that cell happens to be a muscle cell or a xylem cell or a neuron. A bacterium is also a cell, with similar attributes, though much simpler. Likewise, an archeon: a bit more complex than a bacterium, also with no cell nucleus, but capable of metabolism and reproduction. A virus is none of this.

Saying just what a virus *is* has been challenging enough that definitions have changed over the past 120-some years. Martinus Beijerinck, a Dutch botanist who studied tobacco mosaic virus, speculated in 1898 that it was an infectious liquid. For a time, a virus was defined mainly by the size of its particles, as a thing smaller than a bacterium, too small to be caught by a tiny-holed ceramic filter, but which could still cause disease. Still later, a virus came to be understood as a submicroscopic agent, bearing only a very small genome, that replicates inside living cells—which was correct, but only a first step toward comprehending these things.

"I shall defend a paradoxical viewpoint," wrote the French micro-biologist André Lwoff in "The Concept of Virus," an influential essay published in 1957, "namely that *viruses are viruses*." Not a very helpful definition, just another way of saying "unique unto themselves." He was just clearing his throat before beginning a lengthy disquisition.

Lwoff knew that viruses are easier to describe than to define. He knew that each viral particle consists of a stretch of genetic instructions (written either in DNA or RNA) packaged inside a protein capsule, known as a capsid. The capsid, in some cases, is surrounded by a membranous en-velope (like the caramel on a caramel apple), which protects it and helps it catch hold of a cell. A virus can copy itself only by entering a cell, or at least injecting its genome, and commandeering the 3-D printing machin-ery that turns genetic information into proteins.

If the host cell is unlucky, many new virions are manufactured, they come busting out, and the cell is left as wreckage. That sort of damage—such as what SARS-CoV-2 causes in the epithelial cells of the human airway—is partly how a virus becomes a pathogen.

But if the host cell is lucky, maybe the virus simply settles into this cozy outpost, either going dormant or back-engineering its genome into the host's genome, and bides its time. This second trick is what retrovi-ruses, such as HIV-1, do. It carries many implications for the mixing of genomes, for evolution, even for our sense of identity as humans. Eight percent of the human genome consists of viral DNA that has been inserted into our lineage, over millions of years, in this way. That's a very different take on the trope of "I, Virus." Both you and I, as well as Tony Fauci and everyone else, are 8 percent viral in our genomes. And that doesn't even count the viruses of the human microbiome, carried in our bellies and on our skin and elsewhere, but not in our genomes like the retroviral DNA.

The notion of viruses as malign without exception, that they always and only do harm, is not unique to nonscientists. The eminent British bi- ologist Peter Medawar, in a popular 1983 book coauthored with his wife, Jean, asserted that "No virus is *known* to do good: It has been well said that a virus is 'a piece of bad news wrapped up in a protein.'" They had it wrong, and so did a lot of scientists at the time, because 1983 was a little too early for finding viruses in genomes and discerning their func- tion. That remains a view still embraced, understandably, by anyone whose knowledge of viruses is limited to such bad news as COVID-19 and AIDS and the flu. But today many viruses are known to do good. What's wrapped up in the protein capsid is a genetic dispatch—a message in a bottle—and that might turn out to be good news or bad, depending.

Where did the first viruses come from? To answer that requires squinting back almost four billion years, to the time when life on Earth was just emerging from an inchoate cookery of long molecules, simpler organic compounds, and energy.

Let's say some of the long molecules (probably RNA) started to rep- licate. Serving as templates of selfness, pulling in small molecules from their environment to fit where appropriate, they made copies of them- selves. Darwinian natural selection would have begun there, as those molecules—the first genomes—reproduced, mutated, and evolved. Groping for competitive edge, some may have found or created protec- tion within membranes and walls, leading to the first cells. These cells gave rise to offspring by fission, splitting in two. They split in a broader sense too, diverging to become Bacteria and Archaea, two of the three domains of cellular life. The third, Eukarya, arose sometime later. That one includes us and all other creatures (animals, plants, fungi, certain microbes such as amoebas and diatoms) composed of cells with complex internal anatomy, such as a nucleus neatly holding the genome. Those are the three great limbs on the tree of life as presently drawn: Bacteria, Archaea, Eukarya.

Wait, what about viruses? Where do they fit? Are they a fourth major limb? Or are they a sort of mistletoe, a parasitic attachment wafted in from elsewhere? Most versions of the tree omit viruses entirely, because to place them anywhere is to take a position on an issue even more complex than the tree of life.

One school of thought holds that viruses shouldn't be included on the tree because they aren't alive. That's a circular and irresolvable argument, hinging on how you define "alive." More fruitful is to grant viruses inclusion within the big tent called Life and then wonder about how they got in.

There are three leading hypotheses to explain the evolutionary origin of viruses, known to scientists in the field as viruses-first, escape, and reduction. Viruses-first is the idea that viruses came into existence before cells, somehow assembling themselves directly from that primordial cookery of self-replicating molecules. The escape hypothesis posits that genes or stretches of genomes leaked out of cells, became encased within protein capsids, and went rogue, finding a new niche as parasites. The reduction hypothesis suggests that viruses originated when some cells downsized under competitive pressure (it being easier to replicate if you're small and simple), shedding genes until they were reduced to such minimalism that only by parasitizing cells could they replicate and perpetuate their lineages.

Each of the three hypotheses has merits. But in 2003, some new evidence tipped expert opinion toward reduction: the giant virus.

It was found living (or "living," if you prefer) within amoebas. These amoebas had been collected in water taken from a cooling tower in Bradford, England. Inside some of them was this mysterious blob. It was big enough to be seen through a light microscope (viruses supposedly were too small for that, visible only by electron microscope), and it looked like a little bacterium. Scientists tried to detect bacterial genes within it, but found none.

Finally a team of researchers in Marseille, France, invited the thing to infect other amoebas, sequenced its genome, recognized what it was, and named it Mimivirus, because it mimicked bacteria, at least regarding size. In diameter it was huge, bigger than the smallest bacteria. Its genome was also huge for a virus, almost 1.3 million bases long, compared to, say, 13,000 for an influenza virus, 30,000 for a coronavirus, or even 194,000 for smallpox. It was an "impossible" virus: viral in character but too big in scale, like a newly discovered Amazon butterfly with a four-foot wingspan.

Jean-Michel Claverie was a senior member of that Marseille team. The discovery of Mimivirus, Claverie told me, "caused a lot of trouble."

Why? Because sequencing the genome revealed four very unexpected genes—genes for coding enzymes presumed to be uniquely cellular and never before seen in a virus. Those enzymes, he explained, are among the components that translate the genetic code to assemble amino acids into proteins.

"So the question was," Claverie said, "what the hell has a virus the need" for those fancy enzymes, normally active in cells, "while he has the cell at his disposal, okay?" What need indeed? The logical inference is that Mimivirus has them as holdovers because its lineage originated by genomic reduction from a cell.

Mimivirus was no fluke. Similar giant viruses were soon detected in the Sargasso Sea, and the early name became a genus, *Mimivirus*, containing several giants. Then the Marseille team discovered two more behemoths—again, both parasites of amoebas—one taken from shallow marine sediments off the coast of Chile, the other from a pond in Australia. Up to twice as big as a Mimivirus, even more anomalous, these were assigned to a separate group, which Claverie and his colleagues named Pandoravirus, evoking Pandora's box, as they explained in 2013, because of "the surprises expected from their further study."

Claverie's senior coauthor on that paper was Chantal Abergel, a virologist and structural biologist and also his wife. Of the Pandoravirus group, Abergel told me, with a weary laugh, "They were highly challenging. They are my babies." She explained how difficult it had been to tell what they were, these creatures—so different from cells, different also from classical viruses, carrying many genes that resembled nothing ever before seen. "All of that makes them fascinating but also mysterious." For a while she called them NLF: *new life form.* But from observing that they didn't replicate by fission, as bacteria and archaeons do, she and her colleagues realized they were viruses—the largest and most perplexing ones found so far.

These discoveries suggested to the Marseille group a bold variant of the reduction hypothesis. Maybe viruses did originate by reducing from ancient cells, but cells of a sort no longer present on Earth. That is, not from either the Bacteria or the Archaea domains, or even from whatever cellular ancestor those two shared, but from Microbe Lineage X, still another domain of life that went extinct . . . except for its remnant form, viruses. That's a little like the paleontologist's cheerful reminder: *Dinosaurs didn't go completely extinct; they're still here, but we call them birds.* Aber-gel and Claverie don't speak of Microbe Lineage X. They refer instead, in their papers, to a kind of "ancestral protocell" that might have been differ-ent from—and in competition with—the universal common ancestor of all cells known today. Maybe these protocells lost that competition and were excluded from all the niches available for free-living things. They may have survived as parasites on other cells, downsized their genomes, and become what we call viruses. From that vanished cellular realm, maybe only viruses remain, like the crows in the trees, with their deep genetic remnants of *Tyrannosaurus rex.*

30

"The Proximal Origin of SARS-CoV-2," by Kristian Andersen and his collaborators, appeared in the journal *Nature Medicine* just six days after the WHO declared officially what any sensible person could see: that the COVID-19 crisis was a pandemic. By then, China had reported 81,116 confirmed cases, Italy stood second in tribulation with 27,980 confirmed cases and 2,503 deaths, and the United States had counted a "mere" 3,503 confirmed cases, fifty-eight of those fatal. The numbers of unconfirmed cases were undoubtedly much higher, but by how much was unknowable because diagnostic testing was sparse and, in some places, most glaringly the U.S., dysfunctional. Barbados had reported its first two cases, in peo-ple recently arrived from the United States. Ethiopia had five cases and Uz-bekistan had four. Detected or undetected, the virus was getting around.

"Our analyses clearly show," Andersen and his coauthors wrote, "that SARS-CoV-2 is not a laboratory construct or a purposefully manipulated virus." Their logic and evidence covered those two features of the genome that seemed anomalous at first look, the receptor-binding domain and the furin cleavage site. They had recognized, thanks to Matt Wong, that very similar RBDs exist in the wild, notably among coronaviruses infecting pangolins. Natural selection had designed them. The one in SARS-CoV-2 no longer seemed anomalous.

The furin cleavage site was a little more complicated, because no such mechanism had been found in pangolin coronaviruses, nor in RaTG13, the bat virus that resembled SARS-CoV-2 closely. The authors doubted that the cleavage site could have been engineered or cultured-up in a lab, for several intricate reasons, persuasive to most molecular virologists but not destined to satisfy everyone. There would be critics. There would be outcries that maybe this virus, even if not engineered, had leaked from a lab by some gruesome accident. And there would be cogent responses to those outcries, from Andersen and others. The argument over SARS-CoV-2's origin wasn't settled on March 17, 2020.

Further studies by other teams explored the pangolin connection. That similarity of receptor-binding domains in SARS-CoV-2 and the Guangdong pangolin virus—what did it mean? Had a pangolin been infected by two strains of coronavirus at once, and had they recombined during replication, patching an RBD from one into the other? Had pangolins carried that recombined virus for a long time, centuries or millennia, long enough for a mild mutual accommodation to evolve, so that the pangolins became a true reservoir host, a refuge in which the virus dwelt peaceably and securely? Or had some unfortunate pangolin, or maybe a shipment of pangolins, played an intermediate role recently between the natural reservoir and humans? Or maybe the pangolin virus and RaTG13, that similar bat virus, shared a common ancestor that was a bat virus with the RBD, and the pangolin virus kept the RBD, but RaTG13 lost it by recombination. On such questions, the small flurry of studies published in prominent journals, during the next two months, offered some intriguing data and informed guesses. Those studies also preempted the Wong-Petrosino paper, still just a draft in submission at another journal—a journal I won't embarrass by name—where it had been lost or mislaid or buried in a pile or eaten by somebody's dog.

I've already mentioned the paper from South China Agricultural University, the one that was hyped by that university's president at her press conference in early February. Yes, this team had found a SARS-like coronavirus in pangolins but, no, it was not 99 percent similar to SARS-CoV-2. It was 90 percent similar overall, and 91 percent similar in the spike gene, which included the receptor-binding domain. That much resemblance was interesting but not conclusive. The South China Ag researchers, including Yongyi Shen as a senior author, suggested that SARS-CoV-2 "might have" originated by recombination between such a pangolin virus and one similar to RaTG13.

Shen's group drew their data from lung tissues. They screened samples from four Chinese pangolins (*Manis pentadactyla*, a critically endangered species but still sparsely present in Guangdong and other parts of southern China) and from twenty-five Malayan pangolins. These included the same animals seized by the Guangdong Customs Bureau in March 2019, and sampled previously by Jinping Chen's team, plus some others intercepted in August 2019. Shen's group now found coronavirus RNA, as Chen's had, but only in Malayan pangolins, and only those from the March 2019 seizure. The researchers gave vividness to a very technical report when they noted that the March pangolins, at the Wildlife Rescue Center, "gradually showed signs of respiratory disease, including shortness of breath, emaciation, lack of appetite, inactivity, and crying." Fourteen of them died within six weeks. Pangolins are sensitive, hard to keep alive in captivity under even solicitous care, and the harsh conditions of being trafficked internationally would make them especially susceptible to infection. But what killed those fourteen pangolins? Was it Sendai virus, or a coronavirus, or some other cause unrelated to concerns about human health? We'll probably never know. Later in the paper, deeply buried in a section on methodology, Shen and his coauthors added that the animals "were mostly inactive and sobbing, and eventually died in custody despite exhausting rescue efforts." Sobbing might be taken as a metaphor for respiratory struggle, but then again, sometimes a sob is just a sob.

Three researchers at a government lab in Kunming, down in Yunnan province, also went back to the lung tissue samples from those dead pangolins in Guangdong. This team reexamined the same genomic data that Chen's group had published. They reported, in an April paper, what Matt Wong had noticed back in January: that the receptor-binding domain of the pangolin coronavirus closely matched the one in SARS-CoV-2. This suggested that the pangolin virus, just like SARS-CoV-2, might have been quite capable of latching on to human respiratory cells. The researchers implied, but didn't claim, that the pandemic could have begun from a pangolin. But there was no match to the furin cleavage site. For that absence, these Kunming authors posited a simple explanation: the pangolin virus and SARS-CoV-2 could be descended from a common ancestral virus, and the pangolin lineage may simply have lost the cleavage site in the course of evolution, as some birds (the dodo, the moas of New Zealand, also kiwis, penguins, and ostriches) have lost the power of flight.

Suddenly there was a boomlet in pangolin virology. Jinping Chen's group in Guangzhou jumped back into the conversation, offering further analysis of their own samples, those lung tissues from the March 2019 pangolins. Chen and his colleagues now extracted enough RNA fragments, pooled from three animals, to assemble a complete coronavirus sequence. Yes, they reported, it was strikingly similar both to the human virus SARS-CoV-2 and to the bat virus RaTG13. Yes, the receptor-binding domain matched closely the one in SARS-CoV-2. But no, they wrote, their data did not support the supposition that SARS-CoV-2 had come directly from a pangolin. Probably the story was more complicated, and the pandemic strain arose from one or more recombination events among viruses infecting bats and other wildlife, maybe including pangolins. But two things, to Chen's group, seemed clear. First, there are a lot of coronaviruses potentially dangerous to humans circulating among various wild animals—bats, palm civets, camels, pangolins, who knows what else. Secondly, and for the sake of wildlife conservation as well as human health, it is important to reduce disruptive contact between people and wild animals, either captured or farmed, that risks spillover of such viruses. When you have Malayan pangolins, abducted from elsewhere in southern Asia, trafficked across the border, sobbing out their last breaths at a rescue center in a major Chinese city, something is wrong, and not just for the pangolins.

31

All those studies stood balanced on a relatively narrow pedestal: samples from smuggled pangolins intercepted by Guangdong customs and delivered to the rescue center on March 24, 2019. In the meantime, another study expanded the base of evidence from Guangdong to Guangxi, the province adjacent to the west, which borders Vietnam, and found something even more interesting. This group included a highly re-garded and dauntless disease detective at the University of Hong Kong (HKU) named Yi Guan, along with more than two dozen other Hong Kong and mainland scientists, plus Eddie Holmes.

"So what happened," Holmes told me, was that on January 30, "I get contacted by Tommy Lam." Tommy Tsan-Yuk Lam is a statistical geneti-cist and a bioinformatician, educated in Hong Kong, at Pennsylvania State University, and at Oxford, and presently an associate professor at HKU, though he still looks almost young enough to be a skateboard punk from L.A. "Tommy is my old ex-postdoc," Holmes said. Now he was working with Yi Guan. He told Holmes about a curious project that involved some confiscated pangolins in Guangdong. "They've got this respiratory dis-ease," Holmes remembers him saying. "And guess what. They've got, like, this coronavirus in them." Holmes to me: "And I thought, well, that's ex-traordinary." This was just days after Matt Wong, sobered by the news of Kobe Bryant's death, started looking for deeper meaning.

Lam and Guan, like each of those other teams, had laid hold of data from the Guangdong pangolins; but they had something more. Somehow, they had gotten samples from another set of trafficked pangolins, confis-cated by Guangxi customs about two years earlier. They had extracted RNA. They wanted Holmes's help. "We start analyzing these data and what we see, what's so striking—" Holmes interrupted himself, pausing to be sure I followed.

"They're pangolins from two provinces, right?"

"Yeah," I said.

"It's Guangxi and Guangdong, right? They're both illegally smuggled Malayan pangolins." Two batches of the Malayan kind, and yes "illegally smuggled" is redundant, but he was talking fast, excited again just re-counting this. "They're not from China. They're imported there, okay? They've both got some respiratory disease. And that's described in the paper"—the paper subsequently published, by Yi Guan and Tommy Lam and Holmes and their colleagues, in *Nature.* Yes, okay, I knew the paper, having read it one day before this conversation.

"What's so interesting to me is, they both have coronaviruses that are related to the human strain but are not the *same.* Okay? That's what is so striking."

Guan's team in Hong Kong had received frozen samples of lung, intestine, and blood from eighteen pangolins seized in anti-smuggling operations by Guangxi customs officers. They found coronavirus RNA in six samples, and from those fragments they assembled six genome sequences, which they designated the GX lineage, short for Guangxi. They also took the raw data from the Guangdong pangolins, which Chen's group had made available, as well as extracting new sequence data from other samples from those pangolins, and reprocessed all that into full genomes, using their own tools to their own standards of accuracy. These they designated the GD lineage. Both lineages closely resembled SARS-CoV-2, but not in the same ways at the same points in their genomes. Most notably, the Guangdong lineage had a receptor-binding domain very similar to the one in SARS-CoV-2.

"What are the odds?" Holmes asked. What are the odds that you would sample two sets of pangolins in two provinces, both sets illegally imported, each set infected with a coronavirus that happened to be similar, but in unique ways, to a virus lately emerged in humans?

Let me guess, I thought: the odds are low.

"That is absolutely weird to me," Holmes said. "It's amazing to me." It seemed also a bit ominous, even to me, as I scrambled to follow his logic. Why? Because it suggested the presence of many more pangolin coronaviruses than we had imagined, diverse and widespread and at least some of them menacing to humans; or else it reflected a continuous cascade of coronaviruses from bat reservoirs into pangolin intermediaries; or maybe both. "Can't rule out either one of those explanations," Holmes said.

And if you put the two-lineages evidence together with the RBD evidence, he added, it implies that our knowledge of the coronaviruses dwelling in wildlife is "minuscule." Between the closest known bat viruses, such as RaTG13, and the SARS-CoV-2 genome, there still lies a relatively big evolutionary gap. "What's in that gap? I don't know." In what other wildlife might coronaviruses be lurking and recombining and coming closer to people? "Raccoon dogs? Bamboo rats? Who the hell knows? Right? But until we go there"—go to the field, go to the caves and the forests, go to the farms where wildlife is legally bred for food, go to the depots from which contraband animals are trafficked, go to the open markets and the black markets where such creatures are sold—"until we go there and sample them, we're never going to know. That's the critical thing, to re-solve the origins."

"Okay," I said.

"And pangolins hint at that."

Tommy Lam became first author on the *Nature* paper. The discovery of these multiple lineages of coronavirus so similar to SARS-CoV-2, he and his colleagues concluded, "suggests that pangolins should be consid-ered as possible hosts in the emergence of novel coronaviruses." That was the scientific takeaway. The recommended action, the paper's punch line, was that these animals "should be removed from wet markets to prevent zoonotic transmission." No pangolins breathing on the pork. No pango-lins weeping on the shrimp. In a wet market, a crowded place full of meat and poultry and fish and wildlife, full of cages and knives and splashes and funky air, full of people talking and shouting and coughing, full of animals sobbing, you could get much more than you bargained for.

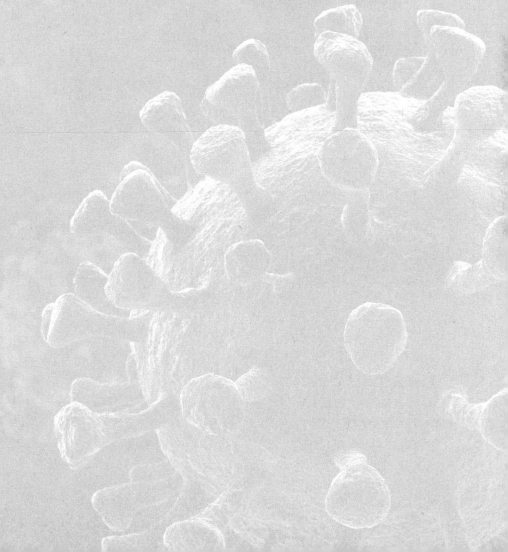

IV

MARKET
DYNAMICS

32

The Huanan Seafood Wholesale Market wasn't the largest emporium of fresh foodstuffs in the city of Wuhan. But it became the most infamous, beginning on December 31, 2019, when the Wuhan Municipal Health Commission made its announcement about twenty-seven hospitalized cases of the mysterious pneumonia, including seven in serious condition, all linked to the market. Past tense is appropriate—Huanan "wasn't" the largest—because the place has ceased to exist as it did. It's still closed and vacant, at last report, with a high blue fence surrounding it at ground level; it may never reopen. On the second floor of the building, which stands at the corner of New China Road and Development Road in central Wuhan, are some shops selling eyeglasses, access to which is controlled by security guards. The downstairs, with its dark, narrow alleys of shuttered stalls and drainage gutters, its lingering smells of disinfectant and rotten meat, is off limits to the public. You might get a guided tour, but only if you belong to a privileged group of visitors, such as the members of the WHO-convened Global Study of the Origins of SARS-CoV-2, a team of seventeen Chinese and seventeen international scientists, who made an inspection visit to the Huanan market on the afternoon of January 31, 2021. That's how Marion Koopmans happened to see it.

Koopmans is head of the Department of Viroscience at Erasmus Medical Centre in Rotterdam, and an expert on zoonotic viruses. She led the group that first traced the MERS virus to camels. On the WHO-China study of SARS-CoV-2 origins, she headed the international team's subgroup on molecular epidemiology. She is a forceful, direct person with a hip, tousled, silver-gray haircut. She and the other team members received briefings beforehand on what they would see and what they wouldn't see at the market—they wouldn't see the products with which this market once teemed, nor the people who bought and sold them—as well as information on work done in the interim, by Chinese scientists, to identify those products and their sources. This visit, remember, came more than a year after the market closed. The weird bouquet of disinfected rot was still notable, still strong, because the closure was done so abruptly and so firmly that many of the products, including meat and entire carcasses, were left behind. So were tools and machinery. No recovery of those items by their owners was allowed. One day a neutron bomb hits a crowded, sloppy market: life gone, structures intact. In this instance, the neutron bomb was a virus.

"There was a complete map of the whole market," Koopmans told me. This map, supplied in their briefings, indicated "where were the cases, what did all the stalls sell." She meant the cases of COVID-19, those early pneumonia patients, most of whom were vendors or suppliers, working in the market or visiting regularly, not customers. The map showed where they had stood, what living and dead animals they had offered; separate information told where those animals had originated. The focus was on products considered risky for viral transport: wildlife, including farm-raised wildlife such as bamboo rats and porcupines, and (because the Chinese team insisted on considering this hypothetical avenue) frozen fish. "You can walk back into twenty different countries along that supply chain," Koopmans said. Wuhan is a city of eleven million people, the largest in central China, and an important nexus for international travel and trade. "You can walk back also with the wild animals, meat on the market, to farming systems in provinces," she said, "where we know there's bat coronaviruses. SARS-like bat coronaviruses." These were leads to be followed. This study mission in early 2021 was supposed to be phase 1, a brisk month of preliminary work—scouting the terrain, testing hypotheses against what data were available, framing a plan to gather more data—of what the WHO conceived as a two-phase study. How the Chinese side would view continuation to a second phase was a different question.

The team members also saw studies tracing the human cases, Koopmans said. That involved molecular epidemiology: the remit of her own subgroup. "It's clear that this started at least early December," she told me, "and by mid-December it really exploded."

Molecular epidemiology for early COVID-19 entailed comparing genomic sequences, then building trees of ancestry and descent, to chart where the virus was, and when, and how it traveled through chains of transmission. The viral sequences came both from human samples, collected case by case, and from environmental samples, the swabbing of surfaces and other modes of collection at the market, done by the China CDC on January 1, 2020, and thereafter. Koopmans and her colleagues saw twenty-five complete viral genomes and three partial sequences from sampled humans. All those came from people sampled during the second half of December 2019. For the first half of December, nothing. "There's no samples testing from that period of time," she told me. So molecular epidemiology couldn't speak on what happened in the market, or around it, as the outbreak began. Or at least Marion Koopmans and her group couldn't speak on it, given the data they had seen.

33

Through the early weeks of the outbreak, as it became a pandemic, the narrative continued to center on the market. Among the first Chinese studies published internationally was one in *The Lancet*, the same British journal that published K.Y. Yuen's report of the family cluster in Shenzhen and the ominous hint of asymptomatic spread. This study, from a group including medical staff at Jinyintan Hospital in Wuhan, where many of the December cases got treatment, appeared online the same day as Yuen's, January 24, 2020. Its first author was Chaolin Huang, vice director of the hospital. Clearly, editors at *The Lancet* grasped the importance of the news out of China and welcomed these bursts of scientific illumination.

Huang's study, coauthored by more than two dozen medical doctors and scientists in Wuhan and Beijing, focused on the clinical aspects of the first forty-one cases admitted to Jinyintan Hospital. What were their ages? How many carried other medical problems—we all know the word "comorbidities" now—such as diabetes, hypertension, heart disease? What were their symptoms? How many feverish, how many coughing, how many with difficulty breathing? What did the lab workups say about their blood? What did the chest CTs reveal about their lungs? One other parameter was notable: How many of them had been directly exposed to the

Huanan Seafood Wholesale Market? The rough answer to that question was, most of them. The report spoke of "the shared history of exposure to Huanan," and implicit in that phrase was an inference: exposure to Huanan meant exposure to the animals sold there. These people, including the forty-one pneumonia patients, weren't just strolling around a building at the corner of New China Road and Development. Some of them, very possibly, were handling and cleaning up after and even butchering wild-caught or farmed wildlife such as civets, raccoon dogs, bamboo rats, Malayan porcupines, and other animals. Furthermore, those visitors who didn't handle animals still breathed the same musty air of the alleys and shops as those who did. When the Huang study appeared, that inference drove the international news coverage.

The Guardian, for instance, a percipient British daily, immediately put up a story about outcries for banning of wildlife markets worldwide, provoked by the coronavirus outbreak and events at the Huanan market, "which has been closed down as the source of the infection." That was accurate but misleading. Yes, the market had been closed *as* the source of the new virus in humans. But *was* it the source?

One day later, on January 25, 2020, an American physician and infectious disease expert named Daniel R. Lucey, who had begun posting news and comments on the outbreak, drew attention to the Huang study. Lucey noticed something between the lines and in the fine print. Concerning that "shared history of exposure" to the market, he saw the obvious fact that such exposure wasn't *universally* shared. Yes, twenty-seven of the forty-one earliest patients at Jinyintan had links to the market; but what about the fourteen who didn't? A bar graph on page three of Huang's paper showed all forty-one cases logged to the date when their symptoms first showed. If you paused to look closely, as Lucey did, you could see that three of the earliest four cases, symptomatic on or before December 10, reported no connection to the market. And the earliest of all forty-one patients, an unidentified person who sickened on December 1 (according to this graph), was a nonmarket case.

Lucey put up a blog post. He framed it as a Q&A with himself, sketching what he called "an evidence-based hypothesis" that the epidemic (it wasn't yet a pandemic) began in November 2019 or earlier, and at some other location, not the Huanan market. His post appeared on a forum

page titled "Science Speaks," maintained by the Infectious Diseases Society of America (IDSA), a large association of doctors, scientists, and
public health professionals. This was already Lucey's update number six
on the new virus. It began by citing the Huang paper and its report of a
case on December 1. That's new information, Lucey noted. Then he asked
himself, and answered:

"Did this earliest of the 41 patients have any exposure to the Huanan
seafood market?"

"No."

"Did any of that patient's family members develop fever or any respiratory symptoms?"

"No."

"Was any explanation given for how this patient became infected?"

"No."

The rhetorical self-questioning went on. Did that first patient have
any link with the forty others? No. When did the next three get sick? Not
for another nine days. Was any explanation offered for the fourteen, all
infected without market exposure? No explanation for thirteen of them.
(One woman was the wife of a market-exposed man.) Do these infections
suggest that person-to-person transmission or animal-to-human spillover
occurred in November or earlier? Yes. Where might those contacts have
occurred? Maybe at another market, or a restaurant, or on a wildlife farm,
or along the routes of wildlife trade. Does this hypothesis say anything
about efforts made to control or contain the virus? Yes: it says that December 2019 was already too late.

Lucey's regular posts on the IDSA website served an attentive audience, but this one echoed loudly around the world, after he sent it to a staff
writer he knew at *Science.* The day the Huang paper appeared, January 24,
was a Friday, and the weekend proved busy. "Every Friday morning, I get
The Lancet on my phone," Lucey told me from his home in Washington,
D.C. *The Lancet* is a weekly journal, published on Fridays (though with
some articles appearing online beforehand), and, with so much to read,
Lucey observes the weekly rhythm. On that Friday morning, as usual, he
scrolled through the medical news. "I looked at the titles of the articles,
and I saw that one." Huang and coauthors, on "Clinical Features of Patients Infected with 2019 Novel Coronavirus . . ." etc., not the sort of fare

you or I might browse with our eye-opener coffee, but Lucey started reading. He peered at the bar graph. "And everything changed."

Lucey wrote his Q&A blog and posted it to the IDSA website, where it
appeared on Saturday morning. He sent the link to his contact at *Science*,
the staff writer Jon Cohen. In late afternoon on Sunday, which was just
before 1:00 p.m. for Cohen, who lives in San Diego, Cohen called him.
Cohen had already emailed the senior author on the Huang paper, Professor Bin Cao, at Capital Medical University in Beijing, for comment on
Lucey's post. Cao's reply was forthright: that he and his coauthors "appreciate the criticism" from Lucey. It now seemed clear, Cao wrote, that the
market was "not the only origin of the virus." He added: "But to be honest,
we still do not know where the virus came from."

"It was memorable," Lucey told me.

He did more homework. "I have a boring life," Lucey said, only half
joking, I think. He lives alone in an apartment near Pennsylvania Avenue
and travels impulsively to volunteer his services as a medical doctor, especially during dangerous disease events in faraway places, such as Liberia
for the Ebola epidemic in 2014, Qatar in 2013 to help with MERS patients,
and some clinical work in both China and Toronto during the original
SARS outbreak of 2003. He spotted the first news from Wuhan early, same
time and same way as Marjorie Pollack and others, from Chinese social
media leaking onto the wider internet. He read about the ophthalmologist
Wenliang Li who had posted on WeChat a warning to his old classmates.
"My hair was on fire, you know, since the night of December 30," Lucey
said. Now he dove deeper, trying to imagine his way back to the start of
the outbreak, searching the web for illuminating bits and pieces. "I have
this approach where I try to get information that is reliable, that is as close
as possible to people who are on the ground, who have direct personal
experience."

He contacted an old friend, in Hong Kong, a microbiologist at HKU,
and he found the website of the Wuhan Municipal Health Commission,
which is written in Chinese but can be read via Google Translate. Where
was this market, this city, and what happens there? Lucey wanted to
know. "It turns out Wuhan is the center for high-speed trains. It's the hub
for high-speed trains everywhere in China," he told me. "So I made a big

picture." He meant that literally: he printed out a map of China, with red lines indicating high-speed rail routes radiating from Wuhan, and at the FedEx store in his neighborhood he had it blown up into a poster. He began carting that poster with him when he gave talks on the China outbreak and what it might bode, including one at a meeting of the U.S. National Academy of Sciences. He would hold up the poster and say, "These red lines represent trains coming out of Wuhan to all parts of China. But when I look at it, what I see those red lines representing . . . are the virus."

Near the end of his IDSA post on the first forty-one cases, Lucey raised a question about looking for evidence of earlier infections, during November 2019 or before. How could such an investigation be done, and *why* should it? Testing archived specimens, of blood or tissue or even swab samples, taken from humans or other animals and saved for other reasons, was the answer to how. Fragments of virus, or antibodies in blood, might still be detected if the samples were properly stored. RNA is more fragile than DNA, but even RNA can last up to a month at room temperature, in the right preservative, and longer still if deep-frozen. Why do such testing? Because positive evidence of the virus elsewhere could help in shutting down other sources or chains of recurrent transmission. Lucey might also have said, but left implicit, that we won't ever know the origin of this virus until we discern just where and when the first human infections occurred.

The location of those very first cases would remain uncertain for at least two years. Even the date and identity of the first *confirmed* case in Wuhan would be a matter of shifting certainties. As later investigations dug deeper and their results were published, the December 1 case recorded by Huang and his coauthors in their bar graph, and of such interest to Daniel Lucey, disappeared from the discussion. That case had evidently been further scrutinized, based on additional information, and its date of onset revised. When the WHO international team reached Wuhan, in January 2021, to work with their Chinese counterparts, the investigators would meet with and interview a forty-one-year-old man, an accountant identified as Mr. Chen, said to be the first confirmed COVID-19 case. Mr. Chen fell sick on December 8, 2019, they were told. Like the case dated to December 1 in the Huang paper (the one which disappeared or was tossed out of the dataset), Mr. Chen reported no links to the Huanan market. He shopped at a large supermarket.

This uncertainty over the first confirmed case would remain like a burl on the trunk of a tree, growing larger and more contorted, over the next two years. I'll return to it, when the attention of others returns to it, near the end of this book.

Meanwhile, people had begun dying. The first recorded fatality was a sixty-one-year-old man, a regular customer at the Huanan market. As the Huang paper reported, on January 24, "The number of deaths is rising quickly." That day it stood at twenty-four.

34

If the virus was circulating in humans before December 1, 2019, and beyond the dank alleys of the Huanan market, where would that be? The logical guesses were greater Wuhan, or elsewhere in Hubei province, or maybe somewhere between Wuhan and the caves of Yunnan, where similar viruses abide within roosting bats, and people interact with those bats at their risk. A less logical guess was northern Italy. But several studies have suggested that possibility.

During late autumn 2019, a group of scientists led by Elisabetta Tanzi, an expert on viral diseases at the University of Milan, investigated what seemed to be an outbreak of measles. They saw thirty-nine suspected cases in patients who later tested negative—for measles, anyway. Each patient was sampled by oropharyngeal swab (a gentle dab to the back of the throat, not the kind that goes way up your nose and seems to tweak your brain), and the swab specimens stored. Months passed, the pandemic began, and it occurred to these scientists to retest those measles swabs for SARS-CoV-2. They found one positive specimen, taken from a four-year-old boy who lived near Milan. He had begun coughing on November 21. He got worse and, a week later, with vomiting and respiratory troubles, he was brought to an emergency room, and then developed a measles-like rash. But it wasn't measles. According to the PCR testing of his swab, as eventually reported by Tanzi and her colleagues, it was SARS-CoV-2. Italy's first recognized case of COVID-19 occurred three months later.

This study met with skepticism and dismissal (contaminations can cause false positives), but the Milan group later doubled down on it, in collaboration with researchers in Rome and elsewhere, presenting evidence for Covid in eleven Italian patients sampled before the pandemic became apparent in Italy. These had all been suspected measles cases, nine of them sampled during 2019, and one of them, an eight-month-old baby, testing positive for viral RNA in a urine sample collected on September 12. Five other patients reportedly tested positive for SARS-CoV-2 RNA in their urine; in the rest, it was respiratory samples that came up positive. Neither the baby nor any of the others reported recent travel to China.

Another puzzling account of early SARS-CoV-2 infection came from France. The country's first confirmed patient was a thirty-one-year-old Chinese male from Wuhan, a tourist, who arrived in Paris on January 19, started feeling flulike illness, and tested positive for Covid five days later—again that resonant date, January 24, when much happened. Three days before leaving home, afflicted by an attack of gout, the man had visited a hospital in Wuhan, which may have been where he got infected. At the Paris hospital, his respiratory symptoms worsened and, after four days, he was transferred to an ICU. He got a blast of remdesivir, the broad-spectrum antiviral drug, and then maintenance treatment, and he survived. But his illness trajectory isn't the point here. The point is that SARS-CoV-2, according to one later study, had actually entered France long before he did.

A group of French researchers, including some at a different Paris hospital, reported finding SARS-CoV-2 infection in a patient treated in an ICU during December 2019. They detected the case by retrospective screening of samples taken from patients who had been admitted for influenza-like illness, based on symptoms, but then tested negative for influenza virus. Once the pandemic got rolling, and the nefarious subtlety of SARS-CoV-2 began to be appreciated, it occurred to these researchers, as to others, that COVID-19 might explain illnesses that were otherwise unexplained. They went back to samples that had been deep-frozen, chose fourteen, thawed them, and ran them through PCR tests targeting SARS-CoV-2 genes. They got one positive. That sample came from a forty-two-year-old man, born in Algeria, longtime resident in France, who had walked into a medical ward, coughing and feverish, on December 27. He walked out again after two days of treatment, leaving behind a frozen sample containing what the researchers later judged to be SARS-CoV-2. This ICU admission was three days, remember, before Marjorie Pollack got her first alert about a strange pneumonia in Wuhan, three days before Daniel Lucey's hair caught fire.

Then came Brazil. The city of Florianópolis straddles a mainland spit and a scenic subtropical island along the coast of Santa Catarina state, about four hundred miles south of São Paulo. It's a getaway locale for celebrities, people such as Neymar and Ronaldo (if you don't know who they are, it's because you think of "football" as a game played in body armor with a prolate spheroid), who reportedly have homes there. Florianópolis has been called "the best place to live in Brazil," although not because housing prices are a bargain. Besides offering lifestyles for the rich and famous, it thrives on information technology enterprises and tourism. There are beaches. There are stately colonial-era churches and elderly fig trees and women selling handmade lace in the streets, plus an abundance of bars and restaurants. There is an old public market. There is sun. The airport is not huge but connects well with the world through São Paulo, Rio, and Buenos Aires. People go there from everywhere. A team of researchers looked at the sewage of Florianópolis, archived from October to December 2019, and found what seemed to be SARS-CoV-2.

Who knew that sewage is archived?

Wastewater microbiologists knew, and they study such archived samples of raw urban sewage to discern community patterns and trends of infection by gut bacteria and other microbes. You can't tell anything about a single individual with municipal sewage, but you can learn whether an infectious bug is present in town, and even at roughly what level of prevalence. A group of Brazilian and Spanish microbiologists applied their methods to Florianópolis wastewater samples that had been set aside on six different dates, beginning in October 2019, and stored frozen. This sewage came from a system that served five thousand residents in the central city. October 30 was negative. November 6 was negative. November 27 was positive. Even in Wuhan, researchers had found no evidence of human infection as early as that. This was ninety-one days before the first recognized case in Brazil. The study appeared in a peer-reviewed journal, *Science of the Total Environment.* It raised eyebrows, and not just among scholars of sewage. The senior author was David Rodríguez-Lázaro, a microbiologist at the University of Burgos, in northern Spain.

"When we got the wastewater," Rodríguez-Lázaro told me, "you're right, very controversial."

The study began, before the pandemic, with a different focus: food-borne pathogens. Rodríguez-Lázaro happened to be in Brazil in October 2019 giving a lecture and conferring with collaborators. They formulated a plan to do the wastewater testing, mainly for gut viruses, and he flew back to Spain. Then came COVID-19. "We decided, okay, why not to check the presence of the virus, of the SARS-CoV-2?" Having found it in the November 27 sample, by what they considered punctilious execution of an extremely reliable methodology, they nonetheless had trouble publishing their paper. One journal editor turned them down because, though he was interested, he couldn't find any other scientists willing to do the peer review. He tried fourteen. Such reluctance seemed to stem from the case of another startling report, also from Spanish scientists, claiming to have found SARS-CoV-2 in Barcelona wastewater as early as March 12, 2019, ten months before the pandemic. That claim didn't make it past the preprint stage. Twitter lit up with criticisms, as Twitter often does, even among scientists who use that mode of mutual alert and goading. The claim about March 12 disappeared from the Barcelona group's published paper, but it seemed to foul the well (speaking of wastewater) for the Rodríguez-Lázaro group's assertion about November. They tried two more journals and got two more rejections, based on peer reviewers who wanted more data or suspected their result was bogus, possibly because of a laboratory contamination. Finally they published the paper, and David Rodríguez-Lázaro went back to his "normal life," he told me, using wastewater data to study foodborne infections, and in particular the underappreciated problem of antimicrobial resistance among bacteria.

"It will kill us slowly," he said, about the resistant-bacteria problem. "Not quickly, as SARS-CoV-2."

35

All of this seemed perplexing, and contradictory of two widely accepted premises: that the virus had entered humans from an animal in the Huanan market, and that this spillover occurred no earlier than November 2019, resulting in the outbreak of forty-one cases. To complicate things further, a group of scientists in Boston analyzed satellite images of Wuhan, archived from before the pandemic, and reported a big increase in hospital occupancy starting in August 2019, as deduced from crowded parking lots at the hospitals. These scientists also analyzed symptom-related internet searches around that time, through the Chinese technology company Baidu, and found that "cough" and "diarrhea" were trending. Their study was another preprint, and though it went up on a Harvard University website, it took immediate flak for its assumptions and methodology, and it seems never to have progressed to publication.

Lessons? First, you can find riches of anomalous information, a bounty of enticing clues, a great range of startling scenarios about SARS-CoV-2 and its provenance, some peculiar coincidences, and a lot of pseudoscientific bullshit, if you own a computer and are capable of typing a few words into a search engine. Second, our knowledge of this novel virus is still provisional, unfolding daily like a *Rafflesia* blossom in time-lapse photography. So it's important to apply basic tools of critical thinking—such as dispassion, scrutiny of sourcing, humility in the face of uncertainty, and parsimony—to what we hear, what we read, whom we trust, and what we think we know.

Among the scientists I trust highly is Michael Worobey, a Canadian-born, Oxford-educated evolutionary virologist at the University of Arizona. I've followed Worobey's work for about a dozen years, ever since coming across his studies of the origin and diversification of HIV-1, the more virulent of the two types of HIV and the one mainly responsible for AIDS. It was the research of Worobey and his colleagues, paired with the work of a German-born scientist named Beatrice Hahn and her colleagues, that placed the start of that pandemic in space and time. They did it by studying viral genomes, their rate of evolution, their degree of divergence from one another, their pattern of relatedness as portrayed in a family tree. That's the discipline known as molecular phylogenetics.

The trunk of this family tree represented viruses ancestral to a lineage known as the simian immunodeficiency viruses (SIVs). The SIVs were discovered in the early years of AIDS research by other scientists, notably Phyllis Kanki and Max Essex, two veterinarians at the Harvard School of Public Health. Such viruses infect dozens of different kinds of African primates, mostly monkeys but also chimps. One branch of the tree led to the chimp SIV, designated SIV*cpz*, meaning "simian immunodeficiency virus of chimpanzees." From SIV*cpz* to HIV-1 was only a small evolutionary shift, like the divergence of a twig, which occurred when the virus passed (during a bloody hunting or butchering incident, presumably) from a chimp into a human. Hahn's group discovered where that shift happened: in the southeastern corner of Cameroon or thereabouts. Worobey's group illuminated when: back around 1908, give or take a margin of error. Those findings, unexpected as they were when published between 2005 and 2008, have stood up well.

Michael Worobey is rigorous, smart, and judicious. There's also a quietly dauntless streak in him, exemplified by an experience I heard about when I first interviewed him years ago. As a young scientist, Worobey flew into a war zone in the Democratic Republic of the Congo, with the great English biologist William Hamilton, to gather field data that might cast light on the origin of HIV-1. Hamilton was seeking proof or disproof of a very controversial hypothesis known as OPV, standing for Oral Polio Vaccine, which blamed the AIDS pandemic on a case of vaccine contamination. In search of such evidence, he wanted to screen Congolese chimpanzee feces for signs of SIV*cpz*, the immediate progenitor of HIV-1, because such a finding might fit with the OPV hypothesis. Worobey, then a young doctoral student at Oxford, where Hamilton held a prestigious professorship, was less enamored of the OPV story but keen to find any new data that could help illuminate where, when, and how HIV-1 got from chimpanzees into humans—including data that might support the OPV hypothesis. So in early 2020, Hamilton, Worobey, and Worobey's friend Jeff Joy flew into Kisangani, a city along the northern bend of the Congo River, a diamond-trading center, where conflict between Ugandan and Rwandan troops amid the Second Congo War had been killing civilians as well as soldiers, and from which a short drive would take the three men to chimpanzee habitat. The war had shut down scheduled airline travel. Hamilton, Worobey, and Joy came in from Entebbe, Uganda, as Worobey recalled it to me in 2011, sharing a small plane with a diamond dealer.

The OPV idea, which had been researched and promoted by several journalists, one of whom captured Hamilton's interest, was that HIV-1 entered humans as a contaminant of an oral polio vaccine, a vaccine known as CHAT, developed by Hilary Koprowski, a Polish virologist working in Philadelphia, and administered at his direction to hundreds of thousands of people, including children, in northeastern Congo during the late 1950s. It was an incendiary accusation, assembled from a factual core (Koprowski did develop a vaccine and test it in Africa), various threads of circumstantial evidence and suppositional narrative, plus some mistakes of detail, and not supported by molecular data. It depended on the fact that Koprowski's was a live-virus vaccine, made with attenuated virus, rather than an inactivated polio vaccine (IPV), like the one developed by Jonas Salk, which contained only virus killed by formaldehyde. A virus is attenuated by passaging it repeatedly through nonhuman cells in a laboratory, causing it to accumulate mutations that render it harmless to people but still alarming to human immune systems. The live-virus approach allowed for an oral vaccine, such as Koprowski's or the one developed by Albert Sabin, which could be administered from an eyedropper onto the tongue or, better still, in a vaccine-soaked sugar cube. That represented an important improvement over Salk's injectable vaccine, which any kid (such as me) who lined up in school to be vaccinated, either by needle (in the late 1950s) or by sugar cube (in the early 1960s), could appreciate. One supposition of the OPV hypothesis was that Koprowski had attenuated the polio virus, not in monkey cells, the usual procedure, but in chimpanzee cells. Another supposition was that those cells happened to be contaminated with SIVcpz, the progenitor of HIV-1. If that much was correct, the progenitor virus might still be out there, forty or fifty years later, circulating among chimpanzees in northeastern Congo. And if it was still out there, then screening chimpanzee feces might detect it. That seems to be what Hamilton hoped, anyway.

In Kisangani, they checked in with the local rebel commander, a leader of Rwanda-backed forces, who wanted regime change in Kinshasa. That commander controlled most of the city. But the city straddles the river, and on the other bank were the immediate enemy, Uganda-backed forces, who also wanted regime change in Kinshasa. It was a complicated war. "We got into the forest as quickly as we could," Worobey told me. They hired local guides, hiked until they could hear the hoots of a chimpanzee group, set up camp, and their guides then went out early each morning to where the chimps had nested, to collect "basically their morning dump and their morning pee." Worobey and Joy then bottled the samples with a solution that stabilized RNA. They collected thirty-four fecal samples and some urine.

Analyzed months later, the fecal samples tested negative for SIV*cpz*. Two of the urine specimens did contain antibodies suggesting past infection with somesuch virus, but still later results from a subsequent solo expedition by Worobey showed that it wasn't the progenitor of HIV-1. Chimpanzees still occupy habitat, at least in patches, from Senegal all the way to the east shore of Lake Tanganyika, one side of Africa almost to the other, and the geographical separation that has yielded different chimp subspecies has also yielded different strains of virus. SIV*cpz* in eastern Congo, where the Koprowski vaccine went into people, was not the SIV*cpz* that became HIV-1. But that answer came too late to satisfy William Hamilton's curiosity. By then he was dead.

Hamilton caught malaria during the forest fieldwork with Worobey and Joy. By the time they left Kisangani on the only available plane, to Kigali, capital of Rwanda, he was very sick. They flew to Entebbe, where a doctor confirmed that it was falciparum malaria, the most lethal kind, and gave him some medicine. Then they bounced to Nairobi, and finally up to Heathrow Airport in London. At Heathrow there was a baggage emergency, a cruel overlay to the medical emergency—their precious specimens, packed in a cooler, didn't show up. Hamilton, still in poor condition, went to his sister's house in the city. Staff at the baggage office told Worobey they had located the cooler, offloaded by mistake in Nairobi, and that it would arrive on a later flight. Next morning, Worobey called the sister. She didn't know who he was, and she reacted sternly. "Who is this? Why are you calling?" He learned then that Hamilton had gone to a hospital, worsening further, and hemorrhaged. "His whole blood volume, almost," Worobey told me. Either the large doses of ibuprofen that Hamilton had been taking, or some other convergence of factors, including very bad luck, had peeled open his gut. Hard hit with that knowledge,

Worobey went back to the airport. But what showed up on the second flight was wrong: a different cooler, containing sandwiches. Worobey, exhausted and frustrated, got emotional with the airline.

"I was actually crying," he told me. "Bill was, like, dying. It was clear that he was dying, and it was just a bit too much on top of everything." A series of surgeries, plus sequential transfusions, amounting to his body's entire blood capacity twice over, weren't enough to save Hamilton. "I think it took seven weeks for him to die."

To say that the OPV hypothesis cost William Hamilton his life would be unfair to Hamilton. His dedication to science, his determination to address a troubling hypothesis with empirical data, is what cost him his life. The specimens, once recovered, proved inconclusive, but Michael Worobey and Beatrice Hahn, along with other scientists who have studied the evolution and phylogeny of the HIV viruses, now consider the OPV hypothesis strongly refuted. The AIDS pandemic did not start with a contaminated vaccine. What does that have to do with *this* pandemic? The common element for Michael Worobey, I think, is a steely attentiveness to the value of genomic data and molecular phylogenetics, as distinct from other forms of narrative, toward understanding how the hell and where the hell SARS-CoV-2 originated.

36

COVID-19 is the first pandemic for which molecular phylogenetics, on a vast scale, has been done while the catastrophe unfolded. That deployment of new technology is a milestone event as important as the invention of war photography by Mathew Brady at the First Battle of Bull Run.

In 2003, scientists sequenced the SARS virus genome from a single case among the Toronto patients, got another sequence from another victim (a heroic doctor, Carlo Urbani, who died in Bangkok while responding to the outbreak), and that much was notable: two sequences. It was too early, the methods were too laborious, the tools too primitive, for massive sequencing.

Continuous improvements in the speed, reliability, and affordability of automated sequencing machines vastly improved the utility of molecular phylogenetics amid the urgencies of a disease event. During the 2013–2016 Ebola epidemic in West Africa, scientists were able to sequence and analyze more than sixteen hundred viral samples from patients, which helped much in tracing how that virus spread. In the five years thereafter, sequencing capacity increased by orders of magnitude, and it has played a huge role in efforts to understand COVID-19. By April 2021, more than a million sequences of SARS-CoV-2 had been deposited in GISAID—an initiative founded in 2008 as a way of sharing influenza genome data—and within six months after that, the number was above 3.6 million. As of early 2022, GISAID holds and shares more than eight million SARS-CoV-2 sequences, with more added continuously, and the turnaround time, from sequencing to uploading a sequence to its availability for other scientists, is measured in days, not months. This has allowed scientists to identify major lineages and new variants as they emerge, to gauge which variants are spreading aggressively, and to draw phylogenetic trees illuminating the when, where, and how of transmission. Michael Worobey, not surprisingly, joined the effort to trace and understand the evolutionary dynamics of SARS-CoV-2 using such data.

Worobey and a group of colleagues wanted to pinpoint the first arrival and track the earliest spread of SARS-CoV-2 in Europe and in North America. They suspected that the answers might be different from what other studies proposed. They knew that the first confirmed case in the United States was detected in Snohomish County, Washington, on January 19, 2020, in that man who flew home from a family visit in Wuhan. They knew that some evidence pointed toward the Snohomish man as perhaps America's Patient Zero, and that he might have infected others, who infected others, in chains of cryptic transmission during late January and early February of 2020, from which the virus spread to California and British Columbia and Connecticut and onward, making the Seattle area the epicenter of the North American epidemic. The Snohomish man's viral genome sequence, designated WA1, as in "Washington case 1," became an object of close scrutiny.

They knew also that the first European case was in a woman who lived in Shanghai, got infected there by her parents when they visited from

Wuhan, then flew for business to Munich and connected to a town nearby, where she infected a man, one of her fellow employees, at an automotive supply company called Webasto, a maker of sunroofs. That man tested positive on January 27, by which time the woman—who has been called Europe's Patient Zero, though she only bounced into Europe long enough to be a transmitter—had flown back to Shanghai, where she worsened and was hospitalized. The German man was hospitalized too, in isolation, and his virus sampled and sequenced. That sequence, labeled BavPat1, as in "Bavarian Patient 1," is almost famous. It differs by just one nucleotide, among nearly thirty thousand, from the founding virus of the lineage that swept across Europe and the United Kingdom in the early months of the pandemic and became known as lineage B.1. (The B in that B.1 did not stand for Bavaria, however; more about lineage identification and naming, below. It becomes important with the rise of notorious variants.) Worobey's group also knew of a study implying that the German case of January 27 seeded the Italian outbreak that flared in March, from which the virus spread also to France, Mexico, and the United States, causing the first wave of terrible mayhem, overtaxed hospitals, and dead bodies stored in cooler trucks for lack of mortuary capacity. From that earlier study, a broader narrative took hold—of Webasto as the source of the European and American epidemics—that reached even the automotive trade press. It was a terrible burden to put on a sunroof manufacturer, and Webasto denied it.

Knowing these basics, Worobey and his collaborators scrutinized more than five hundred genomes, from the U.S. and twenty-seven other countries, to discern whose virus went where. They generated trees. They performed computer simulations, based on the data they had, of how transmissions may have occurred to produce the trees of relatedness they saw. They drew inferences. The Snohomish case probably did *not* spark the outbreaks in California, Connecticut, and elsewhere. Rather, that case was likely a dead end, with no onward transmission, thanks to quick, firm containment measures taken in Washington. And the Bavarian case probably did *not* spark the outbreak in Italy or anywhere else. It led to about fifteen other cases, after which the spread was contained. Both the WA1 strain and the BavPat1 strain of the virus were successfully

squelched, in the judgment of Worobey and his colleagues. Then they
drew conclusions.

> The public health response to the WA1 case in Washington state
> and the particularly impressive response to an early outbreak in
> Germany delayed local COVID-19 outbreaks by a few weeks and
> bought crucial time for U.S. and European cities, as well as those
> in other countries, to prepare for the virus when it finally did
> arrive.

A few weeks might seem a small difference in response time, but it wasn't.
"The value of detecting cases early, before they have bloomed into an outbreak, cannot be overstated in a pandemic situation." And buying time
can be crucial but equally crucial, they knew, was how the time would
be used.

37

Santa Clara County in northern California, encompassing the city of
San Jose and Silicon Valley, was one of the first areas in the United
States to be hit. Sara Cody could see it coming. Cody, a physician and
epidemiologist with degrees from Stanford and the Yale School of Medicine, and experience doing outbreak investigations for the CDC, served
as health officer and public health director for Santa Clara County. "Forty
percent of the people that live in our county were born outside of the
U.S.," she told me. Practicing public health in such a place, at the south
end of San Francisco Bay, always felt like a globalized challenge. "There's
lots and lots and lots of travel, for personal and professional reasons, and
I think that is probably the reason why we tend to see—as infections are
emerging—we tend to see them here first."

It was her husband, a Stanford professor who works on health policy and infectious disease modeling, and a news junkie, who first drew

her attention to the novel virus. "Hey, have you seen these reports out of Wuhan?" She dialed in and began to watch with increased concern after that case turned up in Snohomish. Over the long weekend of the Martin Luther King Jr. holiday, Cody spent much of her time on conference calls, including one with the secretary of health for Washington state. "I remember being really struck that they had activated hundreds of people for one case. Hundreds!" She started getting questions from the local medical community. So her department moved to its Incident Command Structure, a standardized system for coping with emergencies. The county's public health lab was ready and waiting. But then came several weeks of "this incredibly painful, ridiculous process" during which they had to ship samples by World Courier to the CDC in Atlanta for testing there, deal with packaging and tracking numbers, then wait a few days, "just to figure out whether your patient had this new coronavirus." In the meantime, she needed to decide: Do we isolate that patient, on no evidence, or allow continuing community contacts, inviting the virus to spread?

Cody thought the problem would disappear when the CDC sent them test kits and her team could do testing themselves. Then the CDC test kits arrived, in early February, and they didn't work. The assistant lab director, a young Navy veteran named Brandon Bonin, trained in forensic DNA methods, and in charge of the lab until a new director could be hired, stayed up all night working through the test protocol, to his dissatisfaction. He kept getting nonsense results: positive readings from a virus-free control sample of water.

"The problem was that the assay wasn't reliable," Bonin told me. The assay was liquid containing customized molecular probes to detect specific portions of the virus—in this case, three probes, targeting three regions of the capsid protein. For one of those regions, the probe was dysfunctional and therefore the assay was inconsistent and completely unpredictable. It would show a positive result for plain water in one trial, then negative the next time Bonin tried. "It was all over the place." He alerted Cody, and they learned that other laboratories were having the same trouble. At a time of dire national need—those crucial few weeks, about which Worobey and his colleagues later wrote—the CDC had sent them junk.

The days crept on, during which Cody and her team remained ignorant of SARS-CoV-2 infections in Santa Clara County. They had no functional CDC test kits and, although academic labs around the country had developed tests independently, the U.S. Food and Drug Administration would not approve use of those tests. "It was really critical time that was completely lost," Cody told me. "We were just flying blind, week after week after week."

"'Cause you couldn't test," I said.

"We couldn't test! Don't look, won't find." Venting frustration, she repeated that axiom: *"Don't look, won't find."* In addition, the CDC kept advising that, when local health departments *could* test, they should focus only on people with a travel history, or people with severe symptoms, or people known to have been exposed to confirmed cases. This left important questions about the virus unanswered. What was the incubation period, between infection and when symptoms appear? What percentage of people in a population were asymptomatically infected? What percentage would become infected within a period of time? These are basic parameters of infectious disease epidemiology, and the data weren't being gathered.

"It was spooky," Cody said. "I just remember February being spooky. Like, you know, all these bad things are happening, but you have no way to see it. You can't tell." Finally, in late February, they heard from the CDC, whose updated advisory on the delicate process of testing patient samples for a deadly virus, as Cody recollected, was "You know what, just run it with the two probes that you have. Skip the third probe. Two is good enough." Good enough for government work, under the CDC leadership at the time.

Two weeks later, Santa Clara County recorded its first recognized death from COVID-19. "I believe it was March 9," Cody told me. Her memory had that right. "It was the same day that I issued my first Health Officer Order to prohibit gatherings of greater than a thousand people." She remembered that day, a Monday, because she had spent a weekend evening with her friend Greta Hansen, second in command at the County Counsel's office, sharing dinner and margaritas, with their husbands and (but not the margaritas) their dogs, while conferring about whether and how she could issue any such order. Hansen said yes and helped her draft it.

That order was controversial, not least because it would affect home games of the county's professional hockey franchise, the San Jose Sharks. Three days later, though, hockey attendance at the Shark Tank became moot when the National Hockey League suspended the season for all teams. On March 13, Cody issued a more stringent order: no gatherings of more than a hundred people. Over the next forty-eight hours, the case count almost doubled in her county. On March 16, feeling "exquisitely uncomfortable" to have and exercise such authority, Cody led six Bay Area counties in ordering residents to shelter in place.

One victim in Santa Clara County was Patricia Dowd, a fifty-seven-year-old auditor, who died in her San Jose kitchen on February 6 and was found by her daughter slumped at the breakfast bar. Dowd had been suffering flulike symptoms. Her infection was not linked to COVID-19 at the time, because of the absence of local testing capacity and the advisory declaration about who could be tested. Her death seemed mystifying, possibly caused by a heart attack, and only clarified months later when tissue samples tested positive for SARS-CoV-2. Patricia Dowd was probably the first American to die of COVID-19. By March 16, when Sara Cody and five other county health officers issued their shelter-in-place order, the American death toll stood at ninety-six.

38

Other officials were watching other numbers. On February 24, 2020, the Dow Jones Industrial Average fell 1,032 points.

In Washington, D.C., some presidential aides blamed that on Peter Navarro, a pugnacious economist who served as director of the White House National Trade Council, a directorship and a council invented for him at the behest of Donald Trump. Eventually that "council," which consisted largely of Navarro and one staffer, was absorbed by another council, but Navarro retained the title "assistant to the president," which is a big deal for those who trade business cards inside the Beltway. Navarro had come to prominence, and to Trump's attention, by publishing fervidly anti-China books, such as *Death by China* and *The Coming China Wars*, arguing that the U.S. should prepare itself for economic warfare, if not other sorts, with that country; presumably someone read these books to Trump or, more likely, summarized them for him. Trump liked Navarro, for his views and his style, and (as reported by two *Washington Post* reporters, Yasmeen Abutaleb and Damian Paletta, in their book *Nightmare Scenario*) that gave Navarro latitude to be a loose cannon even looser than some others in the Trump administration. Trump enjoyed having such abrasive, plainspoken advisors around him, bashing one another and acting a little "crazy," according to Abutaleb and Paletta. Late in that month, though, Navarro made himself inconvenient to Trump's dismissive line on the coronavirus—that it was no significant problem, that it would recede as spring arrived, and then vanish entirely. On February 23, as Trump was preparing for a fast trip to India, to be feted there at a huge rally, Navarro appeared on the program *Sunday Morning Futures*, one of the Fox News forums.

Asked about the economic impact of the coronavirus, Navarro told the host, Maria Bartiromo, "my job at the White House during this crisis is to review the supply chains we need to treat corona." Face masks, remdesivir—too much of America's manufacturing capacity for such things had been "offshored," he said. Bartiromo wanted him to talk about how those shortages might affect earnings—by which she presumably meant corporate revenue, not wages—but he returned to the supply chain, the restrictions on importing PPE, and said that "in crises like this, we have no allies." After a few further kicks at China, including a complaint that they control the World Health Organization through their "proxy," the Ethiopian director-general, which accounted for why the U.S. was suffering problems with the coronavirus, Navarro closed the show by saying, "This, again, is a crisis." Back in the White House, according to Abutaleb and Paletta, some aides were aghast. Did Navarro just call the coronavirus situation a "crisis" three times in ten minutes, on *Fox*?

The next day, while Trump and his entourage were in Ahmedabad, the stock market took its dive, with the Dow closing at 27,961, down more than a thousand points. During dinner after the event, some of Trump's aides saw the financial news on their phones. Abutaleb and Paletta don't name these aides, simply reporting, "They knew Trump would have a fit when he found out."

That was Monday, February 24. For the nervous market watchers, it got worse. On Tuesday, from the CDC, a senior official named Nancy Messonnier held an online briefing with reporters. Messonnier was director of the CDC's National Center for Immunization and Respiratory Diseases. She had given other briefings, but this time her tone was fatalistic and dour. "The global novel coronavirus situation is rapidly evolving and expanding," she began. There had been community spread—that is, not just imported cases, but local chains of transmission—in a few countries, such as Italy and Iran, but not yet in the U.S., she claimed. (That was a dubious assertion, given the failure of the CDC's tests; public health officials such as Sara Cody were still operating blindly, due to the lack of accurate, speedy testing, and no one knew whether there was community spread in the U.S. or not.) But community spread would come, Messonnier admitted. It was not a question of *if* but of *when*, she said, "and how many people in this country will have severe illness." There was no vaccine against the novel virus. No medicines approved to treat it.

Nonpharmaceutical interventions would be the most important tools. (Nonpharmaceutical interventions, or NPIs, is a fancy term for behavioral modifications to slow the spread of a disease, such as school closures, stay-at-home orders, social distancing generally, and wearing masks.) Yes, school closures might be necessary. Mass gatherings might be canceled. People might miss work and lose income. "I understand this whole situation may seem overwhelming and that disruption to everyday life may be severe." Call your children's school and ask about their plans, Messonnier said. Talk to your kids, like I did this morning with mine. (It started to sound almost like a briefing for the Cuban Missile Crisis.) COVID-19 is coming. "People are concerned about this situation. I would say, rightfully so. I'm concerned about the situation. CDC is concerned about the situation." Now is the time for everyone to prepare. "I also want to acknowledge the importance of uncertainty," she said in conclusion. "During an outbreak with a new virus, there is a lot of uncertainty."

Investors do not love uncertainty. Politicians do not love uncertainty. Not even gamblers, except the loopier sort, love uncertainty. Poker is based on calculation and bluff, hedges against uncertainty. Admitting uncertainty may have been the truest, bravest, most candid thing that Nancy Messonnier said that day—the molecular evolution of an RNA virus embodies more uncertainty than a roulette wheel—but her uncertainty comment was not calculated to reassure. News media reacted. Investors reacted. Aides in the White House reacted, and even a few alert, restless souls aboard Air Force One, returning from India, who saw the news while others slept. That day, the Dow Jones Industrial Average lost another 879 points. Olivia Troye, who served as homeland security and coronavirus advisor to Vice President Mike Pence, made a pithy comment, reported by Abutaleb and Paletta: "People have their televisions on and there were a lot of comments that the stock market is going down the shitter." Donald Trump, awake and angry about Messonnier's comments, began barking by phone at his hapless secretary of health and human services, Alex Azar, who was Messonnier's boss's boss, even before the plane landed.

Meanwhile, people were dying. Two fatalities had occurred in Santa Clara County by February 25—the deaths of Patricia Dowd and another person—though not recognized as Covid-caused until months later. In Italy, of the 323 confirmed cases of COVID-19 by that date, eleven had proven fatal. China's case count had exploded at the start, more than 78,000 by February 25, with 2,715 deaths. Soon afterward, thanks to draconian versions of nonpharmaceutical intervention, China would flatten its curve like a mesa. America's curve over the coming months, into the following year, would resemble the Grand Tetons.

Why? A national ethos of "rugged individualism," by which I mean not real individualism but programmatic self-concern, to the detriment of community weal, was probably a large part of the reason. Cowboys don't wear masks, unless you count the Lone Ranger. Leadership, or rather "leadership," was another factor. Donald Trump's primary concern in those crucial early months of the pandemic, and later, seems to have been getting himself reelected in November, partly on the merits of a robust economy and a rising stock market. But SARS-CoV-2, a virus without intentions, without malice, without anything but Darwinian imperatives to guide it, didn't care about stock indices or elections. It was turning history in a different direction.

39

I almost wrote: *SARS-CoV-2 didn't care about markets*. And that would be correct, because a virus doesn't "care" about anything, except in the most anthropomorphic sense, whereby it attends to those Darwinian imperatives. But markets do remain important to the question of how this virus found its way into humans—and not just one market, the Huanan Seafood Wholesale Market. Among the most peculiar facts revealed by molecular epidemiology regarding this pandemic, the sort of science that Marion Koopmans and Michael Worobey do, is that SARS-CoV-2 emerged in December 2019 as two distinct lineages, which seem to have come from two different sources.

The earliest illumination of this fact was easily overlooked because it appeared in a journal paper sounding dry and arcane, titled: "A Dynamic Nomenclature Proposal for SARS-CoV-2 Lineages to Assist Genomic Epidemiology." I looked at that and thought *viral taxonomy*, *okay*, *but meh*. Looking again, I noticed that the first author was Andrew Rambaut and the group included Eddie Holmes. So I read it and found something interesting. Amid an effort to bring lucid naming to a rapidly expanding body of data about which people were confused, urgently curious, and wrought up, Rambaut and Holmes and their coauthors described the two foundational lineages of SARS-CoV-2 and labeled them simply: A and B. Lineage B, as represented in a sample collected from a Wuhan patient on December 26, 2019, was linked to the Huanan market. That patient was a customer. Lineage A, as seen in a patient sampled on December 30, seemed to come from somewhere else. This resonates with what Daniel Lucey noticed: that fourteen of the forty-one patients first identified did *not* have any known connection to the Huanan market. The December 30 patient carrying lineage A hadn't visited Huanan but *had* visited a different market. Which market? The records don't say. In Wuhan, at that time, three other markets contained shops selling wild animals for food or pets, including the city's largest, Baishazhou. Among the four markets, Baishazhou and Huanan and two others, there were seventeen shops selling wildlife alive. In recent years, those shops had sold in total more than 47,000 wild animals, belonging to thirty-eight species of terrestrial mammals, birds, and reptiles, ranging from Amur hedgehog and Chinese bamboo rat to monocled cobra.

The numbers come from a study led by Zhaomin Zhou, a specialist in the wildlife trade at China West Normal University, in Nanchong, and formerly employed by a forest protection agency in Yunnan. He was "a technician," Zhou told me modestly, but he was a technician with a PhD, assigned "to identify species of animals and/or products derived thereof." His coauthors included several colleagues from Oxford University and Xiao Xiao, an associate professor at a medical university in Wuhan. It was Xiao who did the legwork, discreet surveys of the four wet markets, between May 2017 and November 2019. Xiao presented himself as "an objective observer unconnected to law enforcement," and that was enough to persuade market vendors to speak freely. The original intent of this project had no connection with coronaviruses—to identify the source of a tick-borne disease caused by a different sort of virus—but its relevance to SARS-CoV-2 became very clear by the time the results were published. Zhaomin Zhou and the Oxford collaborators had previously worked on pangolin trafficking. In the new study, they noted that neither pangolins nor bats were sold at those seventeen market shops; but masked palm civets were, raccoon dogs were, as well as other animals (American mink, Siberian weasel, Asian badger) quite capable of carrying a coronavirus. Raccoon dog farming for the fur trade is legal in China, but with fur prices down, those animals are often sold in the live markets for food. Raccoon dogs and Asian badgers went for about $8 per pound, roughly three times the usual price of pork. Hedgehogs were cheap. Some animals came from farming operations, legal or illegal, but Xiao saw gunshot and trap wounds on many of them, indicating illegal capture from the wild.

These were luxury food items, reflecting "the sort of cachet attached to wild animal consumption in parts of the developed world," not subsistence bushmeat. But the clientele was varied, not confined to affluent toffs. The Zhou team had seen, in their own previous investigations, that "a substantial desire to purchase and/or own wildlife products as 'prestige items' still transcends social classes, age groups, education levels and rural versus urban residents, even though this involves breaking the law." Lax enforcement made that not just possible but easy.

One other factor may have exacerbated the risk of spillovers in China, by increasing demand for wildlife meat: a shortage of pork. China is the world's leading consumer of pork, also the world's largest source, producing about 50 percent of the global supply. In 2018, the average consumption in China was seventy-five pounds per person. But in late summer that year an outbreak of African swine fever swept into the country, eventually affecting more than 150 million pigs. The disease is caused by African swine fever virus (ASFV), a DNA virus endemic to sub-Saharan Africa, where its reservoir hosts are bushpigs and warthogs, and ticks vector the virus from one animal to another. With the arrival of European colonizers, bringing domestic swine to Africa, ASFV infected those pigs too. In the twentieth century it got to Europe, then was eradicated, then returned in this century, possibly by way of wild boar imported to southern Belgium from Eastern Europe for the amusement of hunters. Among domestic pigs, the virus is highly virulent, and the strain that reached China, by August 2018, was almost 100 percent lethal.

Within eight months, because China normally produces such a large share of the world's pigs, commodities analysts in the West were cheerily telling their industry clients that ASFV's impact there would "lift all protein boats." A website called RaboResearch, run by the Dutch financial services giant Rabobank, predicted losses in China of 25 percent to 35 percent, and noted that such a wallop to China's pigs and pig farmers, along with shortfalls of pork in Southeast Asia, "will create challenges and opportunities for animal protein exporters." It might also have created opportunities for people who broker bamboo rats and porcupines from one Chinese province to another, but that was off Rabobank's radar screen. By early November 2019, a month before SARS-CoV-2 made itself apparent in Wuhan, pork prices nationwide had risen by 148 percent. From province to province and region to region, though, prices differed, in some provinces double what was being paid elsewhere. Hubei was among the provinces where pork went high. Did that mean people in Hubei were driven to eat less pork and more bamboo rat and more porcupine meat and more muntjac venison and more weasels and squirrels? Possibly. The authors of one study, again a mixed team of Chinese and U.K. researchers, argued that harsh fluctuations in the pork market just before December 2019 "may have increased the transmission of zoonotic pathogens, including severe acute respiratory syndrome-related coronaviruses, from wildlife to humans, wildlife to livestock and non-local animals to local animals." True, it may have. But that paper was a preprint when I read it, not yet peer-reviewed, and neither it nor any others I've found contain data, actual numbers, on rising consumption of wildlife meat in Hubei province just prior to the pandemic.

It's a plausible, dramatic story. "How One Pandemic Led to Another" is the title of that provocative preprint: an African swine fever pandemic in pigs triggering a coronavirus pandemic in people. The causal link is hypothetical, so far without empirical support. It's a narrative.

Then again, it's not the only hypothesis about this virus and its origins that is merely a narrative.

40

Molecular epidemiology, on the other hand, operates in a data-rich context or it doesn't operate at all. It derives conclusions by comparing genomes and fragments of genomes. The complete genome of SARS-CoV-2, as I've mentioned, is almost thirty thousand bases long. If you were to scrutinize 583 genomes of the virus, from 583 different sampled cases, you would be looking at seventeen million points of data. You would want a computer. Also, good eyesight.

That's what Michael Worobey did, along with four colleagues, to estimate the timing of the very first SARS-CoV-2 infection in Hubei, back before it was detected in the Huanan market. Worobey's collaborators on this study included Joel Wertheim, who had been his PhD student in Arizona fifteen years earlier, and Jonathan Pekar, currently a PhD student of Wertheim's, now at UC San Diego—so, a generational string of mentors and mentees that makes Worobey, who seems young, feel a bit old. They used a quiver of conceptual tools and inferences, as well as those 583 genomes from Hubei, all sampled between December 2019 and April 2020, and all available from the database GISAID. They compared those genomes with one another and performed computer simulations of how the genomes could be arranged to represent limbs and branches on a family tree. The simulations differed because they embodied certain assumptions—about the rate of mutation, for instance, and whether some mutations might have mutated back to the original form, which does happen occasionally. There were some inherent uncertainties, some aspects of pure chance; the scientists hoped to learn what was most probable. They drew a number of those family trees. With each simulation, they could infer the point where the big limbs holding all 583 genomes first diverged from the trunk. That point of first branching was important in this sort of analysis: it represented the most recent common ancestor (MRCA) of all the virus strains sampled. Where was it—where in time? Wherever, somewhere back before late December 2019, that point represented *at least* how long SARS-CoV-2 had been circulating in humans.

The first branching point wasn't the ultimate object of this study, though, because it probably didn't represent the earliest presence of SARS-CoV-2 in humans, the "primary" case. The primary case, in a phylogenetic study of a disease outbreak, is the base of the trunk—not to be conflated with the MRCA, the point from which limbs diverge to form the tree's crown. The details of the primary case remain unknowable, while the position of the MRCA is inferred by the study. But those things *do* get conflated, Worobey told me in a recent conversation. Even his colleagues in the field tend to blur the distinction between primary case and MRCA. "It's kind of one of these things that just is constantly forgotten. And sometimes it doesn't matter much, sometimes it matters a lot."

Between the base of the tree and the first major divergence—between the primary case and MRCA—there may have been other limbs, small ones, that didn't receive enough light, never thrived, and hence withered and died. These short lineages of the new virus would have transmitted among a few people, then gone extinct. Such extinct lineages, never sampled, could only be inferred. But to infer them was part of this exercise in probabilities by Worobey and his colleagues, leading to a conclusion: "It is highly probable that SARS-CoV-2 was circulating in Hubei province at low levels in November 2019 and possibly as early as October 2019, but not earlier."

Not earlier: their findings contradicted the notion that SARS-CoV-2 could have existed in wastewater samples in Barcelona as early as March

2019, or in the urine of a baby in Milan in September, and they knew it, stating that those studies were "unlikely to be valid."

But the high extinction rate among viral lineages, which they saw in their simulations, implied that "spillover of SARS-CoV-2-like viruses may be frequent, even if pandemics are rare." The same pattern of multiple introductions of the virus, multiple short chains of transmission dying away before the virus took hold in a place, could have played out elsewhere too—for instance in Santa Clara County, California, during February 2020, yielding the death of Patricia Dowd. The difference was that Hubei's introductions came first.

Such extinct lineages would suggest that the virus was relatively scarce during the early, pre-pandemic period. Little spot fires of infection guttered out. Illnesses resembling flu or common pneumonia were assumed to be flu or common pneumonia. Chains of infection came to dead ends. Other chains might have continued, giving the virus opportunity to evolve and adapt, but there's no persuasive evidence (apart from lineages A and B) of any that survived and proliferated. "We don't know what might have happened in terms of adaptation in humans," Worobey told me. A few key mutations might have made the virus more transmissible. It still needed a bit of luck. Spillovers are frequent, pandemics are rare, as he and his coauthors noted. Most viral lineages in a new kind of host probably do go extinct. But this tree grew a little taller, the trunk grew a little more robust, and then at a critical point—the point that molecular phylogeneticists call MRCA—branches diverged from a sapling, those branches grew into limbs, and from them more branches diverged, ascending into a great crown.

We don't know, as he said. But maybe that's what happened, and one of those early branches led to the Huanan Seafood Wholesale Market.

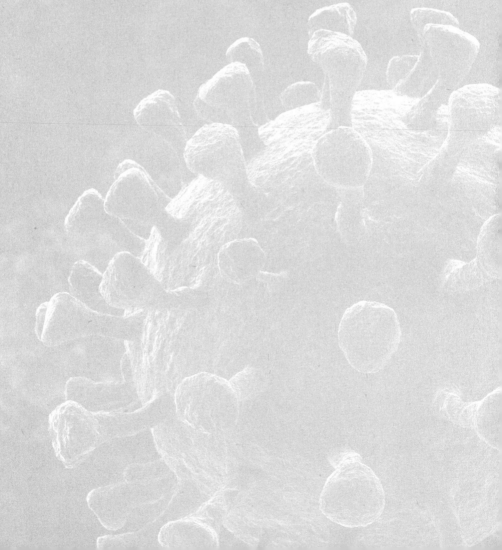

V

VARIABLES AND CONSTANTS

41

Will the virus mutate? people asked.

Yes of course it will mutate, scientists said. Viruses always and continually mutate. The crucial questions are how frequently will it mutate, how abundantly, and how might those mutations be shaped by natural selection into adaptations. Mutations are incremental changes in a genome—one letter here, one letter there—and generally random. Don't worry about mutations alone. Worry about mutations plus Darwin. Worry about how this virus may evolve and adapt. If you hope to prevent it from adapting ever better to the human population, then prevent it from achieving abundant mutation by containing it fast, controlling the outbreak early, taking it seriously, keeping the human case numbers low, applying and adhering to robust nonpharmaceutical interventions until you have vaccines, then get everyone vaccinated, depriving the virus of opportunities to evolve.

But we didn't do that.

42

How frequently would it mutate? That question went to the nature of coronaviruses generally. The precise answer is complex but the simple answer is: enough to be elusively dangerous. Which mutations would be significant, and how might evolution shape them into adaptations? Answers to those would depend on the circumstances in which this virus finds itself. Evolution doesn't carry creatures toward some platonic ideal of perfection. It carries them only toward success within a particular environment at a particular time.

For the first nine months of the pandemic, SARS-CoV-2 seemed to be mutating slowly and evolving little or not at all. Some scientists noted "remarkably low" genetic diversity among the many samples that had been sequenced. One group of researchers, based at the Walter Reed Army Institute of Research, in Silver Spring, Maryland, looked at 27,977 genome sequences from infected people in eighty-four countries and found "little evidence" of natural selection for anything new. The virus, they wrote, "is being transmitted more rapidly than it evolves"—meaning that the discernible mutation rate was less than one nucleotide change per human case. It was passing from person to person, in many instances, without a single copying mistake in its thirty-thousand-letter genome. That's notable constancy, especially for an RNA virus. It didn't mean that the virus might not have mutated at all, within a given person, but that, if there were mutations, they proved unsuccessful in competition with the unmutated strain at getting themselves replicated and transmitted. Evolutionary biologists call that purifying selection. Purify what you've got. If it ain't broke, don't fix it. Maybe this virus didn't *need* to evolve because it was already succeeding so well among humans.

But the constancy, in fact, wasn't that odd among coronaviruses. They differ from most RNA viruses, which are highly variable and therefore highly adaptive, as Eddie Holmes and Donald Burke and others have warned. Coronaviruses lie at one end of the RNA virus spectrum. Their genomes are exceptionally long and their mutation rates are exceptionally low, less than a tenth the rate in other RNA viruses. Those two atypical characteristics are mediated by a nifty mechanism, a special protein called nsp14 (which could stand for "nifty special protein 14," but doesn't). This protein performs a proofreading function, tracking along the genome while it replicates itself, letter by letter, and correcting most mistakes before they can get into new virions. With genomes so long, if they *didn't* have such a mechanism, coronaviruses would accumulate so many errors that they would rattle to pieces, like a Model A Ford in which someone had neglected to tighten the bolts. That outcome is called "error catastrophe." Evolution supplied nsp14, allowing coronaviruses to avoid error catastrophe—by letting them tighten their bolts. A long genome was an advantage, in ways still unmeasured, and the proofreading function of nsp14 made it possible.

So yes, there was a delay at the start of the pandemic, a period of relative stasis, during which no notable or alarming mutations were recognized in SARS-CoV-2. When the first one did appear, it commanded attention. It was a difference of one nucleotide among the thirty thousand, which coded for a different amino acid in one protein, and that change took the label D614G. The amino acid glycine (symbolized by G) had replaced aspartic acid (D) at position number 614. D supplanted at 614 by G, ergo D614G. The protein in question was the spike. Some scientists have proposed that this one-amino change accounts for why COVID-19 patients lose their sense of smell. More importantly, it also seems to have made the virus more transmissible.

The D614G mutation was first observed in China, very early on, and spread by late January 2020 to Europe, where it was noticed in Germany, then Italy. Its spread continued during February and March, across Europe and North America, Australia, and then back to Asia. Tracking its course, performing laboratory experiments to gauge the functional impacts of D614G, working quickly, a team of scientists in the United States and the United Kingdom posted a preprint on April 30. The lead author on that paper was Bette Korber, a veteran computational biologist at the Los Alamos National Laboratory. Korber had worked with Max Essex at Harvard during the early years of molecular epidemiology on AIDS. So this was her second pandemic. As SARS-CoV-2 made its way around the world in March 2020, she got interested in tracing the variants of the virus as they arose.

"I wanted to be able to look at transitions in communities," Korber told me. It was a classic evolutionary perspective—or to be more precise, neoclassical, indebted to the early-twentieth-century biologist R. A. Fisher, the man who mathematized Darwin. Fisher defined evolution as the changes in frequency of alleles (different forms of the same gene) within a population. "Everybody who has ever thought about evolution knows the story about the moths, right? The change in the moths in the trees," Korber said. A famous story, yes: she meant the replacement of light-winged forms of the peppered moth (*Biston betularia*) by dark-winged forms, as the tree trunks of industrialized England became darkened with soot. A mutation for dark wings allowed some moths to escape the notice of predatory birds, and those dark-winged mutants left the most offspring. The change of allele frequency in the moth population—the dark-wing allele or alleles becoming more abundant than the light-winged originals—constituted evolution.

This was different from molecular phylogenetics, as practiced by Michael Worobey and others. Korber was wary of that approach, at this stage, because there was so little variation among genomes during the early months, so few mutations; she didn't trust the family trees. Also, the factor of recombination—the swapping of whole sections from one viral genome to another, which can graft a limb from one phylogenetic tree onto another, confusing things further. Instead, as she said, she wanted to look at changing gene frequencies within communities. Rather than trying to trace where mutations came from, she would measure how much they piled up in this population or that. An individual mutation would often be associated with several other mutations sprinkled through the genome, and any viruses bearing that associated group were known collectively as a variant. If a particular variant piled up inordinately, and consistently replaced the original virus or a preceding variant whenever it entered a new community, you could infer that it was providing evolutionary value. Your vaccines and therapeutic treatments would need to address it.

Korber and her colleagues developed computational tools to apply this approach with SARS-CoV-2: identify a mutation that characterized a new variant, then use the bounteous genomic data available on GISAID to measure how its frequency (its prevalence within a population) might change in different human communities afflicted with COVID-19. They seized on D614G. Other scientists considered that mutation insignificant. It wasn't within the spike protein's receptor-binding domain, crucial for cell entry, and it wasn't in a spot to be targeted by antibodies, so a change there should have no effect on immune response against the virus. Maybe it was a neutral mutation, delivering no evolutionary advantage and carrying no cost. If so, it might continue to appear at low frequency, a random anomaly in some genomes, not many. "But what was happening, and what I could see as soon as we had the tools available," Korber told me, was that in every set of genomes from every population screened for D614G, "no matter a community, state, country, continent, it was increasing rapidly relative to the ancestral strain."

She and her colleagues looked at 997 genome sequences of the virus sampled early, before March 1, 2020. They found D614G in just 10 percent. They looked at 14,951 sequences from March and found the mutation in 67 percent of them. "Our data show that, over the course of 1 month," they wrote, "the variant carrying the D614G Spike mutation became the globally dominant form of SARS-CoV-2." They looked at another 12,194 sequences gathered from April 1 to May 18 and saw that D614G prevalence had increased further, to 78 percent. In lab studies, they tested virus with G614 in its spike against virus with D614, the spike as it was when the virus first emerged, and found that the mutant version was much better at infecting cultured cells. And in clinical data gathered at a hospital in England, on almost a thousand COVID-19 cases, they saw evidence that the mutant virus grew more abundantly in patients than did the virus without that mutation. This was concerning, but with one consolation: the researchers found no link between the D614G mutation and the severity of disease. But because the mutant virus was much more transmissible, leading to more cases, with bad outcomes at the usual rate, it was still potentially a big new problem for global health.

Bigger problems would come. Korber's group foresaw that, and at the end of their paper they included a figure, labeled Figure 7, showing six other mutations within the spike protein that had cropped up in different parts of the world. These mutations bear watching, the authors warned, and so do others that are bound to appear. Variant strains carrying new mutations could be significant challengers to vaccine efficacy, when vaccines arrive, and to antibody treatments based on the nonmutant virus.

"To this end," they wrote, "we built a data analysis pipeline to enable exploration of potentially interesting mutations on SARS-CoV-2 sequences." A data pipeline—Matt Wong created one, Korber's group created one, and we'll hear of more—is a set of computing elements or steps, each step yielding output that goes as input to the next step. An assembly line in a car factory is a pipeline of mechanical steps with physical inputs. At the end of the pipeline, presto, you have a new Chevrolet—or a conceptual result. If the inputs are SARS-CoV-2 genome sequences with this or that mutation, spreading at this or that rate, in this country or that one over there, the outputs from the pipeline could be valuable to vaccine and treatment designers. "The speed with which the G614 variant became the dominant form globally," Korber's team wrote, using the shorthand term for D614G, "suggests a need for continued vigilance."

When we spoke, Korber called my attention to that figure, showing the D614G strain as just one among seven variants of interest. The point, she said, was not just D614G; the point was larger. "I considered it a paper documenting the utility of this kind of approach, as much as the variant itself. And the reviewer was kind of wanting me to take out that last figure that talked about tracking variants."

"Mm hm," I said astutely.

"But the editor let me keep it in." The paper was to appear in *Cell*, a highly regarded journal. The peer review would have been focused and stringent. "They said, 'This is a distraction. You know, we don't need to have this distraction in the paper.'" The reviewer or reviewers wanted Figure 7, with its look forward to other variants, gone. But she fought for it, successfully.

"I said, 'No, this is really central. This is coming! We have to be prepared!'"

43

The virus was coming and, as Korber well understood, the virus was not a constant thing. With any virus, but especially an RNA virus, even one with a nifty special protein for proofreading, the cardinal constant is change. Small changes, such as the single-letter mistake that created the D614G mutation, occasionally have sizable impact. And small changes may accumulate in a sprinkling within a genome, each one adding just a touch of evolutionary advantage, or else simply neutral but lucky, moving as a freeloader with other mutations that confer the advantage. Moving where? Through genome replications into the future. When there is such a sprinkling of mutations, and they spread as a package seen in many viral samples, we call that a variant. Nowadays it's a household world. Breathless media reports make each new variant sound scary. People grow cynical: *What's the latest scariant?* If the variant is exceptionally successful, binding especially well to cells, spreading quickly and widely, the WHO designates that a variant of concern (VOC). A lineage is something else: a branch, short or long, successful or less so, on the virus's family tree. A successful lineage, whether it represents a new variant or not, can travel. It can ride airplanes. This is where molecular phylogenetics meets geography.

Of the two major lineages of SARS-CoV-2 that Andrew Rambaut and his colleagues identified, calling them A and B, lineage B is what first arrived and flourished in Italy. Early reports suggested that it was brought from Germany—from that auto supply business in Bavaria—but, as I've mentioned, Worobey and his colleagues found contrary evidence that it came directly from Wuhan. On January 31, 2020, two Chinese tourists in Rome tested positive. One week later, an Italian man just back from Wuhan, on a special flight repatriating Italian nationals, was hospitalized as a confirmed case.

Then came a two-week lull, during which doctors were directed (by Italian protocols, at the advice of the WHO) not to test for COVID-19 in patients with no travel link to China. One doctor, in a town called Codogno southeast of Milan, had a patient suffering severe pneumonia, a thirty-eight-year-old man who was unresponsive to normal treatments, so she decided to ignore that advice and get him tested. The man was positive, becoming Italy's first confirmed case of locally acquired SARS-CoV-2. He had arrived at the Codogno hospital from a smaller town nearby, Castiglione d'Adda, a little place with a medieval castle not far from the Adda River, and how the man got infected there was a mystery.

The next day, February 21, health authorities announced a cluster of sixteen cases in Lombardy, the booming northern region that encompasses Milan and eleven other provinces, including Lodi, within which lie Codogno and Castiglione d'Adda. None of the sixteen people had lately returned from China, but one of them had met with a friend who had. Two more days passed, plus an anxious ministerial meeting at Italy's disaster relief agency, and then the government of Prime Minister Giuseppe Conte ordered a lockdown for part of Lodi province, ten towns with a population totaling fifty thousand people, declaring the area a "red zone." Schools and nonessential shops had to close. Conte sent troops to seal the borders. It seemed drastic at the time—Italy had only just recorded its first Covid fatality, a seventy-eight-year-old man in Padua—but in fact it was too little too late.

From that point, northern Italy suffered an explosion of cases and a horrific human toll. After two weeks, Conte ordered all Lombardy and the rest of the north into quarantine. One day later, he extended that to all of Italy, putting sixty million people into lockdown. Businesses closed. Restaurants and bars went dark. Italians hunkered in their homes. By March 13, Lombardy alone had tallied 15,113 cases and 1,016 deaths. A week later, at the peak of this wave, the country at large was confirming six thousand new cases per day.

"That's about forty thousand people being hospitalized at the end of March," Marino Gatto told me. Gatto is Professor Emeritus of Ecology at the Polytechnic University of Milan, trained in engineering and mathematical modeling and for much of his career a specialist in disease ecology. "The hospitals were flooded, and certainly many people did not die in the hospital," he said. "Many people probably died at home. They couldn't even be hospitalized." Testing was insufficient, swabs for testing were in short supply, and among those tested and found positive, the case fatality rate was catastrophic, roughly eight times the rate in Germany or South Korea. Was that because of industrial air pollution in the north, causing respiratory stress that left lungs especially vulnerable? Was it because of cigarettes? Was it because of multigenerational households, with fragile grandparents closely exposed to asymptomatic but infectious youngsters? Was it because Italians hug one another more often than other nationalities? The variables were many and the agonies were befuddling. Why, people wondered, was this gracious country suffering so wretchedly? *Someone ask the Pope: Holy Father, does God now hate Italy? This is our thanks for all those beautiful pious paintings?* By June 3, 2020, when the peak had passed and the government lifted the last restrictions, 33,694 Italians were dead.

The north had suffered first and worst. "Why? It's very simple," Gatto said. "Because it is industrialized." The industries meant air pollution, yes, but also international business travel. "A lot of companies with a lot of connection with outside, with Germany, France, *China itself*, and so on." Lombardy had been seeded early with the virus, possibly through one of the three international airports near Milan, by bad luck and fateful connectedness, and for weeks preceding that first recognized case in Codogno, the virus had quietly spread.

Among the Lombardy provinces hardest hit was Bergamo, northeast of Milan, with its factories, its mineral processing, its Orio al Serio Airport, and a vibrant provincial capital, Bergamo city, in the foothills of the Bergamo Alps, encompassing an old walled town on a hilltop, attractive to tourists, and a modern commercial center below, rich with restaurants and museums and arts venues. As of February 25, the province reported eighteen cases, many fewer than the 125 in Lodi province, but an event had occurred the previous week that probably enflamed transmission of the virus throughout Bergamo.

The city's professional football team, Atalanta, had qualified for the playoffs of the Champions League, and on February 19 played the biggest match in its history, against the Spanish team Valencia. For such major matches, Bergamo's "home" stadium was a grand arena in Milan. So on February 19, forty thousand loyal Atalanta fans traveled by bus, train, and car to Milan for the match, which Atalanta won, 4–1, amid much celebratory hugging and kissing and screaming. Many others, who hadn't gotten tickets, gathered in homes and bars for watch parties. "Unfortunately, we couldn't have known," the mayor of Bergamo said. "No one knew the virus was already here." Soon after the match, COVID-19 began its sharp rise. By March 25, the province registered seven thousand confirmed cases and a thousand deaths. The cemetery in Bergamo city couldn't keep up with the burials, so military trucks took bodies elsewhere to be cremated. The main Bergamo hospital also couldn't keep up; an emergency field hospital, quickly established in a trade fair center, added another 142 beds to local capacity; and still there was a shortage, with Bergamo patients transported to Milan and filling up the teaching hospital at the University of Milan, the Luigi Sacco.

A young doctor named Gabriele Pagani worked at Sacco, doing a residency on infectious disease. "We were overwhelmed with all the patients from outside," he told me. "We just had to close every outpatient clinic." They also shut down the surgical work. Milan itself hadn't yet been hit hard, that would come soon, and in the meantime Sacco Hospital absorbed the caseload from Bergamo and a couple other hotspots. Within the hospital, staff were shifted and reassigned to meet the emergency needs. Pagani was sent to a low-intensity ward handling elderly patients without severe Covid, and quarantine cases, and bureaucratic tasks. "We called it the Beauty Farm," he said. "Psychologically it was nice because you actually saw people getting out of the hospital." On the other hand, it felt peripheral to the real battle, which was trying to save patients *in extremis*. "I used the spare time to do research," Pagani told me. He was in the last stage of compiling a multiyear study of dengue fever in Italy. That sense of becalmment changed suddenly when he ran into the head of the Biomedical and Clinical Sciences Department, Massimo Galli. Pagani had known Professor Galli, and had been mentored by the older man since his undergraduate years, during which he did clinical work at Sacco. Now the busy professor walked into a room, saw Pagani, and said, "I got a big opportunity for you. Come with me."

Pagani is a lively thirtyish man with a dark blond ponytail and a scruffy Bob Dylan beard. He had planned for an overseas interlude as part of his residency, doing epidemiology and public health in Madagascar. But with enough of its own disease challenges already, including recurrent outbreaks of bubonic plague, Madagascar went into COVID-19 lockdown even before Italy, and all flights to the island had been canceled. Professor Galli, aware of this, directed Pagani instead to a much closer site of exigency: Castiglione d'Adda, the little castle town southeast of Milan, from which had come Italy's first confirmed Covid case.

The town, with a population of less than five thousand souls and no hospital, constrained within the red zone, in lockdown, was suffering badly and alone. It had endured many dozens of cases since that first, and by the end of March, roughly one percent of its total population would be dead of COVID-19. Galli dispatched Pagani there to organize an epidemiological study by blood screening. The point was to learn how many people were infected—what percent of the populace, young or old, smokers or nonsmokers, hospitalized or not—and whether the town might be approaching a point when, because most people had already experienced a minor or asymptomatic infection, and recovered, the situation could get no worse. It might even get much better. The town might achieve that mysterious state known by the misnomer "herd immunity."

Galli contacted the mayor of Castiglione d'Adda for approval and co-operation. Pagani then found himself charged with organizing this ambitious study, entrusted to him by Galli, who had a thousand other matters on his mind.

"I never did anything like that," Pagani told me. It was a difficult task, and all the more difficult because he couldn't get adequate supplies, he couldn't get enough PPE, he had to raise the money from grants and donations. But he assembled a team of good partners and advisors, and they got it done. They took finger-prick blood and swab samples from more than four thousand willing participants, almost the entire town, an amazing degree of cooperation, and found that 22 percent carried antibodies against SARS-CoV-2. The surprising thing about that result, which was also the worrying thing, was that the number was so low. "This is a lower-than-expected prevalence," as Pagani and his collaborators noted in their published report, "considering that it was recorded in one of the most severely-affected areas in Italy." It seemed ominous, suggesting that "a large part of the population remains susceptible to the infection."

The death count was another troubling aspect of these numbers. At least forty-seven town residents died of COVID-19 during the first three months of 2020. If the infection rate was surprisingly low, that meant the case-fatality rate in Castiglione d'Adda was high: about 5 percent. "There's a lot of elderly people," Pagani said.

Pagani was susceptible too, but youthful and lucky. He had caught his own case of COVID-19 before the Castiglione d'Adda study began. His symptoms weren't severe, just aches, weariness, a terrible pain in his hips, bad sleep, and in the morning after one restless night, he opened a can of coffee and couldn't smell it. That tipped him off: this wasn't a cold or the flu. He quarantined for a few days, staying clear of his girlfriend and his parents, but he infected his cat.

She was a four-year-old female shorthair named Zika. "Like the virus," yes, he confirmed. She started sneezing. No other signs of malaise, and her appetite held. When Pagani mentioned it at the hospital, a friendly virologist offered to isolate Zika's virus and sequence the genome. "Doctors are nerds and nerds love cats," he explained to me. (Maybe, but only an infectious disease nerd doctor would name his cat Zika.) Pagani swabbed her.

He gave his own swab sample for virus isolation and his was sequenced too. The two genomes were 99.9 percent identical, so Zika had come by it honestly. Both genomes belonged to lineage B.

Meanwhile, though Pagani and Zika recovered, people were dying in Italy and around the world. By June 7, when Pagani's field study in Castiglione d'Adda concluded, the national case count was 235,035. Almost 34,000 Italians had died. The global death total was 423,442. And all that was just the first wave, with worse to come.

44

Lineage B became a limb with many branches. The first and most consequential was B.1, detected there in Lombardy at the start of the outbreak, and probably having originated there. It differed from the ancestral lineage B by two mutations, including the one you've heard about, D614G. By early March, B.1 had infected people in Central Europe, the Netherlands, the United Kingdom, and the United States. For one notable instance, it went to New York City, initiating (or helping initiate) the outbreak that progressed to a horrific overload of the city's health care system and drew the gruesome, sad media coverage that we all saw. That crisis reached its crescendo in early April, and by May 10 the state of New York had more than 330,000 cases, most of those in New York City, accounting for 8 percent of the world total.

The B.1 lineage soon gave rise to a branch of its own, designated lineage B.1.1. This one turned up simultaneously in Denmark, Germany, the U.K., the U.S., and there in Lombardy, the simultaneity reflecting a relative shortage of sequencing and making it difficult to say just where B.1.1 started. Arguably, that doesn't matter. What matters is that from this lineage sprang at least seven other branches, of which the most notable was B.1.1.7. These labels seem like numerological befoggery, I know, but they reflect the fact that, as case numbers rose, as mutations added variety to genomes, this virus seized its opportunities, exploring many evolutionary options. We'll move beyond these letter-number jumbles, shortly, to some handier monikers. B.1.1.7, for instance, became the U.K. variant. Later, with more data available, more variants emerging, and for ease of public discourse, that received an even simpler name: the Alpha variant.

Alpha, Beta, Gamma, Delta: the four horsemen of the Covid apocalypse in its second phase.

That phase began in autumn 2020, and it dispelled all uncertainty—still lingering around the D614G mutation—as to whether this virus would evolve. The earliest evidence of the B.1.1.7 variant came from Kent, a county in southeast England bordering Greater London. It turned up in two patient samples collected there on September 20, 2020. But its significance was only recognized later, and thanks to an almost accidental convergence of two kinds of data—epidemiological, charting a sudden, aggressive spread in cases of COVID-19, and genomic, which revealed that the spreading virus contained a sizable, puzzling cluster of new mutations. Andrew Rambaut, at the University of Edinburgh, noticed this cluster by its genomics while he was looking for something more general: mutations in the spike protein, including D614G, brought to attention by Bette Korber's team, and another suspicious mutation that was advancing in South Africa.

"We kept an eye on these things," Rambaut told me. The United Kingdom at that point was sequencing more SARS-CoV-2 genomes than any other country, thanks to foresightful establishment of a consortium, known as COG-UK, with £20 million in government and private support. COG stood for COVID-19 Genomics. The consortium included more than a dozen universities, four government agencies, and the Wellcome Sanger Institute, a research center funded by the Wellcome Trust and named for Fred Sanger, the father of genomic sequencing. COG-UK encompasses sixteen sequencing hubs, scattered across the country and pooling their genomes and analyses, amounting to the largest and most useful body of genomic tracking done by any nation in the world. The executive director is Sharon Peacock, a professor of microbiology at the University of Cambridge with a particular prowess for organization. The effort began in March 2020, with cryptic emails from Peacock to some of her colleagues.

"I didn't say what it was about," Peacock told me. "I just said, 'Can you call me?' I got five different people on the phone, and I said, 'Look, what do you think about setting up a national sequencing capability?'"

Her idea aligned with the thinking of Sir Patrick Vallance, the government's chief science advisor, and they arranged a day-long meeting of potential partners, which yielded agreement on a framework and a funding proposal. The money was promptly committed, mostly from the government's COVID-19 response fund and the Wellcome Sanger Institute, and it started to move on April 1, an impressive feat of collective mobilization and acceleration. Nothing like this happened in, oh, for instance, the United States of America.

"There were doubters who made their views known to us and others," Peacock wrote in a short historical essay on the COG-UK Consortium website. These were people "who felt we were wasting our time. Coronaviruses do not mutate as frequently as some other viruses including influenza and HIV." Peacock's own research concerned mainly bacteria, but she understood viruses and she understood planning. "Why bother getting ahead of the worry curve?" was the attitude of the doubters, she wrote, to which she gave her answer. "We took the view that waiting until the worst happens, only to realise that one is totally unprepared, is not where we collectively wanted to find ourselves." At the end of a year, the U.K. found itself with 259,502 sequences of SARS-CoV-2, more than 40 percent of the world's total, and therefrom a considerable capacity to detect important changes and trends, such as the rise of B.1.1.7.

Sharon Peacock is a person of strong will and hungry mind, and always has been. "I grew up in a working-class family, and nobody had been to university before," she told me. At the age of eleven she failed a crucial exam, which could have led to an academic secondary education and possibly university, but instead diverted her toward domestic skills training (cooking and sewing and basic arithmetic); and so she left school at age sixteen and went to work at a corner grocery near her home. She rather liked working in the shop, but she noticed an alternate opportunity on a bulletin board one day and trained as a dental nurse. Further training as a medical nurse followed. It was during that training, in a hospital environment, that she had a new thought: "I really love this, but I want to be a doctor." She possessed almost no academic qualifications in science, but that could be fixed. She attended college part-time to study math, physics, biology, and chemistry. She tried to get into a medical school and was rejected, evidently because of suspicion toward a female of by now slightly nontraditional age. "So I had to, sort of, ring up the university one day in a fit of bravery, and say, 'You've rejected me twice, but actually I'd quite like you to consider me.'" She got an inteview. She won a place in the class, at Southampton University, where she trained as a physician. She became a member of the Royal College of Physicians, then specialized in microbiology and infectious diseases. "I came a bit late to medicine and then I came, you know, obviously, a bit later to science." She did a PhD in microbiology at Oxford, with a fellowship from the Wellcome Trust, then ran a bacterial diseases program in Thailand for seven years at the Mahidol-Oxford Tropical Medicine Research Unit, in Bangkok. She returned to professorships in the U.K., eventually at the University of Cambridge, plus work as an advisor to Wellcome's program on drug-resistant infections.

So the Wellcome Trust knew Sharon Peacock well—her hungry mind and her tensile will—by the time she went to them, amid the consternation of the early pandemic, with a proposal for a sequencing consortium. The U.K. government knew her too. *By all means, Dr. Peacock, here's £20 million.* As of autumn 2020, COG-UK had generated its first tens of thousands of sequences, with more coming quickly. The country had too many cases of COVID-19, but its ratio of sequences-gathered to cases-identified was high.

It was that body of data, plus sequences available from international databases such as GISAID, that allowed Andrew Rambaut and his lab group, a posse of smart young PhD students and postdocs with keen bioinformatics skills, to scrutinize genomes from all over the country and the world for anomalies. Rambaut looked for D614G and also for the suspicious South African mutation, designated N501Y, and found the latter among some genomes from Wales. Hmm, curious, but those genomes had no other concerning mutations. His young colleagues assigned nicknames, within their informal lab culture, to the mutations. D614G was known as "Doug." If the genome lacked that mutation, retaining the original form, it was "Douglas," as in Doug-less. N501Y became "Nelly."

Rambaut looked further and found Nelly in a cluster of genomes from southeast England—and in those genomes, Nelly was not alone. She was accompanied in genome after genome by a gaggle of other changes, an unusually large number, more than twenty mutations and deletions, seventeen of which were significant. "It was quite extraordinary," he told me. "I had not seen anything like it." Eventually the gaggle got their own collective nickname: "the Elephants." Furthermore, several of those mutations, including Nelly, lay in the spike protein of the virus, possibly affecting its capacity to attack cells. He'd seen nothing like it, he revised himself, except in the occasional case of chronic infection, when a virus lingers in one person and keeps mutating. That thought, occurring to him as he ruminated on the gaggle, would lead to others. One of his PhD students, Verity Hill, an Oxford grad in biology with digital mapping skills, recalled to me later how Rambaut's thinking developed. "Andrew is saying, 'Maybe we should look at this whole herd of Elephants, and not just this one Elephant named Nelly.'"

A week or so into December, Rambaut joined a regular conference call with scientists from Public Health England (PHE), a government agency, and the talk turned to an unusual, fast-growing cluster of COVID-19 cases in Kent and East London. The epidemiologists had seen it in their data. This was early December, and such an aggressive advance seemed weird, given that England had been in lockdown for most of November. But, combined with Rambaut's genomic data on the mutations, it was more than weird; it was ominous.

Epidemiology plus genomics raised this concern to the ministry level. The cabinet held an emergency meeting. On December 19, 2020, Prime Minister Boris Johnson announced another lockdown order, this time for London and most of southeast England. The new variant, he warned, based on preliminary data from the scientists, could be 70 percent more transmissible than the old virus. Christmas gatherings, beyond a single household, were canceled. By then Verity Hill had left Edinburgh—by luck, it was just before the border closure announcement—and returned to her parents' house in Aylesbury, north of London.

"I woke up on Sunday morning with two messages from one of my collaborators," she told me. "He was like, 'Can you make a bunch of maps to track the spread of this across the country?'"

45

When the virus first reached the United Kingdom, back in January 2020, Verity Hill didn't foresee being scientifically involved. She had a PhD in progress, focused on the genomics of the 2013–2016 Ebola outbreak in West Africa, with particular attention to Sierra Leone. She was halfway through her third year, building computer models of a sort known as SkyGrid, useful for tracking the population dynamics of a virus but too abstruse for comprehension by the likes of me and maybe you. She had scarcely heard about the atypical pneumonia cases in China, and paid middling attention until a day in late January, when her boss, Rambaut, arrived at the lab one morning with the front page of a tabloid, *The Scottish Sun*, bearing a bold headline, 5 IN KILLER SNAKE FLU TESTS, and taped it to the computer screen of one of her fellow students. "Killer Snake Flu" is what the *Sun* had taken to calling COVID-19, based on that journal paper just published by a team in China, touting snakes as possible intermediate hosts of the novel virus. This latest update in the *Sun*, on January 24, trumpeted that the "snake flu" virus had reached Scotland and the government was "closely monitoring" its spread.

The snake idea didn't take hold but the virus did, in Scotland and elsewhere throughout the U.K. By March 23, the country had 6,020 cases, and Boris Johnson reluctantly ordered a lockdown, telling Britons to "stay at home." The Rambaut lab shifted to working remotely with a focus on SARS-CoV-2. Verity Hill proposed to Rambaut that she might build a SkyGrid model of the new virus. It was her forte, and she wanted to be useful, "'cause I was, like, pretty excited to get involved," she told me. "This is what I sort of imagined would happen at some point during my PhD." Instead of a SkyGrid, though, she helped create—with a fellow grad student in the lab, Áine O'Toole—a new kind of genomic pipeline that they called CIVET, a simplified tool by which doctors in hospitals or scientists in universities could place a given genomic sequence within a tree of its relatives. CIVET stood for cluster investigation and virus epidemiology tool, a coy acronym harking back to the intermediate host of the original SARS virus.

O'Toole, one year ahead of Hill in the PhD program and poised to start writing her dissertation, was experienced at such tasks, having already created a SARS-CoV-2 pipeline, at Rambaut's request, that she named PANGOLIN. The genomes had begun to flood in, thanks to that national consortium, COG-UK, and those data would be just a chaotic jumble unless they could be placed in family trees by some automated process. Rambaut offhandedly said, by O'Toole's recollection, "It would be really good if we had some sort of tool to assign these things." O'Toole had written pipelines but never one quite like this. She googled for guidance, she noodled for inspiration, she stayed up late one night, "and the next morning there was PANGOLIN," she told me, speaking from an attic room atop her parents' house in Dublin, to which she had retreated to ride out the latest lockdown in Edinburgh and elsewhere. She wore a nose ring and a print dress.

"Why did you call it PANGOLIN?" I asked. Tribute to another host animal, a gentle and beleaguered creature—yes, that part I got. But I wondered how she made it work as an acronym.

"Phylogenetic Assignment of Named Global Outbreak Lineages," she said.

"Whoa."

PANGOLIN became one of the definitive tools for placing SARS-CoV-2 genomes into lineages and showing their place on the big family tree. What it did, which CIVET had not, was to make this service easily available to researchers and public health people around the world. It was a piece of software you could download and run on your own computer, with your own data, or you could access it on the web. "Today I was working with a guy I know, Kefentse, he's in Botswana," O'Toole told me—about Kefentse Arnold Tumedi, a fellow PhD candidate, at the University of Botswana. "They're doing a sequencing run, and they have ten samples," she said. "They haven't put their data anywhere, but they want to check what their data is." Where might those ten samples fit within the scope of SARS-CoV-2 evolution? "They can run it through PANGOLIN. Literally drag and drop the file, and it will produce a report, and tell them what lineages—what lineage—each sample is."

Pangolins may be endangered in the wild, but O'Toole's PANGOLIN thrived. In fact, it just passed a milestone, she told me with justifiable pride: at the time we spoke, the web version had assigned more than a half million sequences to their positions of relatedness by shared mutations. People could see which lineages were present in their communities, which lineages were spreading geographically or increasing in prevalence. "Everyone has run PANGOLIN on their sequences," she said. "It's just sort of incredible." The era of SARS-CoV-2 variants was dawning, and Áine O'Toole, along with Verity Hill and their colleagues in the Rambaut lab, were well positioned and equipped to see its first glimmers.

She and Hill both paused their PhD efforts for six months to work on SARS-CoV-2 as employees of COG-UK. The genomes accumulated and were sorted into lineages. Then came December, and Rambaut spotted the Elephants in Kent, galloping toward London.

On the afternoon of December 18, 2020, Rambaut posted a paper on Virological, with a group of coauthors, alerting the attentive to "an emergent SARS-CoV-2 lineage" that had been "growing rapidly over the past 4 weeks" as it advanced northwestward, from towns and villages in the green countryside of Kent into the metropolis. By now it had also appeared in Scotland, Wales, and four other countries. This was not just a matter of one notable mutation but of many, the paper said: fourteen amino acids switched, and three deletions, a concatenation of changes "unprecedented" among genomes seen during the pandemic. Eight of those changes occurred in the spike protein. One of them was Nelly. Rambaut and his coauthors, following the PANGOLIN nomenclature system suggested months earlier, labeled the lineage B.1.1.7.

How and where did it arise? Multiple mutations accumulating over a short timespan—that had been seen, though not to this degree, in patients with weakened immune systems (for instance, cancer patients getting chemo) who suffer a prolonged infection with a virus. The longer it stays in the body of one person, feebly challenged by a weakened immune system, the more likely a single virus can stack up mutations and then jump to the next person, bearing them all. Rambaut's group hypothesized that this sort of event, relatively rare but not improbable as the number of cases grew so large, had produced B.1.1.7. The new virus called for urgent lab research, they said, and "enhanced genomic surveillance worldwide." Next day, Boris Johnson put London and the southeast back into lockdown.

Further analysis of the data, further conferrals between Edinburgh and England, soon yielded a published report documenting what PM Johnson had already said: that the B.1.1.7 variant seemed to be more transmissible than the preceding virus, and by as much as 75, not just 70, percent. It had expanded its presence rapidly during October and had continued gaining amid the partial lockdown of November. That pointed toward an unnerving conclusion. Not only was B.1.1.7 more capable of getting from one human into another; it was more capable of doing that while people were social distancing and wearing masks. You could almost say that SARS-CoV-2 was learning about us, even as we learned about it.

46

More bad news for people, more good news for the virus, arrived before the end of December, as though SARS-CoV-2 were bringing to a crescendo its successful first year as a human pathogen. Scientists in South Africa announced their detection of another highly capable, multi-mutation variant. This one carried nine mutations in the spike protein alone, including Doug (D614G), Nelly (N501Y), and another, written as E484K, which by the Rambaut lab's mode of nicknaming became "Eek." Sometimes referred to as "Eek!" This variant, with its Eek and its Nelly and its Doug and other mutations, also belonged to the B.1 lineage and got its four-digit variant code (B.1.351) through the PANGOLIN system. But you can forget about that, and we'll shift now to the simplified, popular naming proposed by the WHO, according to which it is: Beta. The Alpha variant, containing multiple mutations, was first detected in the U.K.; and then Beta, another variant carrying a different constellation of mutations, in South Africa.

The Beta variant was first seen in an area called Nelson Mandela Bay, a municipality containing a million people, on the coast about five hundred miles east of Cape Town. A team of mostly South African researchers, led by a Brazilian named Tulio de Oliveira at the University of KwaZulu-Natal, spotted it, in a sample from October 15, 2020, as another lineage carrying D614G and made suspect by that plus five other spike mutations. Those included N501Y (the familiar Nelly), E484K (Eek), and three other significant mutations in the spike protein. The new variant spread quickly toward Cape Town and within weeks it was the dominant lineage of SARS-CoV-2 throughout both southernmost South African provinces, Eastern Cape and Western Cape. By the end of November, it had added three more mutations to the spike, one of which was K417N (dubbed "Karen"). K417N seemed positioned to help the virus evade immune systems. This suggested that the Beta variant might be capable of reinfecting people after prior infection by the original virus—or maybe even after vaccination. South Africa had already suffered badly enough: 698,000 cases, with almost nineteen thousand deaths.

"We had been through it," Penny Moore told me. She's a research professor at the University of the Witwatersrand, in Johannesburg, and an expert on virus-host dynamics. "We had a hectic, hectic lockdown and then we'd come out the other side." They had begun to relax, "until the numbers started climbing again." Moore and others are acutely aware that new variants, such as Beta but not limited to it, might emerge from immunocompromised people, along the lines suggested by Rambaut's team about Alpha.

"We're very worried about that, particularly in South Africa," she said, "because of the huge HIV prevalence." Amid a population of sixty million, the country contains more than 7.5 million citizens living with HIV. If there's anything that represents an underlying medical condition, making COVID-19 patients potentially more susceptible to severe disease or death, it would be AIDS. In addition, the more immunocompromised people there are, the more hosts in which a virus might linger long enough to generate multiple mutations. It is very, very important to recognize that this last statement is no rationale for blaming victims; it is a description of evolutionary circumstances that could entail additional dangers for everyone, but especially for those with AIDS, and it deserves scrutiny as a consideration to them.

The Beta variant, like the Alpha, spread quickly beyond national borders. By January 7, 2021, just a week into year two of the pandemic, forty-five countries had detected Alpha in their COVID-19 patients, and thirteen had detected Beta. Some of those countries reported that Alpha was not only present; it was increasing in prevalence, as it had throughout the U.K. But the rate of international advance for either variant was unknowable, and their dispersal through London or South African airports was untraceable, except by inference from available sequenced genomes. Few countries had mounted anything like the COG-UK enterprise, but by this point Denmark, Iceland, the Netherlands, and Australia were doing fast, routine genomic sequencing from their COVID-19 cases, and South Africa and Botswana had also begun sequencing. More such data were badly needed but it was a valuable, important start—molecular epidemiology on a global scale at high speed. This tracking of variants was described in another posting to Virological, signed by a long list of coauthors from around the world, including Eddie Holmes in Sydney, Marion Koopmans in Rotterdam, Oliver Pybus at Oxford, and Tulio de Oliveira in Durban. The two lead authors were Áine O'Toole and Verity Hill.

In the United States, despite the available resources and expertise, sequencing of SARS-CoV-2 samples was deplorably thin. "We have enough sequencers to sequence SARS-CoV-2 from every case, 100 times over," Kristian Andersen told Amy Maxmen, a senior reporter for the journal *Nature.* At that point, according to Maxmen, the U.S. ranked behind at least thirty other countries in numbers of sequencing done. There was an American version of the COG-UK consortium, called SPHERES (never mind how that acronym spells out), connecting university and corporate labs with government efforts, which at first lacked the power to deliver money, though it later came on strongly. In San Diego, where sequencing surveillance was quite good, thanks to an early initiative led by Andersen, he and colleagues supplemented that coverage by finding a proxy signal, from the Alpha genome, that could be detected by simpler, cheaper PCR tests. That allowed them to infer things about Alpha in the U.S., as described in a preprint: that the variant had arrived near the end of November 2020, and that by January 2021 it had spread to thirty states. It was at least 35 percent more transmissible than the preceding virus, and its relative frequency in the total SARS-CoV-2 population was doubling every week and a half. It would probably soon become the dominant variant in many states, Andersen and his coauthors warned, "leading to further surges of COVID-19 in the country, unless urgent mitigation efforts are immediately implemented."

There was another "unless," at which they hinted: Alpha might become the nemesis virus in the United States, and around the world, unless some other variant came along that was even worse.

47

A third variant, also menacing, arose in Brazil at almost the same time as Alpha in the U.K. and Beta in South Africa. That coincidence of timing could be more than coincidence. It might suggest that SARS-CoV-2, having reached monumental abundance in a short time (47 million cases worldwide by the start of November 2020) had accumulated mutations enough, genetic variation enough, to erupt with new evolutionary gambits. The virus was seeking pathways to success, just as impounded water seeks any pathway that will allow it to run downhill. This third variant made its debut, during December, in the central Amazonian city of Manaus. There weren't many genomes available for scrutiny in Manaus, but of those that were, suddenly 52 percent of them carried a notable group of mutations. Case numbers in the city rose sharply, and so did hospitalizations.

Some of those mutations were familiar from the other variants, but they seemed to have occurred independently. Nelly was there. Eek was there. Something very similar to Karen was there. (Doug was not there, so this was Dougless.) Altogether the new variant carried seventeen significant changes plus three deletions, and three of those changes fell in the receptor-binding domain, that most crucial part of the spike for latching on to cells. The scientists who found this variant named it P.1 (don't ask me why). For our purposes, it is now Gamma.

Manaus is a busy city that sits at the confluence of the Rio Negro and the mainstem Amazon River, reachable by plane or by boat or, not easily, by a single road running south from Venezuela. In the colonial era, Manaus was a great entrepôt for rubber, and therefore a bolus of wealth (and of poverty) along the great wilderness river. It was styled as "the Paris of the Tropics." Rich people paid for an opera house and poor people paid for a cathedral. It stood on the site of a seventeenth-century fort, built by Portuguese colonizers, and it grew into a regional hub, an enticement for indigenous people who wanted wage work or manufactured supplies, missionaries who wanted souls, and soldiers of fortune as crazed as the one Klaus Kinski portrayed in the movie *Fitzcarraldo*. But in the mid-twentieth century it was declared a Free Trade Zone, to help spur its development, and it became a modern city, with a big river port, a financial district, high-rise hotels and apartments, a vast football arena, a nice beach licked by the black waters of the Rio Negro, an Amazonas Philharmonic, and two million residents. Sadly, those residents took a pummeling by the first wave of COVID-19, even before the Brazil variant emerged.

The virus seems to have first reached Brazil from Italy, by way of four travelers arriving in São Paulo during February 2020, just at the time when Italy's own initial outbreak began in Lombardy. From there it moved north quickly, meeting a disorganized public health response exacerbated by political turmoil in some states (such as Rio de Janeiro, where the governor faced impeachment), acute shortage of resources in others (Amazonas, a huge state with few ICU beds and all of those concentrated in Manaus), prolonged circulation of the virus before the first case was reported (Ceará state, on the north coast), socioeconomic inequities, and a poorly informed president (Jair Bolsonaro) with autocratic tendencies who rejected nonpharmaceutical interventions (masks, social distancing), promoted hydroxychloroquine as a treatment, fired his minister of health in April, lost another minister of health in May, and then appointed to the job an army general with no medical qualifications. On April 28, a day when Brazil's total cases exceeded 73,000 and the death toll passed five thousand, Bolsonaro faced a group of reporters at a barricade, and someone mentioned the numbers. "So what?" he said with a shrug. "I'm sorry. What do you want me to do?" Brazilians were so badly served by their president they might as well have had Trump.

Despite the presidential indifference, leaders in São Paulo and Rio de Janeiro, Brazil's two largest cities, ordered partial lockdown measures in March. Schools and universities closed. Theaters closed. Bars, restaurants, shopping centers, and beaches closed. Public transit was limited. All of that helped, and as a side effect air pollution decreased in the two cities, but it wasn't enough to prevent Brazil becoming, on May 24, the second most Covid-infected country in the world, behind only the U.S. This first wave for Brazil peaked on July 29, with more than 71,000 new cases reported that day.

Manaus was anomalous, as you might expect for an insular city surrounded by river and Amazon forest. The virus arrived there in March 2020 and caused "an explosive epidemic," peaking in May, with more Covid-suspected fatalities than Covid-confirmed diagnoses, indicating that it was being grossly undercounted. Public health researchers therefore gauged it by an indirect measure, "excess mortality," meaning that the count of deaths, explained or unexplained, surpassed average deaths by this much, a margin attributable to the virus. Excess mortality had almost quintupled. Most of it was in people over sixty. Many died at home or out on the public byways. Deaths at home and on the roads reflected something else: COVID-19 was hitting poor people harder than the affluent, and government efforts weren't countering that disparity, even though Brazil has a publicly funded health care system, Sistema Único de Saúde, the largest in the world, that is intended to provide universal free care. Manaus had one ICU bed for every nine thousand people. So much for universal.

The virus scorched its way through the city. If you lived elsewhere in Amazonas state, in a tiny settlement upstream on some tributary, or even in a distant town such as Manicoré, you might have avoided it. But not likely in Manaus. By October, according to a survey of blood samples tallying the presence of antibodies, SARS-CoV-2 had infected 76 percent of the Manaus population. Another way of saying that, in the lingo of epidemiology, is that the "attack rate" was 76 percent. There was room for error in the estimate, but it would take a lot of error to reverse the basic message: the virus was *all over* that place. (Down in São Paulo, around the same time, the estimated attack rate was 29 percent.) And this was before the arrival, remember, of any aggressive new variant.

Then came a decline, and a few months of respite. In early November, though, a second wave started to rise throughout Brazil. That's when Carlos Morel, an eminent scientist and physician in Rio de Janeiro, suffered his own case of COVID-19.

Morel is a senior figure in the realm of infectious diseases and global response, having served on the WHO's executive board, and as director of its research program on diseases of poverty; at home, he has been a member of the cabinet of the minister of health and president of the Oswaldo Cruz Foundation, Brazil's leading institution for research and development on public health. He has focused especially on neglected diseases, such as tuberculosis and Chagas and onchocerciasis, and on the neglected people who suffer most from them. He has also been active in the effort to build a global atlas of animal viruses as a step toward surveilling what could be dangerous to humans.

Morel took COVID-19 seriously, as a citizen as well as a scientist, and spent the months of Brazil's first wave carefully isolating, with his wife and younger son, at their home in Rio. He continued his work, including a paper in progress about the importance of genomic viral surveillance and the biosafety labs in which spooky new viruses are studied. Then, in November 2020, he got unlucky. His wife went to a medical appointment, his son went to a business meeting, and they both caught Covid. The son had a mild case that was soon gone; Morel's wife suffered fever and other symptoms for a week, including loss of her sense of smell, then recovered. After another week, Morel himself began feeling ill.

He moved to another part of the house, but it was too late. His breathing became labored, a strange feeling, and he had to command himself consciously to take breaths. Then came fever. "I was getting worse and worse and worse," Morel told me. On November 25, he said to his family, "Guys, I think this can be serious. I have to go to hospital." At the hospital, his breathing capacity was measured: not good. X-rays showed almost half his lung capacity compromised, and doctors said "Tsk, tsk, tsk" over the films. He was moved into an ICU ward dedicated to COVID-19. When he worsened further, they put him on a breathing machine that involved a mask but no tube down his throat, no general anesthetic—"noninvasive ventilation," was the term. A less radical step than intubation. He was awake and could interact with the machine intelligence of the respirator, commanding it by his rhythmic breathing to supply oxygen as he needed. He spent almost two weeks with this device on his face.

"I had nightmares, I was in hell," Morel told me. "It was as if I would be hearing the ring of the gates of hell, see." Weekends could be especially difficult, because the support team changed and the second team sometimes got busy and neglected him for up to eight hours. He resolved to bear it. He was a doctor, and he had been warned by colleagues about going "on the tube," which he might not come off alive. "Some people who are claustrophobic," Morel said, "they can't support this machine, this mask." He knew of another doctor, president of the National Academy of Medicine, who couldn't abide the mask, took the thing off, had to go on the tube. "And he died."

They put a cannula into one of Morel's arteries, a plastic tube, so that medicine could be easily piped in and blood samples piped out. They catheterized his bladder, of course. His own doctor visited the ICU regularly and looked at the latest X-rays. "In these days, he did not have the courage to tell my wife how bad I was." At one point, the personal doctor advised, "We have to go to a very, very high dose of corticosteroid. Otherwise, we're going to lose this patient." Morel found distraction and strength by talking with his family on his mobile phone, when he was able, and conferring with his coauthor on the viral-surveillance paper. He survived, but it was close. He improved; he went home. Months later, when he accompanied his wife to a routine medical appointment, he encountered one of the hospital doctors in that waiting room. "Oh, you are here!" said this doctor. "Welcome back, resuscitated man!"

Morel was seventy-seven, he told me. If you had been a seventy-seven-year-old farmer or fisherman in Manaus, I asked him, what would have happened?

"I think I would not be here."

The second wave for Manaus came a little later than for the country at large, and when it did come, it was more complicated, due to the prior intensity and the new variant. After peaking in April 2020, the numbers of hospitalizations and deaths remained low until almost the end of the year. Did this reflect herd immunity? For those who embraced that winsome concept—a mystic "immunity" for the "herd" of humans in one place or another—an attack rate of 76 percent seemed sufficient to deliver it. In fact, the threshold for herd immunity against SARS-CoV-2 was supposed, by some scientists, to be about 67 percent. Even you and I, without a master's degree in public health, can see that 76 is larger than 67. But the 67 percent supposition carried a lot of conditional stipulations, and conditions could change, and the concept itself is shimmery; so, for anyone guileless enough to hear "herd immunity" as a promise, disappointment was in store. If the threshold was 67 percent, Manaus should have been protected. It wasn't. Hospitalizations for COVID-19 increased sharply at the end of December 2020, and deaths increased sharply at the start of January 2021.

Several explanations might account for this new surge. Maybe the attack rate during the first wave was overestimated—not 76 percent at all, but considerably lower. Or maybe the individual immunity of recovered patients, the protection supplied by their antibodies, had waned over the months as those antibody levels declined. Or maybe first-wave antibodies weren't effective against the new variant, Gamma. Or maybe Gamma was just so much more transmissible that it could run rampant among those Manaus residents not previously infected, even if they did constitute only 24 percent of the population. In other words, maybe the herd immunity threshold was a higher bar than supposed. Another possibility, the least cheerful, was that all four of those factors were involved.

Gamma did get around—it was roughly twice as transmissible, according to one study, as previous lineages. And it showed signs that it could evade the protective function of antibodies. Worse still, it seemed much more likely to kill the people it infected. These inferences came from Manaus during the early weeks of 2021, when the city's health care system was again pushed to its limits. And then the Gamma variant did what subtle viruses do. It rode airplanes. By January 2, 2021, it was in Japan.

48

But the Gamma variant, as matters developed, wouldn't be the worst Covid worry in Japan, nor in many other places to which new forms of the virus dispersed. The spread of Gamma would soon be preempted— so would the spread of Alpha and Beta—by something that was happening in India: the rise of another variant, more menacing than all previous.

This one first appeared in samples from October 2020. The samples came from patients in Maharashtra, a large state in western and central

India, of which the capital is Mumbai. Another of Maharashtra's big cities is Pune, site of the National Institute of Virology (NIV). India by that point had established its own sequencing consortium, a counterpart to COG-UK called INSACOG (Indian SARS-CoV-2 Genomics Consortium), and one of its partner laboratories was nested within the NIV in Pune. When scientists there noticed a sudden "spurt" in the number of Covid cases in Maharashtra, they started sequencing genomes with particular attention. They screened the viral genomes from almost six hundred patients and found a jumble of lineages, including Alpha and almost four dozen others. But one lineage, a new one to everybody, stood out. It accounted for nearly half of the entire group. The researchers ran that genome through Áine O'Toole's PANGOLIN, which placed it on the SARS-CoV-2 tree of life and assigned it a label: B.1.617. What made it notable, besides its abrupt rise to predominance among the COVID-19 cases in Maharashtra, was the presence of multiple mutations in its spike gene, three of which raised suspicion. One of those resembled Eek. Another lay in the furin cleavage site. The third was a novel tweak to the receptor-binding domain, written as L452R, nicknamed "Lazer."

That mutation, Lazer, had appeared elsewhere and before—in Los Angeles. It evidently helped drive a surge throughout southern California during late 2020, and was part of the virus that infected gorillas at the San Diego Zoo. The gorillas (about whom, more below) survived.

The recurrence of these three mutations in the new India variant suggested convergent evolution, not transmission somehow from California to Maharashtra. That is, the similar changes had arisen independently and been perpetuated, in this virus and that one, because of their adaptive value. The familiar Doug (D614G) was also present, though that one might have arrived earlier on chains of international transmission. Those four mutations in combination—so the Pune scientists warned—could make this variant better able to grab human airway cells, better able to enter them once grabbed, and better able to evade antibodies in anyone who had already been infected or vaccinated.

By the middle of February, this new variant accounted for 60 percent of Covid cases in Maharashtra. That was just a modest start. Mumbai is well connected to the world, and before the end of the month, B.1.617 was in the U.K., the U.S., and Singapore. It reached Finland in March, Fiji in April. In the meantime, it continued to mutate and diversify, sprouting new branches. Canada's first cases were confirmed on April 21, one in Quebec, thirty-nine more in British Columbia, suggesting that it had come from two directions. The same day, April 21, Áine O'Toole posted an alert stating that, because PANGOLIN was now inundated with more than six hundred different genomes within the new variant, she had extended the labeling to include three sub-lineages, simply numbered 1, 2, and 3. Sub-lineage B.1.617.2 accounted for about ninety of those sampled genomes. But it was increasing its share of the action quickly. Within two months, it became the dominant variant in the U.K., surpassing Alpha and both other versions of B.1.617. The World Health Organization and the U.K.'s health agency both reclassified it on their alert list from "variant of interest" (VOI) to "variant of concern" (VOC). And then, on May 31, the WHO announced its new, streamlined naming system, for greater clarity and convenience, whereby the major members of the SARS-CoV-2 rogues' gallery would be labeled with Greek letters. That's when B.1.617.2 became known as Delta.

As Delta grew predominant in one country after another, through the summer of 2021, scientists learned more about it. A team based at Cambridge, with partners in India and elsewhere, reported a cluster of breakthrough cases (infections despite vaccination) among vaccinated health care workers in Delhi. Sequencing showed that most of them were infected with Delta. A group in Seattle studied that singular mutation Lazer—the one that turned up in a Los Angeles surge and in San Diego's gorillas, then turned up again in Maharashtra—giving it special scrutiny because of its position within a very crucial part of the receptor-binding domain. They found it in more than a dozen distinct lineages, scattered across countries and continents, where it seemed to have been acquired independently. And it was growing more widespread with each passing day. That might reflect adaptive evolution of the virus "to either the epidemiological containment measures extensively introduced in the fall of 2020," they wrote, "or a growing proportion of the population with immunity to the original viral variants." In plainer words: Lazer might be a nifty trick, tried by multiple variants but perfected by Delta, that allows the virus to circumvent not just our immune systems but also our lockdowns.

It seems circular but inevitable that Delta also afflicted China, more than fifteen months after the country had passed its initial peak and, with draconian and efficient measures, suppressed case numbers almost to zero. On May 20, 2021, a seventy-five-year-old woman walked into a hospital in Guangzhou with a sore throat and a slight fever. She had been feeling a bit crummy for two days. Early next morning, her throat swabs confirmed her as positive for SARS-CoV-2. It took slightly longer to sequence her virus, and in the meantime, four contacts also tested positive: her husband, a waiter who had served her in a restaurant, another customer in that restaurant, and the customer's grandson. They all had Delta. The China CDC, which gives us that much, doesn't say how the woman became infected.

The four contacts were promptly taken by ambulance for isolation and treatment in a quarantine hospital. Too late. More cases appeared, reflecting five steps of transmission. Within the next month, a total of 167 cases, not counting the first woman, were detected in Guangzhou and three other Guangdong cities. All of them carried Delta, their viral genomes forming a neat tree of relatedness tracing back to the seventy-five-year-old woman. And then that was it (for a while). The Guangzhou outbreak ended (though Delta would reappear in another province). A team of scientists in Guangzhou studied these 168 cases, based on the clinical and genomic data, and noticed a few things.

Delta seemed to take hold in its victims faster, with just four days on average between exposure and a positive test, rather than six days for the original virus. Delta also replicated more quickly and abundantly, yielding "viral loads" more than a thousand times greater than the earlier strains. Holy hell, that's quite a bit, but more precisely what they said was "1,260 times higher." Oh, and these big viral loads were piling up early, right about the time a person might test positive, suggesting an increased likelihood, with Delta, for early and asymptomatic transmission.

Delta was on the move. In June it would surge in South Africa. In July it would surge in Turkey. In August it would surge in the American South and in Japan. In September it would surge in Alaska, and reappear as a small outbreak in Fujian province, China, just up the coast from Guangdong. In October it would surge in Montana, where I live. The tourists would go home after Labor Day, thanks g'bye, the snows would begin, and Delta would remain, filling our hospitals.

And after Delta, we knew, would come something else. The Greek alphabet contains twenty-four letters; at that point, the WHO list of variants only went up to *mu*. A virus will always and continually mutate, as I've noted, and the more individuals it infects, the more mutations it will produce. The more mutations, the more chances to improve its Darwinian success. Natural selection will act on it, eliminating waste, eliminating ineptitude, carving variation like a block of Carrara marble at the hands of Michelangelo, finding the beautiful shapes, preserving the fittest. Evolution will happen. That's not a variable, it's a constant.

VI

FOUR KINDS
OF MAGIC

49

One kind of magic: wish it away.

On February 27, 2020, Donald Trump hosted a "roundtable" at the White House with certain Black leaders, in observance of the end of Black History Month. They met in the Cabinet Room. C-SPAN was there. The visitors introduced themselves, in series around the table, and made appreciative statements. Then Trump spoke for half an hour. After some rambling remarks, including a good deal of self-congratulation, he drifted to the subject of the coronavirus. "We have done an incredible job. We're going to continue," he said. "It's going to disappear. One day—it's like a miracle—it will disappear."

That didn't happen.

50

Second kind of magic: herd immunity.

The notion of herd immunity, in its foggiest form, traces back to veterinary medicine of the late nineteenth and early twentieth centuries, among farmers and their veterinarians who dealt with real herds, flocks, and droves. A man named Daniel Elmer Salmon, identified as America's first veterinary doctor, director of the U.S. Bureau of Animal Industry, used the phrase in an 1894 report on the care and feeding of livestock, notably hogs. Breed them wisely, give them a varied diet, keep the sties clean, and your swine will acquire "powers of resistance" against disease. He didn't need theory; experience proved it. "These facts show something besides individual immunity," Salmon wrote. "They demonstrate the possibility of obtaining herd immunity." He didn't explain those ineffable "powers of resistance" or define that mystical entity, herd immunity.

Whatever it was, it wasn't unique to swine, nor were the problems it addressed. There was a cattle malady called "abortion disease," now known as brucellosis, caused by a bacterium passed from cow to cow. It causes spontaneous abortion of calves—representing serious hits to a producer's bottom line. Some stockowners thought that the remedy was bringing in new heifers to replace the aborting cows, but a 1917 installment of *Farmers' Bulletin*, written by two experts from the U.S. Department of Agriculture, Adolph Eichhorn and George M. Potter, explained how that was wrong. Keep your infected cows, raise the calves that do survive, don't bring in new animals, and ride it out. This worked in our monitored herd, Eichhorn and Potter reported, with abortions declining over nine years to almost nil. "Thus a herd immunity seems to have developed as the result both of keeping the aborting cows and raising the calves." Abortion disease was causing losses reckoned at $20 million annually, so this was urgent and pragmatic advice.

The concept of herd immunity, still blurry, slipped across into human medicine and epidemiology during the 1920s. It became loosely mathematized, got a little laboratory testing at the expense of mice, took on a variety of meanings and nuances as vaccine science progressed, and played an important role in the eradication of smallpox, certified in 1980. For a variety of reasons, it hasn't worked as well toward the eradication of polio or measles, and against the influenza viruses, with their variety and changeability, their attacks constantly renewed by spillover of new strains from wild aquatic birds, it has been useless. But it burst clumsily into the global conversation about COVID-19 during the early months of the pandemic—most pointedly on the morning of March 13, 2020, when Sir Patrick Vallance, chief scientific advisor to Prime Minister Boris Johnson's government, mentioned it to an interviewer on BBC Radio. "Our aim," Vallance said, is to "try and reduce the peak—not suppress it completely," and he added, "to build up some degree of herd immunity whilst protecting the most vulnerable."

Vallance was on the spot that morning because of what the PM said one day before. In a news conference from Downing Street, Johnson had conceded that COVID-19 was now a pandemic, and he announced what he called "a clear plan" for dealing with it. The goal of the plan was to "stretch the peak of the disease over a longer period," so that people and institutions would be better able to cope. The means? The means were advice, not mandates—in notable contrast to what was happening on the Continent. Italy's prime minister had just decreed a national lockdown; France's president had ordered the schools and universities closed. Johnson made earnest suggestions. If you were coughing or had a fever, you should stay home. If you were over seventy, you should eschew cruise vacations. If you were a schoolchild, you should decline to go on international school trips; just keep showing up in your classrooms. And whoever you were: Wash your hands. That was it. Next morning, Patrick Vallance found himself tasked to explain: *Why the devil does the "clear plan" seem so passive? What's the point?* His ill-advised answer: herd immunity.

Evidently, he didn't regret those words as soon as he spoke them, not quite, because he reverted to the phrase again that same morning, on Sky News television. Of the virus, Vallance said: "We want to suppress it, not get rid of it completely—which you can't do anyway." The goal, again, was to lower the peak of infections, flatten the curve, stretch the impact over time, "and also allow enough of us, who are going to get mild illness, to become immune to this. To help with the, sort of, whole population response, which would protect everybody."

Now the Sky News host, goading him politely, added the missing phrase. "That herd immunity," the host said. "In terms of building up a herd immunity in the U.K., what sort of percentage of people need to have contracted the virus?"

"Probably about 60 percent," Vallance said.

"*Sixty* percent?"

"Sixty percent is the sort of figure you need to get herd immunity."

Viewers watched the interviewer do a little math in his head: 60 percent of the national population, which is 67 million, times a case fatality rate of, say, one-half percent to one percent. "That's an awful lot of people dying in this country," the man said.

"This is a nasty disease," Vallance agreed.

51

How did Patrick Vallance come to that magical figure, 60 percent? The mathematized concept of herd immunity begins with two scientists named Kermack and McKendrick back in 1927.

William Ogilvy Kermack was a Scottish statistician and chemist who blinded himself in a laboratory accident with caustic alkali, and then turned more fully to math, which unlike lab chemistry he could do in his head. Anderson G. McKendrick was a medical doctor who had worked in the Indian Medical Service, under the British imperial government, and therefore knew a thing or three about infectious tropical diseases. Together, in Edinburgh, they wrote a paper titled "A Contribution to the Mathematical Theory of Epidemics." From the nimbus of their differential calculus emerged a simplified schema, describing the dynamics of an epidemic. It's called an SIR model. As a new infectious disease passes into and through a population, each living individual belongs to one of three categories: susceptible (S), infected (I), or recovered (R). Susceptibles become infected; the infected recover—or die. (If they die, they disappear from this calculation.) The number in each category changes with each click of time, and the flux of the whole system therefore changes too. That's the basic conceptual tool. The SIR model (as in Susceptible → Infected → Recovered) was useful when Kermack and McKendrick proposed it and is still useful today.

To this, the two men added one valuable insight about how epidemics end. The cascade of infections doesn't necessarily proceed until the susceptibles category is at zero. It might terminate earlier—at a point when susceptibles are still scattered sparsely through the population, but the virus or other pathogen *can't find them*. Let me say that differently: Kermack and McKendrick alerted us that a forest fire doesn't necessarily end with every tree burned to ash. It might flicker out while some clumps of combustible trees and brush remain untouched, if those clumps happen not to be reached by the sparks. Maybe the unburned trees are just lucky: standing amid a small green meadow, upwind of the last embers.

Thirty years later, another British mathematician with experience in tropical diseases, George Macdonald, contributed an element to epidemic modeling that also remains essential today: the concept of basic reproduction rate. That's the number of cases in a naïve population directly resultant from a first case. How many people, on average, does an infected person infect? Other modelers since Macdonald have designated that entity R_0 (spoken as R-naught), now known as the basic reproduction *number*, not rate. It's the measure of how infectious a given pathogen is amid any fresh population of susceptible individuals. If the first infected case triggers three other infections, and each of those triggers three more, and so on, with the average remaining at three, then R_0 for that pathogen equals three.

The number differs widely from one bug to another. Viruses passed by airborne transmission, such as measles, tend to have high numbers for R_0. The rabies virus in humans, by contrast, transmitted via saliva when a rabid animal bites a person, has a very low R_0, because the unfortunate person dying of rabies rarely bites and infects another human. George Macdonald studied malaria, a very complicated disease caused by a protozoan that passes from host to host in the bites of mosquitoes, and the R_0 for that microbe depends on several factors, such as the density of mosquitoes, their longevity, how many times each mosquito may bite, and how long an infected human remains infectious to other mosquitoes. The reproduction number could be wildly various, Macdonald noted, ranging from 1.0 upward to who knows what. If there were long-lived and voracious mosquitoes biting people with long-enduring active infections, it might be 735. No wonder Macdonald and his public health colleagues couldn't eradicate malaria, and seven decades later the disease still kills almost a half million children each year.

Respiratory viruses are simpler, which is not to say simple. The reproduction number can change over time, as circumstances for the virus or the host population change. A mutation in the virus enhances transmission? The value of R goes up. A government message advises social distancing, and people comply? Transmission is interrupted, impeded, reduced, and R goes down. That changeability is important for understanding what herd immunity is and is not, if and when it arrives.

Here's why I'm torturing you with these math details, which seem dry but are actually very wet. The number R_0 plays a big role in determining the supposed threshold at which herd immunity takes effect. That threshold is a percentage level—*this* much of the population has passed from susceptible to infected. And what does the threshold mark? It marks the point at which R has fallen to 1.0. Transmission has slowed because susceptibles have become fewer, and therefore harder for the virus to locate. Chains of infection come to dead ends. R keeps falling. From then on, so long as the reproduction number remains below 1.0, the outbreak subsides. All things being equal, it will decline to a tolerably low level of endemic disease or, best-case scenario, the virus will disappear entirely from that population because it can't find anyone new to infect. That's how smallpox eradication was achieved. The last active smallpox strains in the last few human cases were not killed by antiviral drugs. They died of isolation and loneliness, without leaving offspring.

All things are never equal, of course. Smallpox was special. It produced visible symptoms in each infected person, so cases could be identified and isolated while health workers vaccinated everyone around them, stopping viral spread by containment; and it dwelt within no other reservoir than humans, so it couldn't reemerge from a nonhuman animal host. Polio eradication has been harder. That virus too inhabits no other host, but many polio infections are asymptomatic, making them harder to identify and isolate. Eradication of SARS-CoV-2 will almost certainly be impossible, again because it transmits from asymptomatic carriers, among other reasons. Even herd immunity on a global scale against SARS-CoV-2 will be difficult, tentative, and intermittent at best, not final. Herd immunity is a local phenomenon—localized by city or by nation or by island—turning itself on and off like your furnace in response to a thermostat. So long as R remains greater than 1.0, the disease outbreak will expand through the population. If it falls below 1.0, the outbreak will wind down, and eventually end (like smallpox) or become an endemic disease (like measles), circulating sporadically within limited regions or populations. The equation used to calculate herd immunity is very simple, so simple that even I can understand it and you can too: threshold = $1 - 1/R_0$.

He prints equation. Eyes roll back in heads. But no, wait, look how easy this is. If the reproduction number is three, meaning three secondary cases infected by each primary case, you need nothing more than grade-school fractions to work it out: 1 minus 1/3 equals what? Two thirds, right? So, in that case, the threshold for herd immunity is two thirds: 67 percent of the population.

This is what lay behind Patrick Vallance's passing statements on that very uncomfortable morning of March 13, 2020. When he said "about 60 percent" infection of the U.K.'s population would bring herd immunity, he was assuming a reproduction number for the virus of "about" 2.5. You can trace backward and see that. You can do the math on your phone, as I just did.

$$\text{herd immunity threshold} = 1 - 1/R_0$$
$$\text{that is, 1 minus 1-divided-by-2.5}$$
$$\text{okay, 1 divided by 2.5} = .4$$
$$\text{and 1 minus .4} = .6$$
$$\text{therefore}$$
$$\text{herd immunity threshold, when } R_0 \text{ is 2.5,}$$
$$\text{equals 60 percent}$$

The reality is not nearly so exact. Certain assumptions are built into this clean, easy calculation, and those assumptions never quite do justice to actual viruses, people, or circumstances. One assumption is that, when an individual recovers, that person is left with full and permanent protection (such as antibodies, which can block viruses, and T cells, which help destroy them) against being reinfected. Another assumption is that the virus does not evolve to evade such protections. Both those assumptions, applied to SARS-CoV-2, represent optimism, not confirmed knowledge.

A third assumption is homogeneous mixing, random interactions, among all members of the population in question. That maximizes the unlikelihood of continuous chains of transmission as the number of susceptibles falls. If the susceptibles are distributed heterogeneously, in pockets—for instance, ethnic neighborhoods with high interaction among the neighbors, low interaction with outsiders, and perhaps a cultural stricture against vaccination—then those pockets will enjoy no herd immunity, even if the rest of the population does. A fourth assumption: no newcomers to the population, no addition of susceptibles. (Ecologists would say, "no immigrants," but in our current, ugly political climate that can be mistaken for an echo of xenophobia.) This crucial insight traces back, as I've mentioned, to the veterinary advisors Eichhorn and Potter, discussing their monitored cattle herd in that 1917 report on brucellosis. "During the years that the herd was being replenished by purchase," they wrote, "abortions were frequent, but that practice was discontinued." That practice—meaning the purchase of new heifers to supplement those lost. Instead, the heifer calves born within the herd, despite the presence of the bacterium, were raised to become the new bearing cows. Eichhorn and Potter didn't say so, but their successful mothers must have had some degree of natural resistance to the bacterium, and they passed that along. It worked: herd immunity developed as an aggregate of individual immunities plus falling R_0, and abortions declined. "Therefore it seems safest for a herd owner to raise his own calves and avoid bringing in new infection." You would call that a closed population.

Translated to a human epidemic: if travelers continue to arrive, and some of them are susceptible, that undermines the clean math of the SIR model and refreshes the possibility of chains of infection, bringing the reproduction number back up, therefore raising the herd immunity threshold. And there will *always* be travelers. A human population is never, apart from the total population of planet Earth, closed.

The implications of these two factors—homogeneous versus heterogeneous interactions, closed versus open populations—are well illustrated by the history of measles in Rhode Island.

52

Measles virus was zoonotic in origin, having diverged from the cattle virus that caused a disease called rinderpest. Here's our punishment (among others) for domesticating bovids: measles got into humans from our cattle and has been with us for roughly two thousand years. Rinderpest has now been eradicated but the measles virus in humans, not. The reason for that, I think, is that it has been easier for us to contain and control bovine behavior than our own.

Measles has remained a serious disease, quite capable of killing unvaccinated children, notwithstanding the efficacy of measles vaccines.

We tend to forget about this virus (another single-stranded RNA virus), at least in high-income countries, because most people have been vaccinated in childhood and outbreaks are rare. Not everywhere: in the Democratic Republic of the Congo, with low measles vaccine coverage, roughly 310,000 measles cases occurred in 2019, with six thousand deaths, mainly of young children. That's tragic, given that we've had a licensed vaccine since 1963. In the United States, mass vaccination of schoolchildren, starting soon after the vaccine became available, nearly eliminated the problem of childhood measles. In the state of Rhode Island, that program began on January 23, 1966, with a statewide event advertised as End Measles Sunday.

Despite a bad snowstorm, more than 31,000 children were vaccinated that day. A week later, "mop-up" clinics added another few thousand children. Measles almost disappeared in the state. Before the program, in 1965, there had been nearly four thousand cases during that year. In 1966, cases totaled less than a hundred. But the population of Rhode Island was not homogeneous and not closed.

The neighborhood of Fox Point, within the city of Providence, contained a strong concentration of ethnic Portuguese residents. Roughly 60 percent of Fox Pointers either were immigrants from Portuguese territories or descendants from Portuguese ancestry. Fox Point lies on a tip of land at the confluence of the Providence River and the Seekonk, so it's somewhat isolated by geography; culture and language added to that separateness. The residents centered their social activities around a single church. Many of the families had immigrated recently, after End Measles Sunday, and missed that opportunity to get their kids immunized. Their attitudes toward vaccination may have leaned toward reluctance, skepticism, or mistrust. Into this community, on September 15, 1968, came a three-year-old boy, returning with his family from a visit to Portugal. He had measles.

Two weeks later, the boy's sister broke out with measles. Other cases soon appeared among students at the sister's school and preschoolers in the neighborhood. Another neighborhood, East Providence, just across the Seekonk River, was closely linked to Fox Point by a bridge; it also contained people with Portuguese backgrounds, and some East Providence kids interacted with Fox Point kids. East Providence became another outbreak center. By the end of the year, Providence contained ninety-one measles cases among children, only three of whom came from non-Portuguese ancestry. None of those children had been vaccinated. Fortunately, none of them died.

Rhode Island at large may have achieved herd immunity, but Fox Point and East Providence had not. Many things have changed since 1969, and SARS-CoV-2 is not measles virus (which is extraordinarily transmissible, with an R_0 reckoned at twelve to eighteen), but the limitations of the concept remain. People cluster. People travel. The human population is a very rambunctious herd—not like a hundred dairy cows in a barn, or even a thousand Herefords alone on the high plains of northern Montana.

Let's be optimists for a moment. Let's accept all the assumptions and posit that herd immunity regarding COVID-19, in one country or another, however temporary, is attainable. What does it look like? What happens when we reach that magical threshold? This leads back again to the SIR models of Kermack and McKendrick.

Imagine a population of one hundred people, none of them ever exposed to SARS-CoV-2. You have one hundred susceptibles (S). The virus arrives, people catch it, they transmit it to others, until sixty of your susceptibles have become infected (I). Two of them die, so those two don't make it into the recovered (R) category. (Small note: the recovered category symbol R is not to be confused with the R that's the reproduction number. It *is* confusing that they both carry the same letter, I agree, but that's not my fault.) The reproduction number, R, has now fallen below 1.0, but not to zero, so some unlucky people continue to get infected. But the outbreak is in decline. The likelihood of any of the remaining susceptible people getting infected is lower. Therefore, by the logic of Patrick Vallance and others, your herd of people has reached "herd immunity," whatever that phrase may mean. What *does* it mean? What do they win?

What they win is *not* a magical immunity for the remaining susceptible people. What they win is a reduced likelihood that those people will be exposed to the virus, because so many other people have been infected and recovered, or so many others have been vaccinated, or a combination of both. What they win, at best, is a slowly declining rate of infection among the remaining susceptibles of the population. Any one of those susceptible people, though, can still become infected and die. Herd immunity is like "immunity" from lightning if you wander out on a golf course during a thunderstorm: it'll probably hit the other guy or a tree.

53

T hird kind of magic: drug therapies.

Hydroxychloroquine has been around for a long time, but it entered the headlines in late March 2020, after Fox News host Laura Ingraham, with help from Larry Ellison, the cofounder of Oracle, and possibly also Elon Musk, the entrepreneur and spaceman, planted that word like an earwig in Donald Trump's brain. Trump, as you may have heard, is not a scientific sophisticate. He would become slightly confused some months later when discussing what he called "herd mentality," by which he seems to have meant herd immunity (although no one can know for sure). But hydroxychloroquine is just a drug, a tablet, a thing that you eat, and therefore simpler.

Hydroxychloroquine is a derivative of chloroquine, and both substances have been routinely prescribed for prevention and treatment of malaria. Sometimes they are also used against rheumatoid arthritis and lupus. Chloroquine can be toxic—an overdose can cause severe reactions, even death—and the addition of the hydroxy bit seems to temper that problem. The usage against malaria traces back to natural quinine, an active ingredient of the bark of *Cinchona* trees, traditionally used by indigenous peoples in Peru as a remedy for fevers. In the seventeenth century, it was brought in bark or powder form to Europe, where it was known as *quina* and put to the same use. A synthetic version was created in 1934 by chemists at the Bayer laboratories in Germany and deployed to North Africa along with Rommel's troops. After the war, manufactured chloroquine became a standard prophylaxis against malaria.

That utility lasted about fifty years. I took it as a preventative myself, back in the 1980s, whenever I traveled in malarial zones. It seems to have worked, because there was no shortage of mosquitoes and I never caught that particular illness. But don't take it for too many months in a row, I was told, or it'll wreck your liver. It functions, so far as scientists can tell, by blocking a metabolic step, as the malarial parasites gobble up hemoglobin and replicate within a host's red blood cells, so that the invaders die of poisoning in their own waste products. But then the parasites, with their high rates of replication and a reasonable amount of mutation, evolved resistance to chloroquine. Certain areas of the tropics became danger zones for chloroquine-resistant malaria, and if you went there, you were advised to take a different drug. Some of those concoctions were even more odious than chloroquine, but side effects such as headache, nausea, vomiting, hives, nervousness, and ghoulish nightmares (like the one with the giant snakes in the walls, harrowing even to me, and I like snakes) were preferable to malaria, at least until we had something better.

The notion that chloroquine might be effective against viral infections wasn't new in 2020. As the drug went out of favor for antimalarial application, it drew attention as a possible weapon against viruses, including SARS-CoV in 2003. A group of researchers based in Italy and Belgium argued, based on a review of published studies, that it could inhibit cell entry and replication by several kinds of virus, including coronaviruses and HIV. In 2005 another team, including three of the most revered virologists at Special Pathogens Branch during the good old days of the CDC (Pierre Rollin, Thomas Ksiazek, and Stuart Nichol), confirmed that it hindered replication of SARS-CoV in cultured cells, yes, and reduced infection from one cell to another. That action, they wrote, "suggests a possible prophylactic and therapeutic use." Two years later, a trio of French scientists proposed, based on a review of several dozen laboratory studies, that chloroquine and hydroxychloroquine could be valuable for treating a long list of bacterial, fungal, and viral infections, most notably a pathogen called *Coxiella burnetii*, a tough intracellular bacterium that causes a disease called Q fever, and including among the viruses again SARS-CoV. Apart from some efforts on Q fever patients, these were almost all *in vitro* studies, meaning "in glass": drugs versus viruses versus cultured cells in petri dishes. Still, they showed that the idea of chloroquine against a coronavirus wasn't crazy. It was merely untested by systematic clinical trials among humans.

Trump, soon after receiving the inspiration from his guiding spirits, in March 2020, began speaking and tweeting about hydroxychloroquine as a treatment for COVID-19. At first, he was commendably tentative. On March 19, 2020, during a news conference, he mentioned both hydroxychloroquine and chloroquine, noting correctly that they were government-approved antimalaria drugs and adding his hope that they might be helpful in the pandemic as well. "If they work, your numbers are going to come down very rapidly. So we'll see what happens," Trump said. "But there's a real chance that they might"—chloroquine or hydroxychloroquine—"they might work."

Tony Fauci wasn't on the podium that day, standing with Trump, Vice President Pence, Deborah Birx (Trump's coronavirus response coordinator), and others. But the following day, he was asked whether there was evidence to support use of hydroxychloroquine against Covid.

"The answer is no," Fauci said.

By then, several small studies had been published, mainly by scientists in China. One group found that chloroquine did impede entry of SARS-CoV-2 into lab-cultured cells—a lineage of cells derived from monkey kidneys. Another group reported that clinical trials in ten Chinese hospitals showed positive effects, but that paper offered no data. A French team tested hydroxychloroquine on twenty-six patients, dropped six of those patients from the study for various reasons (including nausea and death), and saw decreased viral load in fourteen of the remaining twenty. Fauci was evidently unimpressed. At a meeting of the White House Coronavirus Task Force, at which hydroxychloroquine was much extolled by Peter Navarro, Trump's excitable economic advisor, Fauci called the evidence "anecdotal."

The senior author on the French paper was an eminent and controversial microbiologist, a prideful contrarian named Didier Raoult. Among Raoult's more solid accomplishments was sharing leadership, along with Jean-Michel Claverie, of the group that discovered that first giant virus, Mimivirus, lurking inside an amoeba. Raoult runs a large and well-funded research institute in Marseille, and puts his name as coauthor to more than a hundred scientific papers in a typical year. He's a blusterous leader and a contentious scientist, by several accounts, with a formidable scowl and the stringy, shoulder-length hair of an elderly druid. He has touted hydroxychloroquine for a long time; he was one of the three authors of the Q fever paper in 2007, to which the drug's antiviral potential was an addendum. Now, during the early months of the pandemic, Raoult went all in on hydroxychloroquine, as described in a profile of the scientist by Scott Sayare, in *The New York Times Magazine*. The fact that other scientists might be skeptical about the drug was, to Raoult, only a goad.

"I've spent my life being 'against,'" he told Sayare. He found conflict thrilling. He was very sure of his own vaunting claims and sure also that other scientists, present and past, including Charles Darwin, were guilty of blindered thinking, doltishness, and hubris. Raoult considered the tree of life as implied by Darwinian evolutionary theory "entirely false," he told Sayare, and scoffed that Darwin himself "wrote nothing but inanities." He posted a YouTube video titled "Coronavirus: Game Over!" and declared that COVID-19 is "probably the easiest respiratory infection to treat of all." Hydroxychloroquine was so good, he warned, that pharmacies would soon be sold out—and he was right at least on the second part of that self-fulfilling prophecy.

Back in the White House, Donald Trump listened to people who listened to people who listened to Didier Raoult. Trump increasingly liked what he heard of hydroxychloroquine—"I'm a big fan"—and began pressuring his secretary of health and human services, the pliable Alex Azar, and his head of the Food and Drug Administration, the somewhat less pliable Stephen Hahn, to approve anti-Covid use of hydroxychloroquine. Despite some resistance within the FDA, on March 28 that agency issued an emergency use authorization, allowing doctors to prescribe, and health care providers to administer, the drug against COVID-19. Clinical trials began, and results started to roll in. Those results did not support the bullish enthusiasm of Didier Raoult.

On June 15, the FDA revoked its emergency use authorization, declaring that, on the basis of emerging scientific data, chloroquine and hydroxychloroquine "are unlikely to be effective in treating COVID-19." In light of serious cardiac problems and other potential side effects, said the agency's press release, the benefits (if any) no longer outweighed the risks.

Meanwhile, people were dying. The Covid death toll in France, as of June 15, 2020, was 29,411. In the United States, following the grim April surge in New York and elsewhere, it was 120,780. If hydroxychloroquine wasn't the answer, what was?

Remdesivir is an antiviral with a very different backstory. It was created by the company Gilead Sciences, with funding from the U.S. government, through an effort begun in 2009 and focused on hepatitis C virus and respiratory syncytial virus (RSV). It delivered marginal value against hepatitis C, some against RSV, but not enough to raise expectations, and it was set aside. At that point it was just considered a candidate drug, known by the label GS-5734, with GS presumably standing for Gilead Sciences. Several years later, as emerging viruses became an increasing concern, Gilead launched an effort, in partnership with the CDC and the U.S. Army Medical Research Institute of Infectious Diseases (USAMRIID, the place where Don Burke began his Army career, a facility famed for its work on badass viruses), screening roughly a thousand candidate compounds to see what might be effective against these scary new viruses such as SARS-CoV, MERS-CoV, Ebola, and Zika. This was done in cell cultures. GS-5734 showed promise with Ebola in particular, so it was tested against Ebola infection in rhesus monkeys. The drug was delivered directly into a vein, and at low dosage it helped somewhat. At higher dosage, all of the test animals survived (at least until some of them, grown moribund, were euthanized). So remdesivir was tried on people, under a "compassionate use" rule, at the end of the 2013–2016 West Africa epidemic of Ebola. It seemed to save the lives of a thirty-nine-year-old woman and a newborn baby. But when the use expanded to a larger trial, on more than six hundred cases, it did poorly: 53 percent of the treated patients still died.

In a 2017 study, though, remdesivir showed itself effective against a broad range of coronaviruses in human cell cultures. That work was done at the University of North Carolina, by researchers quite aware that SARS and MERS probably wouldn't be the end of the coronavirus story. Their drug testing, they noted, might also be relevant to "circulating zoonotic strains with pandemic potential."

Then the novel coronavirus emerged. In January 2020, Gilead provided remdesivir to the China CDC for lab testing against this new threat. The testing was done by scientists at the Wuhan Institute of Virology, with two Beijing-based partners. One of the Wuhan collaborators was Zhengli Shi. The group found remdesivir "highly effective" against SARS-CoV-2 in laboratory cell cultures. Then again, they tested chloroquine in the same set of experiments, and found it highly efficacious also. This is how science proceeds, from limited and provisional results toward more experiments, toward more observations, toward inferences that are less provisional. The action of remdesivir against SARS-CoV-2 in human bodies, and of chloroquine likewise, were two unknowns to be addressed independently.

The first major clinical trial, with patients randomly assigned to receive either remdesivir or a placebo, was conducted at ten hospitals in Wuhan during February and March 2020. The drug went into veins of 158 patients, and the results were disappointing. The researchers concluded that remdesivir "did not significantly improve" symptoms, time to recovery for those who recovered, or mortality, above the outcomes of the patients who got the placebo. And this well-organized trial ended early, before reaching the hoped-for number of enrolled patients, for a reason that may seem, to those of us in the rest of the world, enviable and weird: Wuhan ran out of Covid patients. The strict lockdown of that city, the testing and tracing and isolating, all the nonpharmaceutical interventions, had made pharmaceutical interventions unnecessary. On the day that the Wuhan remdesivir study was published, April 29, 2020, China at large had a total of four new cases and no new deaths.

Ivermectin presents a different conundrum, because that drug is cheap and easily obtained, therefore tantalizing to desperate or scared people, and the evidence about its effects is mixed. You can find studies reporting that it has high value against SARS-CoV-2 and reduces mortality, other studies concluding that it has no significant value, and review articles arguing both sides: on the one hand, that an aggregate view of multiple studies strongly supports ivermectin utility against COVID-19, and on the other hand, that many of the pro-ivermectin studies are flawed and falsified. (One of the most positive studies in the positive-leaning review, which was posted as a preprint and contributed heavily to the positive average, was later withdrawn after criticism of its legitimacy and honesty.) You can read anecdotal accounts of ivermectin misadventure, such as one from the CDC:

An adult drank an injectable ivermectin formulation intended for use in cattle in an attempt to prevent COVID-19 infection. This patient presented to a hospital with confusion, drowsiness, visual hallucinations, tachypnea, and tremors. The patient recovered after being hospitalized for nine days.

Tachypnea is fast, shallow breathing. Confusion is a symptom we all face during Covid, but ivermectin can evidently exacerbate it.

And ivermectin is available over the counter, at your local pet store or feed store, or from Amazon with a couple of clicks—in tablet form, chewable tablets, pour-on liquid, or as an apple-flavored paste, all intended for deworming your dog, your horse, your cows, or your goat. That's because it's an important and trusted tool among veterinarians and livestock producers for treatment against lice, mites, and parasitic worms. Then again, the claw hammer is an important and trusted tool among carpenters, but it's not recommended for use in dentistry.

Ivermectin was discovered in 1975, by two biologists who eventually won a Nobel Prize for that work, and the WHO includes it on a list of essential medicines. People get tormented by lice and mites too, and in some parts of the world parasitic worms cause severe, widespread morbidity. Onchocerciasis, also known as river blindness, caused by a kind of roundworm spread by the bites of blackflies, affects about fifteen million people, mostly in sub-Saharan Africa, and causes some degree of vision loss in almost a million. Lymphatic filariasis, commonly called elephantiasis for how it can swell a person's legs by blocking the lymphatic system, caused by roundworms of the filarial group, is carried by mosquitoes. As of 2018, more than 51 million humans were infected. Ivermectin is a blessing to those people, well worthy of a Nobel Prize. Again, I've taken the stuff myself, in small dosage, when I was walking across swamps and forests in the Republic of the Congo and Gabon, being bitten continually by blackflies, and hoping to avoid river blindness. All of us on that trek took it, administered from an eyedropper by our intrepid expedition leader, an ecologist named Mike Fay, a man with long Congo experience, who typically bought his ivermectin supplies from a feed store in America or a boy on a street corner in Brazzaville or Libreville. Even in America, which is somewhat less diversely wormy than Congo forest, but with heartworms that threaten dogs, stomach worms that subtract billions in value from cattle, and roundworms that can grow more than a foot long in people, ivermectin has a market. The Food and Drug Administration has approved it for human use against such invertebrate parasites.

But use against COVID-19 is another matter, discouraged by the WHO, the CDC, and the FDA. One group of authors, having scrutinized a large number of studies for the Cochrane Library, an online service that reviews medical science, judged that "the reliable evidence does not support the use of ivermectin for treatment or prevention of COVID-19," except in well-designed clinical trials. Researchers at the University of Oxford began a huge clinical trial in June 2021, expected to be the world's largest, but that will take time. For now, the concern about ivermectin seems to be, not whether it's safe, after having been so widely used by humans, but whether it works against SARS-CoV-2.

That's not the situation with molnupiravir, a new offering from Merck and its partner on this project, a small Florida-based company called Ridgeback Biotherapeutics. Molnupiravir has the merit of oral delivery, as a prodrug form of an active drug known for short as NHC, a tricky compound with a long history. (A prodrug is a compound that can be taken orally, then is metabolized in the body into an active drug.) Molnupiravir comes as a pill, like hydroxychloroquine and ivermectin, not an intravenous drip, like remdesivir. In an interim analysis of a clinical trial, according to a Merck press release issued in October 2021, molnupiravir showed dramatic results, reducing the risk of hospitalization or death among adults with mild or moderate COVID-19 by about half. (By the time the trial was complete, though, total risk reduction had fallen to a lower level.) It works by causing mutations in SARS-CoV-2, to the point where the virus becomes dysfunctional. Impaired by the drug from replicating its RNA accurately, the virus reaches that threshold of mistakes, in its long thirty-thousand-nucleotide genome, that I've mentioned before: error catastrophe.

Molnupiravir also drags with it a tricky history, and part of that history is evidence that it may cause mutations not just in the virus but also in mammals. Ronald Swanstrom, a biochemist and evolutionary virologist at the University of North Carolina, led a team that did some of that cautionary research. Swanstrom has worked for decades on the evolution of viruses, especially HIV-1, within human hosts; he has coauthored many papers on drug discovery and drug resistance among viruses. And he has seen the merits of molnupiravir as well as its possible dangers. In early 2020, he was among a group of researchers reporting that the drug worked well against three coronaviruses: SARS-CoV, MERS-CoV, and the novel virus by then commanding everyone's attention. This was very good news, both because there were no approved drugs for treating any coronavirus at that point, and because molnupiravir could be taken orally, vastly increasing its potential for controlling COVID-19 cases in their early stages, before people had to be hospitalized. One year later, Swanstrom anchored a study illuminating the downside of molnupiravir: its potential to alter the DNA in mammalian cells—cultured hamster cells in the lab, and maybe also fetal cells in pregnant women or stem cells that produce sperm in men. The drug "has potent antiviral activity" far beyond other drugs of roughly the same sort, Swanstrom and his coauthors wrote, "but is also mutagenic to the host." Such mutations, they added, could produce birth defects or cancer.

Ronald Swanstrom is not categorically opposed to any use of this drug in humans; just concerned and inclined toward caution. It might prove valuable, on balance, for older people but inadvisable for young men and women. "I would probably take molnupiravir (I have age as a risk factor)," Swanstrom told me by email, "but I would want to be part of a cohort that was followed to look for cancer risk." The big problem, he added, is that the level of risk is unknown. It might be inconsequential, it might be significant.

"Our result was not a surprise," he added. It was a confirmation of what had been long known about metabolic pathways in animal cells and the way this drug works. Molnupiravir belongs to a group of molecules known as nucleoside analogs, which resemble the nucleotides from which RNA and DNA genomes are built, and it can therefore interfere with genome replication. It was first synthesized at Emory University in Atlanta in the early 2000s, and researchers there hoped it might be effective against the virus that causes Venezuelan equine encephalitis, a disease that kills horses and sometimes people. Next it was tested against strains of influenza, and it suppressed those viruses too. So it's considered a "broad-spectrum" antiviral drug, showing promise as a weapon in the fight against emerging viruses. It functions by mimicking the RNA nucleotide bases cytosine and uracil, thereby getting itself inserted into the genomic molecule and, because it's not what it pretends to be, creating mutations. In place of a C it can present itself as a U, and do that over and over again. These wholesale mutations proceed to a point where the viral strand is unfunctional: again, error catastrophe. With the drug thus interfering, the virus population crashes. That's a blessed result for the host—except there's also a significant chance, as shown by the work of Swanstrom and his colleagues, that this sneaky molecule will also imper-sonate building blocks of the DNA in host cells, as those cells reproduce, and cause mutations in them as well. Many cells in our bodies don't re-produce very often, but fetal cells do, and so do the stem cells that pro-duce sperm.

"The biochemical pathways and our data say it is a mutagen," Swan-strom told me. "I am disappointed this potential is not a more public part of the discussion. I think there is a concern that talking about it will cause the drug to get tainted."

Big money as well as lives are at stake. Once the pandemic took hold and the need for coronavirus-stopping drugs became acute, Emory li-censed molnupiravir to Ridgeback Biotherapeutics. Ridgeback sought government support for its further development, and that caused con-flict within the U.S. Department of Health and Human Services (HHS) while it was under Donald Trump's control. Nested in that department is the agency known as BARDA (Biomedical Advanced Research and Development Authority), established in 2006 with a mission of develop-ing and procuring weapons against threats to public health, either from acts of bioterrorism or from pandemic disease. Part of BARDA's role is to make grants that help bring drugs and vaccines to market. Emory Uni-versity had already received federal funding to test the drug, and in spring 2020 the founders of Ridgeback, two former investment managers, tried for more money through BARDA. The director of BARDA was a scien-tist named Rick Bright, an immunologist and virologist by training, who had worked at the CDC, in the private sector, and then at BARDA since 2010. As the pandemic began, Bright fell afoul of his boss at HHS, by his account, when he dissented against the broad use of chloroquine and hy-droxychloroquine and felt that political pressure was overriding scientific concerns about those drugs.

Then came the Ridgeback people, asking for cash to advance this other enticing drug, molnupiravir, and using what Bright considered personal connections, over his head and around him, to get what they wanted from his agency. On April 20, 2020, Bright was removed from his directorship of BARDA. On May 5, he filed a 57-page whistleblower complaint to the U.S. Office of Special Counsel. More than a year later, Bright settled his employment claim against HHS, details undisclosed, by which time he had been hired as a senior pandemic response planner at the Rockefeller Foundation. But the ambiguities of molnupiravir's risk/benefit balance have *not* been settled. In October 2021, Merck announced the interim analysis of its clinical trial, reporting that "molnupiravir reduced the risk of hospitalization or death by approximately 50%." But when a study appeared in *The New England Journal of Medicine* presenting the final analysis, the reported margin of efficacy for the whole group of participants was reduced, as I mentioned above: to 30 percent.

More recently, Pfizer too has also gotten FDA approval (in the form of an emergency use authorization, an EUA) for an oral treatment against Covid, this one under the brand name Paxlovid. It's actually a combination of two drugs (both with more tongue-twister names that you needn't contemplate), and though the combination seems highly effective, each of the two component drugs has a downside: difficulty of manufacturing for one, drug-interaction complications for the other. Paxlovid has potential to be important, Ron Swanstrom told me, but not the panacea for a worldwide pandemic—that is, a cheap, oral, nontoxic, test-and-treat drug—that might be desired.

As for molnupiravir and the danger of promoting mutations: it's important to remember, for the sake of context and good sense, that some cancer chemotherapies too are mutagenic. Some of them even involve nucleoside analogs, bringing a chance of altering the DNA in healthy cells. It's a tradeoff. If I were fighting cancer right now, I might well agree to treatment with such a drug, accepting the possibility of a future tumor (caused by a new mutation) for the sake of shrinking or eliminating the tumor already growing within me. And because I am a septuagenarian (like Ronald Swanstrom) and birth defects would not be an issue, I might also be grateful for a course of molnupiravir, if I were struggling against COVID-19, just as I was grateful for a gulp of ivermectin amid the Congo swamps. That's not magical thinking. It's calculated risk.

54

Fourth kind of magic, and the most truly amazing: vaccines. During the last days of December 2019 and the first days of January 2020, as I've mentioned, Tony Fauci tracked the squibs of news out of Wuhan with sedulous attention. He conferred with his leadership team at the Vaccine Research Center, which is part of the vast institute (NIAID) he has run for almost four decades. That team included John Mascola, the VRC director, and Barney Graham, deputy director and chief of the center's Viral Pathogenesis Laboratory. What was the pathogen causing this outbreak of mysterious pneumonias? Unexplained cases of lower airway and lung infection turn up continuously, all over the world, for a variety of interesting and uninteresting reasons; but this was a pattern, and patterns put infectious disease scientists on alert. If the cause was a virus, as seemed likely, what kind? If it was an old virus, it could be identified and the reports would show that. If it was a new virus, what kind? It "smelled," to their keen and experienced noses, like a coronavirus, respiratory and highly transmissible and dangerous. Some rumors out of China suggested it might be SARS-CoV or SARS-like. But you can't build a vaccine from rumors and smell.

Graham wanted to see the genomic sequence, immediately if not sooner, so he could begin activating a new sort of vaccine approach—new to the wider world, anyway, but familiar to him and some others from years of research. Graham was primed and ready. "Because he had been working on a mRNA type vaccine for MERS-CoV and for Nipah virus," Fauci told me. The principles were transferable, from a coronavirus such as MERS-CoV, to a paramyxovirus such as Nipah, to another coronavirus. What Graham needed, for starters, was a particular target: a crucial section of genome. "He was very much involved in the mRNA technology," Fauci said, "because of its easy adaptability. And I remember, we were on a phone call, and he said, 'Let's get the sequence, and we'll really get moving on it.'" If there were whole or partial sequences passing confidentially between labs in China, by that time, as there evidently were, Fauci hadn't seen them. Then, in the wee hours of January 10 to 11, Edinburgh time, Eddie Holmes posted the sequence from Australia.

The Vaccine Research Center is in Bethesda, on Eastern Standard Time, five hours behind Edinburgh. Barney Graham recalls first seeing the sequence at 8:30 or 9:00 that night. It was a Friday. The VRC already had a collaborative relationship with Moderna Therapeutics, an upstart biotech company based in Cambridge, Massachusetts, signed into force several years earlier, for manufacture of vaccines developed by the VRC. That collaboration included the Nipah vaccine project under Graham's guidance. So Graham was in touch with Stéphane Bancel, the mercurial Frenchman who was CEO and part owner of Moderna. They partnered on earlier projects, they had another in the works, and now they agreed to shift that effort to the new virus. "As soon as you send me the sequence," Bancel said, by Graham's recollection, "we will start manufacturing." Four days after its release, they did. Less than nine weeks after that, on March 16, the first event in the first clinical trial of the Moderna-VRC vaccine occurred: a shot delivered into a human arm.

"People say, 'Wow, it's amazing how quick you went,'" Fauci told me. "That was really, really, really rapid. But it reflected a lot of work that had gone on before."

55

The saga of vaccine development against SARS-CoV-2, which un-folded slowly over many years and then culminated with extraordinary quickness, is not one story, with one beginning, one denouement, a handful of characters, and maybe a single hero or several. It's more like the *Mahabharata*: an epic, braided from a thousand threads. Some of those threads come out of Hungary and Germany and the University of Pennsylvania and lead toward the Pfizer vaccine. Some threads emerge from Janssen Vaccines, in the Netherlands, and from Beth Israel Deaconess Medical Center, in Boston, and (money threads) from BARDA, that biomedical development agency in Washington, D.C., from which Rick Bright was sent into exile. Those threads lead toward the Janssen COVID-19 vaccine, better known as the Johnson & Johnson. Some come out of China, in two distinct bundles, passing through Chile, Indonesia, the Philippines, Morocco, Bahrain, and Pakistan, among other places, on the way to the CoronaVac and the Sinopharm BIBP vaccines. Some threads tangle their way from Oxford, England, all the way to Cambridge, England, then onward and outward to the Serum Institute of India, in Pune, and the Vietnamese Ministry of Health, in Hanoi, yielding the Oxford-AstraZeneca vaccine, which may soon be available as a nasal spray. Out of Russia, by way of Abu Dhabi and Italy, wind other threads, at the end of which lies Sputnik V. And then there's the Moderna vaccine. That story has many threads too, one of which begins more than thirty years ago in the laboratory of Barney Graham.

Barney Scott Graham grew up near Paola, Kansas, a farm boy who helped with the cows, hogs, and quarter horses, his father a dentist when not tending livestock, Barney soon a tall young man bright enough to be valedictorian of Paola High School and headed for Rice University in Houston. After college, and then medical school back in Kansas, he found himself, in his early thirties, chief resident for internal medicine at a hospital in Nashville, Tennessee. That didn't exhaust Graham's ambitions to make sick people well and keep others from getting sick, so he got a doctorate in microbiology and immunology and became a professor of medicine. He also did research, beginning in the mid-1980s, focused largely on vaccine efforts against HIV and on a more common, less notorious, but still serious pathogen: respiratory syncytial virus (RSV), which can seriously afflict children and the elderly. RSV gets around, infecting nearly every human child by the age of three, bringing only coldlike symptoms to most but severe respiratory illness to many, causing about three million hospitalizations per year. Graham started work on RSV in 1986, and stayed with it, because there was no vaccine.

He left Nashville in 2000, joining the newly established Vaccine Research Center, within NIAID, which in turn is part of the vast National Institutes of Health (NIH) complex in Bethesda. He was one of the Center's founding investigators. Graham carries a recollection of the VRC creation

story, as recounted to him—recounted to many people, over the years—
by Tony Fauci. Bill Clinton, around the midpoint of his presidency, got
very interested in the imperative of conquering AIDS. Clinton summoned
Fauci for a briefing.

"He had one of those posterboards on an easel in the White House,
in the Oval Office," Graham told me of Fauci's presentation. "Al Gore
and Bill Clinton were sitting there in rapt attention, and Harold Varmus
was sitting back behind." Varmus, a distinguished cancer biologist with
a Nobel Prize and other decorations, was then director of the NIH, and
therefore Fauci's boss. Fauci was using his easel display to instruct Clinton on the complexities of how HIV infects white blood cells. Clinton, to
his credit, sat through it attentively. Then, as he ushered Fauci out of the
Oval Office, by Fauci's account to Graham and others, and by Graham's
to me, Clinton said, "What do you need to really solve this problem once
and for all?"

"Well, we really need a center where we can bring people together
from different disciplines," Fauci said, "and have a focus on HIV vaccine
development." Clinton looked back to his chief of staff, Leon Panetta, and
said, "Leon, just get this done." Or words to that effect. "Make this happen." The VRC was established by executive order in 1997, opened its labs
several years later, and expanded its vaccine research beyond HIV.

Graham, amid other responsibilities at VRC, continued work on respiratory syncytial virus, for which there remained acute need of a vaccine. RSV had a low profile in public awareness, but it was the leading
cause of hospitalization for children under five years old, and it killed
upward of sixty thousand in a typical year, possibly as many as 200,000,
with 99 percent of those in developing countries. A brutal threat to poor
kids in unprivileged circumstances. But this virus presented some special
challenges for creating a vaccine: it attacked victims at such a young age;
it possessed some tricks for immune evasion; it could reinfect a child, or
an adult, after recovery from a first infection; and there was an unfortunate history, dating back to the 1960s, of RSV vaccine efforts gone amiss
during clinical trials, in which a trial vaccine made matters worse, provoking a crisis of inflammation when the virus hit. The polite name for
this dark sort of event was "vaccine-enhanced disease." Although the RSV

trials fiasco occurred while Graham was an adolescent in Paola, by one account it "haunted" him. A vaccine researcher needed to step cautiously with RSV, and Graham did.

His careful steps led him, as others were led, to the idea of using messenger RNA, instead of live attenuated virus, or inactivated (chemically killed) virus, or portions of viral protein, to mobilize the human immune system. The element that provokes an immune response, and against which antibodies and immune cells are targeted, is called an antigen. The ingenious concept of using mRNA was that the antigen would be produced—in mass quantities, from a set of genetic instructions—inside the patient's own body. That idea went back decades, to several sources, among the earliest of whom was Jon A. Wolff, at the University of Wisconsin School of Medicine and Public Health, who published on it in 1990.

Katalin Karikó came to that idea independently, with help from a colleague. Having grown up in Hungary and taken a PhD at the University of Szeged, Karikó went into research, moved to Philadelphia as a postdoc at Temple University, then held a faculty position at the University of Pennsylvania, seeking grant money—tirelessly and without much success—to support a lab of her own. She was deeply committed to a prospect that seemed promising to her and few others: using synthetic mRNA to fight a variety of diseases involving protein deficiency. One of the problems with that approach, which Jon Wolff recognized but didn't solve, was that introduced mRNA degrades quickly in the body, too quickly to be effective in generating immune response. In 1997 Karikó met an immunologist named Drew Weissman, a new colleague at Penn, who had lately arrived from a fellowship at the NIH and was setting up his lab to do vaccine research. They were drastically different personalities, Karikó and Weissman—she tall and sandy-haired and outgoing and forceful, he balding and laconic. But they meshed well as scientific partners, with the goal of creating mRNA vaccines, and together they solved the problem of how to smuggle their customized mRNA through or around the human body's immune defenses and protect it from being degraded before it could function. Eventually, their ideas were acquired by a German-based biotech company called BioNTech, focused chiefly on cancer immunotherapy, and founded by a husband-wife team of physician-scientists. In 2013, Katalin Karikó became a senior vice president of BioNTech. Drew

Weissman remained at the University of Pennsylvania. All this occurred, remember, in a pre-pandemic world among scientists with no notable interest in coronaviruses.

Then the new virus emerged, the globe turned, and BioNTech began work promptly in January 2020 on a vaccine against COVID-19. Within three months, the company signed partnership agreements with Fosun Pharma (based in China) and Pfizer (based in New York) for hundreds of millions of dollars in development and manufacturing support. If you've gotten the Pfizer vaccine, as I have, you can thank a lot of people but probably Katalin Karikó—with her solitary, persistent voice that people once considered crankish—belongs near the top of the list.

56

The Moderna vaccine, quite similar in that it employs the mRNA strategy, has a very different creation story featuring a different set of ambitious entrepreneurs and visionary scientists. This one converges earlier with coronavirus research, and it involves not just mRNA engineering but also a field known as structural biology, wherein molecular biology, biochemistry, and biophysics combine in the elucidation of big molecules (especially proteins), their three-dimensional shapes, and how those shapes affect their functions. Moderna's history loops us back to Barney Graham and his long-standing concern with RSV and all those suffering children. Into his ambit, around 2009, stepped a young postdoctoral fellow named Jason McLellan. Sheer chance is often catalytic in science, and the chance in this case was crowded lab workspaces at the Vaccine Research Center.

"Jason was a postdoc from Peter Kwong's lab," Graham told me. Kwong was a structural biologist at the VRC, working on the intractable problem of an HIV vaccine. "And they ran out of space on the fourth floor. And so he came down"—McLellan did—"to the second floor and wanted to work on something besides HIV that wasn't quite so competitive." Graham had a suggestion: protein structures of respiratory syncytial virus. There was one of special interest, on the surface of the virus, roughly equivalent to part of the spike protein on a coronavirus because it's responsible for fusing with the membrane of a cell under attack, so that the viral genome can gain entry. Scarcely anything was known about this fusion protein of RSV, except that it can take two forms, one before it fuses with the cell membrane and one after. You could think of those as the Before and the After forms, relative to that crucial fusion. Graham and others presumed, from their work on RSV, that the most relevant antigen—the element that elicits protective antibodies—is the Before form. An effective vaccine, therefore, would need to offer something that mimics the Before protein. In the case of an mRNA vaccine, that would mean a small molecule, a free-floating protein, produced in the person's own body from mRNA instructions, in the shape of Before. This antigen mimics a part of the real virus, thereby eliciting the production of antibodies, which attack the real virus when it arrives. If your antigen takes the After form, it will be no good, because that form exists only after fusion with the membrane, when the virus genome has entered the cell—too late.

"So we started picking away at structures on the F protein," Graham said, referring to the alternate forms of that fusion protein. This wasn't easy. It involved growing crystals of the protein, hitting those with X-rays to show the protein shape, or else viewing them by electron microscopy. But it was all within Jason McLellan's wheelhouse. "Finally, in 2012 and '13," Graham said, "we captured the right structure." They saw the Before form of the protein.

But the Before form of that protein is unstable—prone to snap into the After form, like a mousetrap snapping closed. A closed mousetrap catches no mice. A fusion protein in After form doesn't initiate fusion with the cell membrane—nor does it present a good target for the antibodies you want from a vaccine. Your vaccine antibodies need to be preventative—targeted at the Before form to *stop* it from fusing to cells. All of this, remember, was a challenge to efforts at designing a vaccine against RSV, that child-killing virus, years before COVID-19 came along. RSV was a dangerous virus, a dire problem, but it was also a dress rehearsal for SARS-CoV-2.

During 2013, McLellan and Graham and some colleagues, with help from Peter Kwong and his lab people upstairs on the fourth floor, and others from the U.S. and the Netherlands and China, found a way to put a chock in the Before form of that RSV protein so that it wouldn't flip to the other shape.

"We were able to find stabilizing mutations to hold it." Graham was in his home office as we Zoomed, bookshelves converging to a corner behind him. He reached aside for two multicolored plastic models. "To hold it in *this* shape"—his left hand elevated one model, shaped like a fat Christmas tree, its top branches orange and red, with some patches of lavender and green and yellow and blue down below, all representing an intricate, unstable configuration of proteins—"instead of *this* shape," Graham said, and he held up another model.

The model in his right hand was taller and narrower and flat-topped, like a champagne flute. Lavender, green, yellow, blue; no sign of orange and red. "Because *this* protein"—left hand—"flips into *this* protein"—right hand—"spontaneously." I was mesmerized.

"*This* is pre-fusion, *this* is post-fusion," I said, referring to the Before and After, as I pointed to his left, then to his right. Slow learner.

"*This* is pre-fusion, *this* is post-fusion," he confirmed.

All the red and orange stuff at the top, he added, was where the antibodies needed to glom on. That would stop the virus. But that stuff was all obscured on the After form of the protein. And the After was what everyone had been using in their vaccine development efforts. "It had failed every time." Those vaccine candidates were duds. Graham had been working on RSV for years by the time Jason McLellan wandered into his lab. But once he and McLellan and their team managed to stabilize the Before shape of the fusion protein, he said, "it just changed everything."

"That was published at the end of 2013," Graham added, "just as Jason was going to his first faculty position, at Dartmouth."

McLellan became an assistant professor of biochemistry in the Geisel School of Medicine, named for Theodor Geisel, aka Dr. Seuss, a generous Dartmouth alum. He moved to New Hampshire in summer of 2013, McLellan told me, "and the plan was to work on RSV." A busy time for him: settling in a new town (Hanover), beginning to establish his own lab, and publishing two papers, himself as lead author, in *Science*, the second of which won recognition as one of the journal's Top 10 Breakthroughs of the year. That paper, with Barney Graham and Peter Kwong sharing senior position at the end of a long list of coauthors, described how Mc-Lellan and his colleagues created a stabilized version of the F protein of RSV, in its pre-fusion form, that worked well in a vaccine against the virus when tested on mice and monkeys.

Despite the "Breakthroughs" honor, McLellan struggled for grant funding—as newly independent young scientists often do—to pay the salaries of postdocs, support graduate students, and equip and run his lab. The NIH turned down three of his proposals for continued RSV research. Discouraged, at a low point, McLellan spoke with his mentor by phone.

Graham suggested a tactical diversion sideways. Why not try using his structural approach against coronaviruses instead of RSV? A nasty new one had lately emerged on the Arabian Peninsula: the MERS virus. Because its lethality was high and its potential to spread was still unknown, this virus might seem more urgent and promising to NIH grant reviewers. Following that advice, McLellan did get funding to apply his structure-based approach to the spike proteins of coronaviruses, starting with MERS-CoV, in hopes of creating a stabilized effigy of its spike that could be used in vaccines.

"But the protein was hard to work with," McLellan told me. "It was just not a well-behaved protein." Very unstable. Didn't crystallize well for X-ray crystallography. "It was being a pain." He and a group of colleagues struggled with that problem and eventually solved it, partly by using a fancy new technique (cryogenic electron microscopy, or cryo-EM) and partly by honing their methods against a somewhat less difficult coronavirus (HKU1, a common-cold coronavirus discovered by researchers at the University of Hong Kong in 2004). For expertise on the new microscopy, McLellan turned to Andrew Ward, a cryo-EM maven at the Scripps Research Institute in La Jolla.

McLellan knew Ward by email, and was aware that Ward had done sterling cryo-EM work. He asked Ward about helping with the spike of HKU1. Ward, a youngish full professor with a toothy smile and a jaunty attitude, said, "Yeah, let's do it." They would solve the structure, have beers at a conference, present their work—science with the right colleagues could be fun.

"So the three of us, my lab, Andrew's lab, and Barney's lab, started working together," McLellan told me. They succeeded in detecting the structure of the HKU1 spike, and they succeeded also eventually with MERS-CoV. As the colleagues wrote up a paper on the HKU1 work, McLellan and his lab were already busy designing similar modifications—these would be bioengineered mutations—that would jam the MERS-CoV spike in its Before form. And remind me, why did they want to jam the spike in its Before form? *Yes*, because that's the form to use as antigen in a vaccine! That form, above all other forms, will evoke antibodies to stop the virus. They stabilized the MERS-CoV spike by engineering a tiny molecular monkey wrench into its works. They jammed that sucker: stuck in Before form, unable to flip to After.

McLellan and Ward and their groups published the MERS results in a good journal, at a time between MERS outbreaks, when the virus seemed to have disappeared back into the camels where it abides. First authors on the paper were three young colleagues: Jesper Pallesen, Nianshuang Wang, and Kizzmekia Corbett. The methods described here, they noted at the paper's end, can be useful against MERS and . . . adding what seemed like a vague additional promise, "provide an important step in the development of broadly effective coronavirus vaccines." Who knew when those might come in handy? It was 2017.

57

Barney Graham, with his four decades of experience, his long labors against punishing old viruses and alarming new ones, acted on one other key insight during those last few pre-pandemic years: that it would be important not just to create new vaccines, but to create new vaccines quickly. Speedy vaccinology and production would entail not just solving protein structures, designing the right antigens, coding them into synthetic mRNA, and wrapping that mRNA in tiny bubbles of lipid so it could penetrate cells before the body's enzymes ate it apart. It would involve doing all those things *fast*, and running the vaccine through lab animals to test efficacy *fast*, and also through human clinical trials to test safety as well as efficacy *fast*, and then manufacturing the vaccine in millions of doses *fast*. It would even require doing some of those things simultaneously, further compressing the timeline.

So, in 2017, Graham arranged a partnership with that small biotech company, Moderna Therapeutics. Moderna was only seven years old, undercapitalized, not widely esteemed by investors or scientists, not yet having brought any product to market. Graham enlisted the firm to collaborate on a speedy design-to-manufacturing vaccine project, with Nipah virus as the target. Moderna seemed like a good choice as partner, despite its lack of achievements, because it was dedicated wholly to the same idea that intrigued Graham: delivering a vaccine in the form of mRNA instructions to build antigen. The name itself, Moderna, was a portmanteau of "modified" and "RNA," with the added appeal that it contained "modern." Stéphane Bancel, the CEO, was an astute entrepreneur but not a scientist; he had been hired in 2011 for his fundraising élan.

Graham's lab designed the Nipah antigen. Moderna turned that into an mRNA vaccine prototype. Moderna also produced an mRNA candidate vaccine against MERS, based on the stabilized spike protein that Graham's lab designed. This was part of an ambitious larger plan that Graham had been spearheading within the NIAID, to develop vaccine design concepts for at least one virus in each of the twenty-six viral families containing viruses known to infect humans—vaccine concepts that could be generalized to other members of that family. By the end of 2019, the Nipah vaccine was ready for a clinical trial, the first step toward assessing its safety. Then came the novel virus out of China. By January 6 or 7, Graham heard what he calls "a backstory rumor" that it was a coronavirus similar to SARS-CoV. *How* similar? Well, not identical but close enough to be concerning. Graham and Bancel exchanged emails, agreeing at once to redirect their rapid-vaccine project toward this virus.

"He said, 'As soon as you send me the sequence,'" Graham told me, "'we will start manufacturing.'"

Which is why Graham in turn had told Tony Fauci, as Fauci remembered it: "Just get me the sequence. We're all set to go."

By this time, Jason McLellan was an associate professor at the University of Texas at Austin, with a larger lab, better equipment, and his own cryo-EM capacity. On January 6, 2020, he happened to be in Park City, Utah, on a vacation trip with his wife and two kids, getting a pair of new snowboarding boots heat-molded to his feet, when his cell phone rang. McLellan saw it was Graham and assumed that his mentor was calling to offer holiday greetings. But no. "It looks like this virus that's causing the pneumonia clusters is a coronavirus," Graham told him. "Similar to the SARS coronavirus." Graham was putting together a team to dive onto the new virus, determine its spike structure, get that stabilized in Before form, and create a vaccine quickly, "just in case it happened to spread out of China." Are you in? he asked. "Of course," McLellan said. He messaged a grad student in his lab, Daniel Wrapp, the in-house cryo-EM hand, and told Wrapp to get ready. "We've just got to wait for the sequence to be released." He also alerted Nianshuang Wang, whose role would include designing modifications to the new genome that could turn an unstable Before spike into a stable one.

McLellan enjoyed a few more days of snowboarding before his world (and ours) changed. The sequence—the one from Yong-Zhen Zhang and Eddie Holmes, posted on Virological—came out that Friday night. (Graham and McLellan were unaware of the sequences that George Gao and his colleagues had submitted a day earlier to GISAID.) Graham was in his home office. "We started aligning things and talking about it," he told me, "and made some choices on Saturday morning." More choices would come, as they converted the spike region of the sequence into protein reality and considered the options of how to proceed toward a vaccine.

Wang got busy in McLellan's lab on Saturday morning and worked through the weekend, generating possible genomic changes that could jam the spike, as they had jammed the one on the MERS virus. The lab was nearly empty, the solitude allowed Wang to concentrate, and he had enough instant noodles, boiled up in a microwave, to keep him fueled. On Monday he sent a series of sequences to a company that converted them to physical DNA molecules. Receiving the DNA back, Wang worked with Daniel Wrapp to express those genes through cultured cells and harvest the different versions of modified spike protein. "They designed a series of about ten," McLellan told me. In Graham's lab, at the same time, Kizzmekia Corbett and others did parallel work, developing similar options for how the spike of the new coronavirus—the one coded in the genome just released by Zhang and Holmes—could be stabilized in Before form to serve as the antigen in a vaccine. Most of these options included the same molecular monkey wrench they had used on HKU1 and MERS-CoV. On January 23, Wang shipped what he had to Corbett.

The clock was ticking, and Barney Graham was the wise elder who had to place the bet, a large pile of chips representing many millions of public and private dollars, and many lives. He had to choose which option would most likely work. He chose. But Graham, not a vaunting man, shrugs credit toward Stéphane Bancel, Moderna's CEO.

"He trusted our judgment, he trusted me. We were in this together," Graham told me. "He took the risk of manufacturing without any additional experimentation on the sequence that we sent them."

Near the end of our chat, an hour later, I asked Graham the same question I asked Fauci and many other scientists I interviewed: What was the most important decision you made in 2020?

"Probably just the final decision on which sequence to choose and send to Moderna," Graham said. If he had chosen wrong, he said, it would have cost them six to eight weeks. Moderna was a small company and couldn't have borne the lost time and expense, maybe, in competition with giants such as Pfizer. And six to eight weeks lost might have played out as six to eight weeks of more unmitigated pandemic, if no other vaccine effort succeeded.

How did you choose the right sequence? I asked Graham.

"It was just based on my work with these fusion proteins over the last seven, eight, nine years," he said. You tried to envision how that fusion protein would move, he said. The top might be moving, so you didn't want your synthetic version to be too rigid. It's got to remain a good antigen. It's got to instruct the immune system how to meet the real spike of the real virus, in pre-fusion form. "You know, maybe those decisions made a difference, and maybe they didn't. But that's what I kind of sweated over the most, is making that final choice of the sequence."

I started to interrupt with a question, but he added, "Because that final choice was my choice."

It worked. On November 30, 2020, Moderna announced results from its big clinical trial, involving thirty thousand participants. According to the trial, the vaccine was 94.1 percent efficacious against infection, a stunningly good result for a new vaccine. Against severe illness, it showed 100 percent efficacy.

Meanwhile, people were dying. On March 16, 2020, the day Moderna began its Phase 1 trial, the United States recorded "only" twenty-two COVID-19 fatalities, but within a month the daily death count rose a hundredfold. On December 18, the day Moderna's vaccine received emergency use authorization from the U.S. Food and Drug Administration, 3,171 Americans died of Covid, as well as ten thousand other people around the world. But the vaccines were arriving now, for those who would accept them, and for those who could get them.

58

Sadly, getting a vaccine was difficult or impossible for most people in most countries. "Vaccine inequity is the world's biggest obstacle to ending this pandemic and recovering from COVID-19," according to Tedros Adhanom Ghebreyesus, director-general of the WHO. When he said "recovering," he meant something more than medical recovery. "Economically, epidemiologically and morally, it is in all countries' best interest to use the latest available data to make lifesaving vaccines available to all." Implicit in the statement from Tedros were two big ideas. For rich nations to turn their backs on struggling nations, during this crisis, would be unforgivable. And for rich nations to turn their backs on struggling nations would be unwise, because the virus will always come back stronger. There is no lasting herd immunity except for the human herd.

Six months into the vaccination phase of the pandemic, as of June 20, 2021, less than one percent of people in low-income countries had received even a single dose of vaccine. That group included Yemen, Madagascar, and many of the nations of sub-Saharan Africa, from Sierra Leone across to Mozambique. In high-income countries, by contrast, 43 percent of the population had gotten at least a single vaccine shot. Taken by continent, only 2.4 percent of Africans had been vaccinated, compared with about 41 percent of North Americans and 38 percent of Europeans. That disparity was predictable as well as deplorable, and the trend continued. Five months later, near the end of 2021, 50 percent of the world's population had received at least one dose of vaccine, but only 4.1 percent of people in low-income countries, and only 8.5 percent in Afghanistan. The vaccination percentages were far lower—down around one percent—in African countries such as Mali, Chad, and the Democratic Republic of the Congo. (But that situation was complicated by the fact that those countries had also recorded surprisingly low numbers of Covid cases and deaths, so far, for reasons no one fully understood.) In the world's thirty poorest countries, only 2 percent of the population were fully vaccinated.

The mRNA vaccines, delivered so quickly by brilliant and committed (and, some but not all of them, now very wealthy) scientists and science managers are wondrous; but those vaccines don't travel well in hot, poorly resourced countries, because they require refrigerated or even deep-frozen storage of a kind not often available at a remote clinic in Burkina Faso or Chad. They are also relatively expensive by the time they arrive in some countries. Pfizer and BioNTech and Moderna are for-profit companies, after all. The conundrum of intellectual property rights versus public health, common good versus the expectations of venture capitalists at play with the pharmaceutical industry, is a monster I'll only nod at here, forgive me, and keep walking.

Efforts have been made, organizations and programs created, to cope with this vast dilemma. The Coalition for Epidemic Preparedness Innovations (CEPI), which I've mentioned, is a donor entity established in 2017, with vast starter contributions from the Bill & Melinda Gates Foundation, the Wellcome Trust, Norway, Germany, India, and other sources, to support foresightful vaccine research aimed at emerging and neglected viral diseases. GAVI (the Global Alliance for Vaccines and Immunization), founded in 2000 and now formally known as Gavi, the Vaccine Alliance, is an international public-private partnership to increase immunization in low-income countries against infectious diseases of all sorts, from murderous old ones such as yellow fever and polio to murderous new ones such as Ebola. COVAX is something more recent and focused: a global initiative directed by CEPI, GAVI, and the World Health Organization to bring COVID-19 vaccines to low- and middle-income countries on a scope somewhere near commensurate with what's available in high-income countries, not as a mercy mission but as a global health enterprise in the best interests of everyone.

Those were the good intentions, anyway. All three organizations have been criticized during this pandemic for failures to deliver, both in the metaphorical and in the literal sense. Uruguay's ambassador to the United Nations, according to the respected health news website STAT, complained that his country bought vaccines from COVAX, didn't receive them as expected, and that officials at the organization were unreachable or unresponsive. Libya's U.N. ambassador described similar frustrations. Both countries, STAT reported, "gave up on waiting for their COVAX purchases and made their own deals with pharmaceutical companies, effectively paying twice." Somalia received vaccine doses from COVAX, but not the syringes needed to inject them. Pakistan turned away from COVAX to try dealing directly with manufacturers, which proved difficult because wealthy countries, with big orders, stood ahead in line. For COVAX, the supply problem owed partly to a sudden constriction of delivery by one of its major suppliers, the Serum Institute of India, a private company manufacturing the Oxford-AstraZeneca vaccine at a rate of millions of doses daily. When the pandemic surged in India during April 2021, those vaccines became precious for domestic use, and exports almost entirely ceased. Other suppliers charged COVAX higher prices. COVAX was also squeezed when promises from rich countries, to donate almost a billion doses from what those countries had bought or produced, and which might not be needed or accepted by their own populations, went unkept. As of September 24, 2021, according to STAT, only 18 percent of those doses had arrived.

In the meantime, privileged people (such as me) in the United States, having received two doses of the Moderna or the Pfizer, were lining up for booster shots. This drove Tedros to exasperation. He was "appalled," he told a news conference in Geneva. "I will not stay silent when companies and countries that control the global supply of vaccines think the world's poor should be satisfied with leftovers." The situation was chaotic, perverse, and not tractable to easy correction simply by heeding the critiques of Tedros, aid organizations such as Oxfam and Médecins sans Frontières (Doctors without Borders), and keen journalists such as those at STAT.

The individuals who guide COVAX, such as Seth Berkley, CEO of GAVI, and Richard Hatchett, CEO of CEPI, and Tedros himself, are smart and well-meaning people dealing with a viral challenge for which none of the world's institutions, structures, arrangements, or sensibilities were ready.

We have a lot of vaccines, but not enough, and not enough of those we do have are available where most needed. More than a hundred different candidate vaccines have gone into clinical trials, dozens more are in development, embodying various approaches, including viral vector vaccines, protein subunit vaccines, inactivated virus vaccines, attenuated virus vaccines, and others. There's ZyCoV-D from Zydus Cadila in Ahmedabad, Medigen from Taiwan, QazCovid-in from a research institute in Kazakhstan, Soberana 2 out of Cuba, Zifivax from China, Covaxin from Bharat Biotech in Hyderabad, and COVAX-19, also known as SpikoGen, as a joint venture between companies in Iran and Australia. If you're not sure about Sputnik V, you can try Sputnik Light, one dose instead of two, less filling. Pricing is various, degree of public-private funding is various, efficacy is various, levels of vaccine resistance from country to country are various.

Delivery systems are various too: one shot, two shots, a patch on the skin, a nasal inhaler. Some public health experts and virologists emphasize that the cold chain requirements of many vaccines—freezer temperatures for longer storage, refrigeration at least for brief storage—are a severe constraint on getting the whole world, or at least most of it, vaccinated. Ilaria Capua is a professor at the University of Florida, a veterinarian with decades of experience in zoonotic diseases, an expert on avian influenza, and a former member of the Italian Parliament. "*The* thing we have to fix is heat-stable vaccines," Capua told me, meaning those that don't require refrigerated or frozen storage. Needles are another issue. If a reliable heat-stable vaccine could be administered from an adhesive patch on the arm, or a lozenge dissolving under the tongue, "we could be out of this and we could be reaching all the people that we need to reach."

Peter Hotez is a professor at Baylor College of Medicine, a pediatrician, a molecular virologist, a leader in vaccine development, and the author of several books, including *Forgotten People, Forgotten Diseases.* Hotez has become widely known and trusted among Americans because of his tireless availability to news media of every stripe, his willingness to explain complicated things patiently and resist politicized obscurantism, recognizable by his bow ties and by the burly directness and spectacles reminiscent of Teddy Roosevelt without a squint. "Relying on the multinational pharmaceutical companies alone is not adequate," Hotez told me. "We need to build indigenous capacity for making vaccines and monoclonals and therapeutics locally. Right now, no vaccines are made on the African continent." Barely any are made in Latin America, he added, barely any in the Middle East. "That needs to be fixed," he said, and so does the problem of those "onerous freezer requirements," he added. Yes, we need a thermally stable oral vaccine, a nasal spray, a skin patch. "I think it is doable," he added. What's required is time and additional money. In the months after our conversation, his work vindicated that optimism. Hotez and his colleagues at the Texas Children's Hospital Center for Vaccine Development have led efforts toward a vaccine—less famous than the mRNA vaccines, using a different molecular method, and more suited to mass production and use in low-income countries—that has already received emergency use authorization in India and may soon be available elsewhere.

A bottle of pills can go anywhere. You can put a case of them on the back of your motorbike, if you're a physician's assistant in Niger or Afghanistan or Colombia, and get them to an outlying village where the second wave has just begun. You can offer those pills to people who can't pay, and whose governments can't pay, for the development and production of such things—people who may distrust needles, may distrust the modern medicine of outsiders generally, perhaps due to a legacy of racist and imperialist medical "care" and experimentation, and who trust their own traditions, practices, and remedies more—but who now face a nontraditional danger. With luck, you can welcome such people to the prospect of protecting themselves against a novel virus with a novel prophylaxis. That's not magic. That's just science, manufacturing, and humanity.

VII

THE LEOPARDS OF MUMBAI

59

When a novel virus appears suddenly as an infection among humans, one of the first questions is always: Appeared from where? Everything comes from somewhere, including viruses. The origin of an unfamiliar, dangerous virus is a matter of urgent interest for several reasons, including prevention of further such surprises and understanding the biology of the thing. Understanding the biology of the thing, and its evolutionary history, can be crucial to the development of therapeutic drugs and vaccines. But tracing it to that origin is often difficult and takes time.

Infectious disease scientists know that a virus being "new to people" doesn't mean it's new to the world, and that "newly recognized by science" doesn't necessarily mean new to people. Ebola virus gained its first fame during a dramatic outbreak among humans in 1976, centered on a remote mission hospital in northern Zaire (as the Democratic Republic of the Congo was then known). A similar outbreak caused by a closely related virus (Sudan virus) began, by sheer coincidence, at a cotton factory in southern Sudan (now the nation South Sudan) at about the same time. Either or both of those ebolaviruses may have been causing small, mysterious clusters of gruesome illness and death among Central African villagers intermittently for centuries before they attracted outside notice. Scientists speak of "emerging" and "reemerging" viruses—there's even a journal devoted entirely to this field, *Emerging Infectious Diseases*, published by the American CDC—because they know that a virus newly arrived in one kind of host must have emerged from another kind. New viruses in humans generally come from wild animals—more specifically, from mammals and birds—sometimes by way of domesticated animals. There are possible exceptions, involving laboratory-assembled viruses, wholly or partly built from sections of wild viruses (and I'll discuss that topic below). It's also theoretically possible that a new virus could spill into humans from reptiles, amphibians, or even plants, though that has seldom if ever been seen. Some human-infecting viruses, such as West Nile and eastern equine encephalitis virus, have turned up in captive snakes and alligators, but in those cases, the reptiles acquired a virus already known in people. Mosquitoes, ticks, and certain other arthropods deliver pathogens such as the yellow fever virus, African swine fever virus, the dengue viruses, and Zika virus to humans, but those insects and arachnids are vectors, not reservoir hosts, in that they seek out human hosts, rather than spilling their viruses passively. If you assume, as experts in this field do, that a shocking new virus in humans has most likely come from a nonhuman animal—a reservoir host, in which it abided inconspicuously over time—then identifying the host is a priority task. That might be achieved quickly. Or the mystery might remain unsolved for decades.

Machupo virus was first isolated in 1963 from the spleen of a two-year-old boy who died in San Joaquín, a small town in northern Bolivia, while suffering a disease that came to be called Bolivian hemorrhagic fever (BHF). San Joaquín lies within Beni Department, near the Brazilian border, an area where the grasslands of the Moxos Plains give way to the westernmost edges of Amazon rainforest. The disease was first recognized in 1959, afflicting a man who farmed crops on a subsistence plot twelve miles from San Joaquín. His name was Augusto Avaroma. He went down with fever, he vomited, red spots appeared inside his mouth and in his armpits, his nose bled, his wife gave him liquids that he could barely swallow, but after a week the fever broke, and Avaroma survived. Further cases occurred intermittently, most involving other men in rural areas, and the causal agent remained unknown, until an outbreak in 1963–1964, centered on the town of San Joaquín, which eventually tallied 637 cases and 113 deaths. The outbreak caught the attention of scientists at a U.S. disease laboratory in Panama, the Middle America Research Unit (MARU), part of the NIH, which sent a response team to San Joaquín. The team leader was a young physician and virologist named Karl M. Johnson, who would become a legendary figure, a dauntless pioneer, in the field of emerging viral diseases. Bolivian hemorrhagic fever in San Joaquín was like Sherlock's first case.

Johnson focused on finding the responsible pathogen and learning a bit about its ecology, to understand how and why this disease emerged when and where it did. He succeeded, but not before contracting a case of BHF himself, being evacuated to Panama, and nearly dying there. Recovered and back on the mission, he led the team that cultured the novel virus from that human spleen. (All this is told, and much else, in an unpublished book manuscript by Johnson and his wife, Merle, that I've been privileged to read.) They named the virus Machupo, after a river not far from San Joaquín. In 1964, Johnson and his colleagues identified an animal host of the virus: a rodent known as the large vesper mouse (*Calomys callosus*), which favors habitat where forest intergrades into savanna. The virus accumulates in the blood, saliva, and urine of the mouse. The mouse does well near human habitations, as well as forest-grassland edges, and a combination of factors—including an increase in grain farming in the area, possibly also a decrease in the domestic cat population of San Joaquín—brought an explosive abundance of mice in the town, an increase in mouse urine to be swept up with the daily dust, and the outbreak of BHF. Solving the reservoir host mystery enabled a remedy for the outbreak. Johnson's team organized a rodent-trapping campaign in San Joaquín, which eliminated three thousand large vesper mice in two weeks, and they encouraged the importation of cats. BHF never disappeared entirely, but it became susceptible to control efforts once the reservoir dynamics were known.

Notwithstanding their heroic labors, and Johnson's life-threatening illness, and their innovative methods (Johnson imagined into existence the first portable glove box, for safely doing virus-isolation work in the field), the MARU team was lucky in one perverse way, and without that luck they might not have solved the reservoir mystery. The large vesper mice of San Joaquín carried a *lot* of Machupo virus. The prevalence in the population was high. The team live-trapped seventeen mice for their isolation efforts, and they grew the virus from fourteen of them. In later studies, the infected rate among wild-caught vesper mice has ranged from 11 percent up to 80 percent. If a virus infects only a smaller minority (say, 2 or 3 percent) among its reservoir host, as is true with some other viruses, then it's more difficult to find.

Cracking the case of Marburg virus took much longer. That virus made its first known appearance in August 1967, when laboratory workers in the cities of Marburg and Frankfurt, Germany, and Belgrade, Yugoslavia (now Serbia), took delivery of some African monkeys sent live from Uganda for use in medical research. Almost simultaneously, in the three places, outbreaks of a fearful and unidentified hemorrhagic fever began. In Marburg, twenty-three people sickened, mostly workers at a pharmaceutical factory who handled tissues of the monkeys, and five of them died. In Frankfurt, it was six cases and two deaths. In Belgrade, a veterinarian at a vaccine research institute became infected, and then also his wife, after caring for him during his illness. They both survived. All the monkeys came from the same exporter in Uganda, where they had been trapped on islands in Lake Victoria. A novel virus was isolated from blood and tissues of several patients in Marburg and several patients in Frankfurt. Marburg the city, with most of the cases, got the naming "honors" for Marburg the virus.

Those outbreaks stretched over a few months in autumn 1967. Identifying a reservoir host of the virus—it wasn't the African monkeys, they were only intermediaries—took much longer. Forty years passed. In the meantime, three cases of Marburg virus disease occurred in what then was Rhodesia (now Zimbabwe) in 1975, when an Australian student on a hitchhiking vacation fell ill, and then the hitchhiker's girlfriend and a hospital nurse came down with the illness too. Only the young man died. In 1980 there was a single case in Kenya, probably associated with a visit to a certain cave on Mount Elgon, where bats roosted; then two cases in the Soviet Union, resulting from accidents (one of them, a needle-stick injury) at a laboratory; and sizable outbreaks in both the Democratic Republic of the Congo (DRC) and Angola, each accounting for more than a hundred deaths. The reservoir host of the virus remained unknown, but bats were among the suspects.

Caves and mines were also suspect, because it was in or near such places that Marburg infections seemed to happen. In the DRC, for instance, at least 154 cases of Marburg occurred between 1998 and 2000, with a case fatality rate above 80 percent, centered around Durba, a village near several gold mines in the northeastern corner of the country. Most of those cases were in young male miners and their family members. One team of scientists, led by a field-toughened South African virologist named Robert Swanepoel, went to Durba twice during 1999, while the outbreak was sputtering along, and made a broad investigation of local fauna in search of a reservoir host of the virus. They sampled eight kinds of bats, seven kinds of rodents, three shrews, four crabs, a frog, and thousands of arthropods, including cockroaches, crickets, spiders, wasps, bat flies (little wingless insects that parasitize bats), and mites. They found fragments of Marburg virus, using a PCR method, in some of the bats. They found antibodies against Marburg in some of the bats. But those positives only indicate that the animals had been exposed to the virus, not that they carried it chronically, serving as long-term incubators. Swanepoel's team found no evidence whatsoever for Marburg in any of the rodents or spiders or cockroaches. The crabs were clean. The frog was exonerated posthumously. And none of the samples, from bats or anything else, yielded functional virus that could be grown in the lab. That's the gold standard for identifying a reservoir host: isolating live virus. The reservoir of Marburg virus remained obscure.

But the Swanepoel team's effort provided important clues. Most of the positive hits they got for antibodies, and for fragments of virus by PCR, were in samples from just two kinds of bat: the eloquent horseshoe bat (*Rhinolophus eloquens*), a little insectivore, and the Egyptian fruit bat (*Rousettus aegyptiacus*), a sizable creature with a squirrel-like face and strong wings suited to ranging widely for fruit. And one other clue: more than 90 percent of the infected miners worked at one site, the Goroumbwa mine, an underground operation. There were open-pit mines at Durba also, but men who worked in those seldom got infected with Marburg virus.

Then, in 2007, came reports of another cluster, linked to a mine called Kitaka in southwestern Uganda, about 250 miles south of Durba as the bat flies. One signal feature of the Kitaka mine is that it served as a roost site for a huge colony of Egyptian fruit bats.

At the CDC in Atlanta, some scientists within the Special Pathogens Branch followed this series of events and clues with high interest. One of them was Jonathan Towner, a lean and dark-haired molecular virologist with a strong tolerance, like others of his guild, for the physically arduous, ecological side of viral research. Towner and a small group of CDC colleagues pounced on the news from Kitaka as a grim opportunity to further the search for the Marburg virus reservoir. They flew to Uganda, converged with colleagues from Johannesburg (including Robert Swanepoel) and elsewhere, and descended upon the Kitaka mine with traps, nets, PPE, collection vials, and other matériel for investigating a focused hypothesis: that the Egyptian fruit bat might be it.

"More and more epidemiological data was pointing towards a cave-like environment for the source," Towner told me a dozen years ago. "If you're looking in a cavelike environment, you know, 99 percent of the species in the jungle don't go into caves." So your list of possible hosts can be narrowed. "What lives in caves?" Bats do, some rodents do, crickets and spiders do. Then again, Swanepoel had had no luck with crickets and spiders. It turned out that forest cobras (which can grow to ten feet in length) and gigantic African rock pythons (more burly than the cobras, and capable of reaching nineteen-foot lengths) live in caves too, at least in southern Uganda, evidently because of the bounteous supply of toothsome bats, but Towner and his colleagues didn't learn that until they got to Uganda.

The collection effort in Kitaka, as described to me by Towner and one of his colleagues, Brian Amman, a mammal ecologist with the CDC who specialized in bats, was wonderfully hellish. They suited up in Tyvek coveralls, respirator helmets, goggles, boots, and gloves. The mine tunnels were hot and humid, the goggles got foggy, the standing water was dark and you couldn't see how deep, the head room was low, and some of the pinch points from chamber to chamber were tight, especially for Brian Amman, a large man. Ticks abounded, gathering in little cracks near the bat roosts, waiting for their chances to climb aboard some unfortunate chiropteran for a blood meal; human blood might have satisfied them just as well, so you didn't want to jam your hand, at a moment of imbalance, into one of those cracks. Swanepoel's report from Durba hadn't mentioned ticks. Might *they* be carrying Marburg? Amman described some of the niceties of this adventure to me, such as squeezing through a slot into one chamber, a blind room, and finding hundreds of dead bats. Probably these animals died of asphyxiation, after local workers tried to rid this mine of bats using fire and smoke. If the bats had died of Marburg infection, that meant they couldn't be a reservoir host, contrary to the guiding hypothesis; but the litter of carcasses, like a compost of fallen sycamore leaves and poisoned rats, might still have been teeming with the virus. "It was really unnerving," Amman recollected as I sat with him in a clean, comfortable room at the CDC. I've quoted him before, but I find his deadpan calm unforgettable. "I'd probably never do it again."

Twice was enough, after the team's return field trip to Kitaka in 2008. Altogether they captured more than a thousand bats; killed and took tissue samples from 611 of those, and found fragments of Marburg virus in thirty-two of the 611; swabbed and tagged the others, the tagging done for a mark-and-recapture study that would allow them to estimate the total number of bats in Kitaka. They estimated more than a hundred thousand. They found fragments of Marburg virus RNA in 5 percent of the bats they collected. That meant, as they later concluded, the mine contained more than five thousand Marburg-infected bats.

A year after the second field trip, this team published a paper containing some of these details (including the forest cobras) and one headline result: they had isolated live virus from five of the bats. The reservoir host—or anyway, *a* reservoir host—of Marburg virus had been found. It was safe to say, thanks to Towner and Amman and Swanepoel and their colleagues, that the outbreaks at Kitaka and Durba, probably also those cases linked to the cave on Mount Elgon, and possibly too the virus in the African monkeys shipped to Europe in 1967, originated from Egyptian fruit bats. That discovery had taken only forty-one years.

60

On February 6, 2020, just a week after the WHO declared the spreading coronavirus a Public Health Emergency of International Concern (PHEIC, one cautious step short of calling it a pandemic), a day by which China had 31,161 lab-confirmed cases of the new infection and the United States had its first (so far unrecognized) Covid death, two researchers in Wuhan posted a preprint article on a social networking site, titled "The Possible Origins of 2019-nCov Coronavirus." They were calling the virus by its provisional name before it became SARS-CoV-2. The first author was Botao Xiao, a young scientist establishing his own research group at Huazhong University of Science and Technology, after completing a postdoc at the Harvard Medical School. The second author was his wife, Lei Xiao, employed at Tian You Hospital. The article ran barely more than a page and did not claim to report original research; it was an essay or a comment, a legitimate form of scientific publication that generally involves mulling over other people's data. It cited several other papers, including the one by Zhengli Shi's group reporting that the new virus had its "probable" origin in a bat, and the one linking most of the first forty-one Wuhan cases to the Huanan Seafood Wholesale Market. Xiao and Xiao challenged the inference of connection. "The probability was very low for the bats to fly to the market," they wrote. "Was there any other possible pathway?" Yes, they asserted. The new site of the Wuhan Center for Disease Control and Prevention, with its laboratories, stood only 280 meters from the market. Twelve kilometers away (by their measurement, though Google Maps says fifteen) was another set of laboratories, at the Wuhan Institute of Virology. Both contained labs where research on bat coronaviruses was done.

"In summary," Xiao and Xiao wrote, "somebody was entangled with the evolution of 2019-nCoV coronavirus. In addition to origins of natural recombination and intermediate host, the killer coronavirus probably originated from a laboratory in Wuhan." The first of those two sentences was portentous. The second was grammatically incoherent—hard to tell what they intended by "In addition to origins of natural recombination and intermediate host . . ."—but the final clause was daring and clear. Those ten words were the match that lit the tinder of the lab leak hypothesis.

Here's something else important about the Xiao and Xiao preprint: three weeks later, it was withdrawn from the website, and it has never been published. Botao Xiao explained this retraction in an email to *The Wall Street Journal*: "The speculation about the possible origins in the post was based on published papers and media, and was not supported by direct proofs."

This statement, as far as it goes, is clearly true. But did Botao Xiao jump to his retraction, or was he pushed? Had he become embarrassed by the tenuousness of the incendiary accusation, or were he and his wife under pressure—from their institutions of employment, or from higher Chinese officialdom? (Or were they both embarrassed and pressured?) The way you find yourself inclined to interpret this opaque event, gentle reader, probably reflects a predisposition you already hold toward the question of the virus's origin. But maybe, too, your predisposition is not immutable. The Xiao and Xiao episode is another small Rorschach test, probing attitudes about SARS-CoV-2. Some people will look at this ink-blot and see a bat. Others will look at it and see a laboratory.

61

A different idea, also dark, emerged at about the same time: that the virus had been created in a laboratory by some form of genetic engineering. Among the first statements of that theme was a preprint, which I've mentioned already, posted on January 31, 2020. The authors, nine researchers in New Delhi, noted four very short stretches of the spike protein of SARS-CoV-2 and claimed that those four stretches bore an "uncanny similarity" to four stretches of the equivalent protein in the pandemic AIDS virus, HIV-1. This similarity, they claimed, was "unlikely to be fortuitous in nature." It represented an "astonishing" link between the two proteins, found in two viruses that belong to two different virus kingdoms. The senior author on this dramatic paper was Bishwajit Kundu, a professor and protein specialist at the Indian Institute of Technology, in New Delhi.

Kundu and his coauthors called those four stretches "insertions" in the SARS-CoV-2 spike (although not specifying: insertions to *what*), and speculated that, because the "insertions" lay in the receptor-binding domain, the spike protein's grab-ahold spot, they might have increased the virus's capacity for latching on to cells. This "uncanny" and possibly advantageous similarity to HIV-1, the authors wrote, suggested "unconventional evolution" of SARS-CoV-2 that "warrants further investigation." The implication, left unsaid, was that someone had used parts of the HIV-1 genome in designing or enhancing the SARS-CoV-2 genome.

This paper drew negative reactions fast. Other scientists noted at least two big problems. First was the paper's claim that these so-called insertions "are not present in other coronaviruses." That was just wrong. The second problem was calling the match of short sections, between

SARS-CoV-2 and HIV-1, "uncanny." In fact, it was entirely unremarkable. The four stretches amounted to six amino acids at one spot, six others in another spot, and slightly more (eight aminos, twelve aminos) in each of two other spots. SARS-CoV-2 is constructed from roughly ten thousand amino acids, coded by its genome. The alphabet of amino acids in living creatures (and in viruses too, in case you don't want to call them "living") contains only twenty letters, repeated and arranged variously to specify all proteins. To find six ordered letters in one genome of that length, and to find the same six in another, is a coincidence that the odds favor. Finding four stretches of about that length, in two genomes, is not "uncanny." It's humdrum. If you scanned T. S. Eliot's poem *The Waste Land* and Robert Service's poem "The Cremation of Sam McGee," as I have just done, you could find certain words of six, seven, or nine letters included in both, though neither of those poems is as lengthy as a coronavirus genome. (But, alas, you won't find "Lake Lebarge" in the Eliot.) This does not prove that T. S. Eliot was highly influenced by Robert Service. Likewise, if you search the big genome databases for matches to the four "insertions" in the SARS-CoV-2 spike, you'll find those same letter combinations in genomes of mammals, insects, bacteria, and various other viruses, including influenzas and giant viruses. You will also find them, contrary to what Kundu's group wrote, in the genomes of three coronaviruses known from bats.

Absorbing such critiques, Kundu's group promptly took the preprint down. Two days after it went up, it was still findable online but marked WITHDRAWN. The paper's first author, Prashant Pradhan, added a note saying, "It was not our intention to feed into the conspiracy theories and no such claims are made here." According to later reporting in an Indian newspaper, *The Sunday Guardian*, the team offered a revised version to seven journals over the following six months and were refused by them all.

The senior author of the paper, Bishwajit Kundu, was less daunted than Prashant Pradhan. Contacted more than a year later by *The Sunday Guardian*, Kundu said, "We still stand by what we had published." To him the four "insertions" still looked "unusual," though he evidently didn't say why. "We believe it is a laboratory made virus."

During that brief period when the preprint was up, while numerous scientists tore into it on Twitter and elsewhere, Tony Fauci expressed some reactions privately, later revealed among his published emails. To the director of the NIH, Francis Collins, Fauci forwarded an op-ed on origin stories, noting, "The Indian paper is really outlandish." To colleagues within his office, on an email thread requesting his guidance for a possible response, he wrote, "Geeeez."

62

So the Xiao and Xiao preprint, and the Pradhan and Kundu preprint, came and went. Other questioning voices, raised during early months, were more persistent and, some of them, more judicious. They challenged the premise that the virus had evolved naturally and spilled naturally from an animal host into humans. These critics focused especially on three aspects, real or imagined, of the novel virus and its genome. First, the receptor-binding domain—that sticky little patch on the spike protein, which allows the virus to catch hold of ACE2 receptors on cells. Where had it come from? Why was it absent from the closest bat virus known so far, RaTG13? Had it been engineered into SARS-CoV-2? Second, the furin cleavage site—that hinge area between two major parts of the spike, which reacts at the touch of the right protein (furin) to let the parts split (cleave) and enable the viral envelope to fuse with the cell membrane, after which the virus's genome enters the cell. Again, not present in RaTG13. Again, what was its source—an unknown bat virus, a magician's top hat, or a laboratory?

The third point of dissent from the natural origin scenario was that this novel coronavirus seemed, from its earliest appearance in Wuhan, just *too well adapted* to humans—too well for comfort, for coincidence, for a bat virus. Had it somehow been "pre-adapted" for infecting and transmitting among people?

Kristian Andersen and his coauthors of the "Proximal Origin" paper, posted as a preprint on February 16, 2020, and then published in *Nature*

Medicine, had foreseen these arguments, as I've described. They had even shared an initial puzzlement about both the RBD and the furin cleavage site, until Matt Wong alerted them to similar RBDs among wild coronaviruses, and until other factors resolved their unease about the cleavage site. But the doubters remained doubtful, including a number of scientists and more than a few opinionated nonscientists expressing themselves in newspapers and social media.

William R. Gallaher reacted promptly to these statements of suspicion about the cleavage site. Gallaher is an emeritus professor at the Louisiana State University School of Medicine, an expert on the molecular genetics of viruses, and a longtime collaborator with Robert Garry, one of Andersen's coauthors on the "Proximal Origin" paper. He's a distinctive person, highly regarded by peers, a bit of a polymath who publishes novels and poetry as well as scientific papers, and he doesn't shy from pointed disagreement or correcting misstatements, his own or anyone else's. It was Eddie Holmes who alerted me to Gallaher and a comment he posted online—in early February, three days before the "Proximal Origin" paper—about the furin cleavage site. "Very interesting," Holmes said. "Read Bill Gallaher's piece."

It was on the Virological website, easy to find. "I have been privately dealing with rumors and inquiries," Gallaher wrote, about whether the virus "may have a suspicious origin as an engineered, laboratory-generated virus either accidentally or deliberately released" in or near the Huanan market. Much of that suspicion focused on RaTG13, the similar bat virus that lacked any such cleavage site. Might someone have taken the bat virus and added this feature, making an engineered virus, one that was more infectious to humans? "I see no evidence at all to support such a claim," he wrote, and explained why. His reasons were technical, but they included the fact that the RNA code on either side of the cleavage site also differed, by nineteen mutations, from the code seen in RaTG13. It made no sense, if you were engineering a virus, to do that. What the evidence did suggest, Gallaher wrote, is that SARS-CoV-2 inherited its furin cleavage site from some ancestral virus in the distant past. "It is not of suspicious origin," he concluded, and RaTG13 was neither its first cousin nor its laboratory template.

That was just the start of a long thread on Virological, with Gallaher posting and others responding, which reads like a script of molecular virologists discussing SARS-CoV-2 in the privacy of a steam bath. On May 2, almost three months after his first post, Gallaher wrote, "I have found a probable source of the putative insert"—meaning the stretch of twelve RNA letters, coding four amino acids, that constitute the furin cleavage site. That stretch, he reported, was almost identical to a sequence in another bat coronavirus, called HKU9, isolated from a *Rousettus* fruit bat in Guangdong province in 2011. This finding was mildly unexpected, because the coronaviruses most closely resembling SARS-CoV-2 came from horseshoe bats, little insect eaters, not from fruit bats. But the geographical ranges overlapped, Gallaher noted, and fruit bats sometimes share roosting caves with insectivorous bats. They probably share viruses as well as caves.

More important was the mechanism that Gallaher proposed, by which the furin cleavage site from HKU9 got patched into a virus that became SARS-CoV-2. It involved recombination of a particular sort. The first step was that both viruses infected a single host animal and, within that host animal, a single cell. Step two was an accidental event within that cell, while the two viral genomes were replicating themselves, called a "copy-choice error."

When a viral genome copies itself, the original genome serves as a template while a copying gizmo (an enzyme called polymerase) chugs along its length, turning out a second linear strand. Call that strand the Copy. Call the original genome Template A. Ordinarily this process produces a nice, complete twin of Template A. But in some cases, the polymerase hits a bump and jumps off Template A, landing on another viral genome, Template B. Whoops, it has just chosen the wrong template to copy. But it proceeds to copy a stretch of that genome into the same linear strand it was generating. Then it bounces again, back to Template A. The result is a new viral genome, a recombinant, with one stretch of B patched into A. In the case of SARS-CoV-2, Gallaher proposed, that was the furin cleavage site from HKU9, patched into a different bat virus that became the progenitor of SARS-CoV-2. "The only laboratory required," he wrote, "is the natural laboratory of the bat cave with multiple species of bats and bat coronaviruses."

"This is a really nice finding, Bill," posted Andrew Rambaut. It offered an origin for the furin cleavage site, and it eliminated the need for a pangolin or any other creature as intermediate mixing bowl, at least for that feature of the virus. "The sequence analysis all seems very cogent," posted someone else, "especially the discussion on copy-choice errors."

Five days later, Gallaher posted again, offering a further piece: a candidate for what caused the copy-choice error. He had found the speed bump, he wrote, that caused the copying gizmo to jump from one template to another in mid-copy. It was a short palindromic stretch of RNA letters. Remember what a palindrome is? A sequence of letters that reads the same in either direction, front-to-back or back-to-front. A palindrome for Napoleon: ABLE WAS I ERE I SAW ELBA. A palindrome about Ferdinand de Lesseps: A MAN, A PLAN, A CANAL: PANAMA. The palindrome that Gallaher had found in SARS-CoV-2 was shorter. "The entirely natural origin of SARS-CoV-2," he wrote, "is as simple as CAGAC." That palindromic sequence or its near approximation, CAGAT, immediately precedes both the furin cleavage site and the receptor-binding domain. Not everyone on Virological agreed, but Gallaher was confident.

This idea of copy-choice error was echoed by other scientists, including Spyros Lytras, a young Greek PhD student at the University of Glasgow. Born in Athens, Lytras came to Scotland when he was eighteen as an undergrad at the University of Edinburgh, then moved over to Glasgow, where he works with David L. Robertson among other advisors. Robertson is head of Bioinformatics at the MRC-University of Glasgow Centre for Virus Research (CVR) and a coauthor on some of the more interesting SARS-CoV-2 papers by Kristian Andersen and Eddie Holmes. Lytras is a smart young man with a bladelike nose and wide brown eyes, who wears his long hair cheerily and variously dyed (yellow, the week I met him) and falling elegantly to both sides from a midpoint of dark roots. At Edinburgh he studied classic evolutionary biology, he told me when we Zoomed, but he did a summer internship with an evolutionary virologist researching fly viruses, and that drew him toward viral genomics. Several months after William Gallaher's "copy-choice error" post on Virological, Lytras posted on the same site, seconding Gallaher's idea and adding that he had found something intriguing in "an overlooked fragment" of another viral sequence, "providing a clue" about the source from which SARS-CoV-2 might have acquired its furin cleavage site. It was a short sequence within the cleavage site itself, similar between SARS-CoV-2 and another virus, lately found in a bat.

At this point we should stop thinking about the "origin" of SARS-CoV-2, and proceed by thinking about its *origins*, plural. The propensity of coronaviruses for recombination, swapping parts of their genomes with other coronaviruses, and the evidence that SARS-CoV-2 has resulted from such swapping, mean that we're not looking for a single origin; we're looking for several. No matter whether you prefer to believe the virus was engineered and released intentionally, or manipulated in a lab and leaked by mistake, or evolved naturally by the processes available to coronaviruses, it remains clear that SARS-CoV-2 is to some degree a pastiche.

The other virus to which Lytras referred was RmYN02, a new one to this discussion. RmYN02 is another bat-borne coronavirus (the *Rm* stands for *Rhinolophus malayanus*, the Malayan horseshoe bat), collected from bats in Mengla County, Yunnan province (YN), in 2019. To be more precise, it's the assembled genome of a virus, pieced together from overlapping segments extracted from eleven fecal samples from those bats. It's the second of two genomes assembled from those samples, ergo RmYN02. The research team, led by Weifeng Shi, included ten other Chinese scientists from Shandong, Beijing, and Wuhan, plus Alice C. Hughes, a British ecologist who has long worked in China and Southeast Asia, plus Eddie Holmes. In early May, 2020, this group posted a preprint, soon afterward published in the journal *Current Biology*, that presented RmYN02 and described two notable things about it. The first was its similarity to SARS-CoV-2. It shared 93.3 percent of its overall RNA sequence with the Covid virus, though differing markedly in its spike; and throughout much of its genome, it was even closer to SARS-CoV-2, at 97.2 percent. Its second notable aspect was that it carried a stretch of three amino acids, right at the hinge of the spike protein, same spot as the furin cleavage site, that seemed to prefigure such a site in SARS-CoV-2. It suggested—as Weifeng Shi and his coauthors had noted—that such a cleavage site was natural in a bat-borne coronavirus.

That's why Spyros Lytras invoked RmYN02 in his August 2020 note on Virological. It was similar enough to SARS-CoV-2 throughout most of its genome, he wrote, that the two viruses would have shared a common ancestral virus in roughly the 1970s. As their lineages diverged, "these viruses must have co-circulated in bats in the same geographical location, and occasionally co-infected the same individuals." That gave opportunity for recombination between them, by the sort of copy-choice error Gallaher had described. Lytras cited Gallaher's palindrome just before the cleavage site in SARS-CoV-2, which he considered "a very likely culprit" for why the copying gizmo had jumped from one strain to another, grabbing a cleavage site from the RmYN02 lineage and patching it into the SARS-CoV-2 lineage. Lytras also coauthored a paper, with a postdoc named Oscar MacLean and some veteran colleagues, including David Robertson as senior author, that went up as a preprint and eventually into a journal. It proposed that SARS-CoV-2 must have shared an ancestor with RmYN02 that infected some bat around 1976, and that the SARS-CoV-2 progenitor continued its evolution in bats, grabbing its furin cleavage site and its receptor-binding domain from other bat viruses along the way. An intermediate stopover in a pangolin or some other host wasn't impossible, they wrote, but "collectively, our results support the progenitor of SARS-CoV-2 being capable of efficient human-human transmission as a consequence of its adaptive evolutionary history in bats, not humans," and then they added an important tag, "which created a relatively generalist virus."

A relatively generalist virus? That means one possibly capable of infecting not just bats but humans, and not just bats and humans but also mink, ferrets, house cats, lions, tigers, snow leopards, gorillas, hippopotamuses, deer mice, and white-tailed deer. And this is the virus we have.

63

Which takes us back to the early critics' third point of dissent from natural origins: that the virus, from its Wuhan debut, seemed too well adapted for infecting humans. Too well adapted, anyway, to be simply a bat virus on a getaway.

The "too well adapted" theme featured in a preprint, posted in early May 2020, by three scientists who had not previously been coronavirus researchers but who knew plenty about genomes. The first author was Shing

Hei Zhan, a genomic analyst then at the University of British Columbia, who had published diversely on animal and plant genomics. The third author was Alina Chan, a molecular biologist and a postdoc at the Broad Institute, a research center affiliated with MIT and Harvard, in Boston. (The second author, credited for advisory conversations, was Benjamin Deverman, Chan's supervisor at the Broad.) Zhan, Deverman, and Chan noted that the original SARS virus of 2003, SARS-CoV, had acquired several adaptations, during the course of that epidemic, which seemed to improve its human-to-human transmission. "Our observations suggest," they then wrote, "that by the time SARS-CoV-2 was first detected in late 2019, it was already pre-adapted to human transmission to an extent similar to late epidemic SARS-CoV." They called it a "human-adapted" form of virus. Other scientists had already wondered whether SARS-CoV-2 might have adapted better to humans during a period of unrecognized transmission in people before December 2019—even Andersen and his colleagues, in the "Proximal Origin" paper, considered that possibility—but this group of three, among whom Alina Chan was the driving force, took it as a premise that the virus was "human-adapted," then asked where and how that had occurred. Was it a natural process during unnoticed circulation in people? Or had a progenitor virus changed while being studied in a lab?

Chan is Canadian, partly raised in Vancouver, but her parents were computer scientists, mobile professionals, and she spent most of her childhood and middle school years in Singapore. She was there in 2003 when the original SARS struck that city, and she remembered news coverage of the quarantined cases, the patients in ICUs, the advisories to the public about avoiding infection. She remembered also her frustrations and bad treatment in the draconian school system, where she and other students were required to bow (literally) to teachers and subjected to corporal punishment—struck with a large ruler, in front of the class—for poor performance or rebelliousness. "I got whacked so many times," she recalled to a writer for *Boston Magazine*.

A vicious circle: she detested the circumstances, she skipped classes, she hung out at the local arcade, she did badly on tests, she was punished. "They were on the brink of expelling me when I was in middle school," Chan told me. Inside the disaffected girl, though, was a fierce intelligence. Chan relished mental challenge, but of a different sort.

"I was a very disengaged child," she said. "I just loved spending hours thinking about puzzles and, like, how to solve them." A percipient teacher recognized that and saw a way to channel young Alina's potential. "She kind of rescued me from the brink of being expelled from the school," Chan said. "By getting me involved in mathematics and puzzles." Chan's grades went from bad to superlative. She became a math wizard; she attended Mathematics Olympiads. "I don't want to boast too much about it. I'm sure my teachers and my classmates thought I was a pain," she said, and then laughed at herself.

Returning to British Columbia, finishing high school there and starting college, she switched to biology and began studying viruses. "Viruses are kind of like puzzles," she told me. Within not many years, she had a PhD, then a postdoc position at the Harvard Medical School, where her work, in a field called synthetic biology, involved creating human artificial chromosomes. A human artificial chromosome (HAC) is a sprig of DNA that can be inserted into human cells to cope with a congenital disease, such as muscular dystrophy, or into experimental cells and animals to make them models for disease research. Her postdoc time in Boston, though, was not altogether satisfactory. "I had some experiences that really made me consider whether or not I should leave academia." By academia she meant doing science, at the lab bench, in a university or an institute. She politely declined to offer details, saying merely, "I was tested, and the conclusion was, I want to stay in science." Hired on a second postdoc, into Ben Deverman's lab, she found the situation far more congenial. Her work there has involved designing vectors, such as artificial viruses, that can deliver genetic payloads into human cells and lab animals toward the goal of ameliorating congenital diseases in people.

Chan's sense of alarm about SARS-CoV-2 began as it began for other attuned scientists: with online reports, on that last night of December 2019 or the early days of January, and then videos of stricken patients in crowded hospitals. "I really started to feel a bit panicked." Others around her dismissed the possibility of any global risk—a new virus, big deal, it'll fizzle away in China—but she took to stockpiling supplies: soap, hand sanitizer, beans, rice, and large quantities of frozen fish. She also scanned the web for scientific information—especially about interactions between the virus and human cells, her bailiwick—and then also about the genomic diversity of the virus as it spread to more and more people. Genomic diversity: at first, she noticed, there seemed to be very little. "I think this was in March," she told me. "The virus was genetically stable." She paused. "That's when, I think, the light bulb went off, that something was weird here with this virus." It was mutating, as all viruses do, but it didn't seem to be evolving—acquiring fixed changes that would suggest adaptation to a new host. That is, the mutations weren't accumulating in new lineages, she asserted, either by sheer happenstance (which is sometimes enough) or because they conferred new advantages and were being favored by natural selection. The virus wasn't chancing upon better ways to infect people—not yet anyway.

What about D614G? I asked. She knew, of course, what I meant: the early mutation that Bette Korber and her colleagues identified, which spread quickly around the world; the one that the Edinburgh trackers called Doug.

"That 614G mutation had already appeared in January," Chan acknowledged, "so it was within the first three months."

"Yeah. Right."

"But it was a single mutation." The original SARS, back in 2003, she noted, had retained many more early mutations than this one. And that difference led Chan to her controversial inference: that this novel virus didn't *need* to evolve because it was already so well adapted for infecting humans.

Stirred to action, she contacted Shing Hei Zhan, a friend from grad school days at the University of British Columbia and an adept computational biologist, asking him to compare the degree of genetic divergence among early SARS-CoV genomes versus early SARS-CoV-2. Zhan looked at forty-three genomes of the former virus and forty-six of the latter and found that, yes, there had been considerably more early divergence in the original SARS virus than in this one. And there was no evidence of divergence among the SARS-CoV-2 genomes sampled from surfaces at the Huanan market, according to the data they saw. As the writer from *Boston Magazine*, Rowan Jacobsen, described her reaction: "Chan's puzzle detectors pulsed again. 'Shing,' she messaged Zhan, 'this paper is going to be insane.'"

In the preprint, posted on May 2, 2020, Chan and her coauthors declared that SARS-CoV-2, as first detected in late 2019, "was already pre-adapted to human transmission" at a level that SARS-CoV (which was also, to some degree, a generalist virus) only achieved later. What could explain that? they asked. Did the progenitor of SARS-CoV-2 spill from an animal into humans, early in 2019, and circulate unrecognized for months? Or was it already well adapted for humans while dwelling in bats or an intermediate host? A third possibility, they wrote, was that a wild virus became well adapted to humans "while being studied in a laboratory" due to some form of intentional manipulation, such as passaging through cultured human cells. That third scenario, they added, "should be considered, regardless of how likely or unlikely."

The preprint was cautious about positing what happened next. The word "leak" appeared nowhere in its text. But that was the implication: once transformed, the virus might have infected a lab worker, by some accident, and gotten loose.

Has your preprint now been published in a journal? I asked Chan. "No," she told me mildly. "The reaction kind of stalled us a little."

In the *Boston Magazine* story, Jacobsen captured a different tone. Two weeks after Chan posted the preprint, he wrote, a British tabloid took note of it, then *Newsweek*. "And that, Chan says, is when 'shit exploded everywhere.'"

64

Alina Chan is not the most doctrinaire among those who suggest that COVID-19 began with a laboratory leak. She argues that such a leak could have happened and, in her view, probably did, not that the case is proven. But she is among the most tenacious and broadly informed of those criticizing the natural origins hypothesis. After the British tabloid picked up on her thinking and the "shit exploded everywhere," she was drawn into that media furor and began to play a large role in arguing for more investigation. The lab leak hypothesis became a subject of keen interest, fevered discussion, amateurish as well as scientific investigation, and wide coverage, especially on social media but also via newspaper op-eds, television, and magazine stories. That discussion continued throughout the rest of 2020, became further inflamed after the ill-fated and inconclusive WHO-convened Global Study of the Origins mission to Wuhan in early 2021, and came to a sort of crescendo that spring, with an outburst of narratives and speculations in popular media. Some of those stories announced that the lab leak hypothesis had gotten more plausible. Safer to say, it had gotten more popular. The controversy may not have sold a lot of newspapers, as such things were quaintly measured in the past, but it certainly commanded a lot of clicks and eyes.

Chan herself, meanwhile, continued to gather facts, arguments, and forms of evidence (circumstantial and otherwise) supporting her view that a lab origin was plausible and worth investigating. She insisted that the world needs further data, deeper research, and better clarity on the origins of SARS-CoV-2, a point with which few thoughtful observers disagree. Late in 2021, when she learned more about coronavirus research done at the Wuhan Institute of Virology, she shifted to the view that a lab origin is not just plausible but more likely than a natural origin. She also published a book, *Viral: The Search for the Origin of COVID-19*, co-authored with the highly respected and provocative British science writer Matt Ridley. Now a declaration of conflicting interest: I have known Matt Ridley a long time, he has been a friend of mine, and I hope he still is. On some matters we agree to disagree.

The Chan-Ridley book asserts, as the Zhan-Deverman-Chan preprint did, that when the novel virus started sending people to hospitals in Wuhan, it was "very likely already well adapted to its new human hosts." True fact: it was indeed. The key question in that regard is whether it was inexplicably, suspiciously, and uniquely human-adapted, or just well adapted period. Let's add some additional data to the pot in which this question is stirred.

65

Humans aren't the only mammals susceptible to infection by, or testing positive for, SARS-CoV-2. There have been instances among quite a few others. The first to ring alarms internationally was a small dog in Hong Kong. On February 26, 2020, a seventeen-year-old male Pomeranian, with a heart murmur, pulmonary hypertension, renal disease, and other secondary conditions, tested positive for the virus. ProMED relayed this news to its global subscribers two days later. Many of us read it and thought, Hmm, odd. By that time, because the Pomeranian's owner had been sick for two weeks and tested positive herself, the dog was quarantined in a government-run facility. Throughout his quarantine period, the dog "remained bright and alert with no obvious change in clinical condition," by one report, but his clinical condition already wasn't too good. Anyway, he survived to bark again. The second known pet was a young German shepherd, also in Hong Kong, also from a household with a human case.

Next it was cats. A group of scientists in Wuhan began promptly in January 2020, as the outbreak among humans made headlines, testing the blood of domestic felines for signs of the virus. They gathered data through March and posted a preprint on April 3. This team included researchers from a college of veterinary medicine, and maybe they were simply following a hunch. They took blood samples from a total of 102 cats, including abandoned creatures harbored at animal shelters, cats at pet hospitals, and cats from human families in which COVID-19 had struck. (They also looked, for purposes of comparison, at thirty-nine cat samples drawn before the outbreak, all negative.) They found evidence of the virus in fifteen cats and, in eleven of those, strong evidence of antibodies capable of neutralizing the virus. "Our data demonstrated that SARS-CoV-2 has infected cat population in Wuhan during the outbreak," they wrote in the preprint. By the time their study appeared in a journal, other cats elsewhere had become infected.

A cat in Belgium tested positive. A cat in France tested positive. Another study from China, done by experiment at a veterinary institute in Harbin, in the north, showed that cats inoculated with SARS-CoV-2 became infected and could transmit the virus to other cats through the air. A cat in Hong Kong tested positive. A cat in Minnesota, a cat in Russia, two cats in Texas. In Italy, as I've already described, Gabriele Pagani's cat Zika began sneezing, then tested positive, evidently having caught the virus from him. In Germany, a six-year-old female cat at a retirement home in Bavaria tested positive by throat swab after her owner died of COVID-19. In Orange County, New York, just up the Hudson River from New York City, a five-year-old indoor cat started sneezing, coughing, draining from her nose and eyes, about eight days after her person developed similar symptoms. She tested positive.

Domestic cats aren't social creatures in the ecological sense; they don't aggregate in dense populations (except amid the pungent households of obsessive cat hoarders and overgenerous rescuers), so the opportunities for cat-to-cat transmission tend to be low. But many a house cat, once it gets outside, interacts with mice in the barn, the shed, or the backyard. Those mice generally belong to two groups, house mice (*Mus musculus*) and deer mice (several species within the genus *Peromyscus*). Deer mice are well documented as hosts of hantaviruses and the Lyme disease bacterium, and recent laboratory work shows them susceptible to infection with SARS-CoV-2. A mouse can carry the virus for as long as three weeks and transmit it efficiently to other mice. Deer mice are the most abundant (nonhuman) mammals in North America. It may be only a matter of time before SARS-CoV-2 gets into a population of deer mice, from a cat, and begins mouse-to-mouse transmission in the wild. More on this theme, below, when we get to the mink and the white-tailed deer.

Among felids infected with SARS-CoV-2, it hasn't been just the domestic kitties: a tiger named Nadia, at the Bronx Zoo in New York, appeared sick and tested positive for the virus, presumably transmitted by one of her zookeepers. It seems she wasn't alone. According to a statement from the Animal and Plant Health Inspection Service (APHIS, within the U.S. Department of Agriculture), Nadia's testing came after several lions and other tigers at the zoo showed signs of respiratory distress. Within weeks, four more of the Bronx tigers and three lions tested positive. A puma (an American cougar) at a zoo in South Africa tested positive. A female snow leopard and two males, at the Louisville Zoo, in Kentucky, started coughing and wheezing, then tested positive.

In the Netherlands, during the spring of 2020, SARS-CoV-2 began showing up among farmed mink. Those outbreaks carried large economic consequences as well as public health implications, because the mink were held in crowded conditions, raised in the thousands for their fur, and they proved very capable of transmitting the virus, both from mink to mink and possibly (with human help) from farm to farm. The first detected cases occurred on two farms in the province of Noord-Brabant, which is in southern Netherlands along the Belgian border. "The minks showed various symptoms including respiratory problems," according to a statement from the Ministry of Agriculture, Nature and Food Quality. Several roads were closed, and a public health agency advised people not to walk or cycle in the vicinity of those farms. But the virus spread quickly, soon affecting ten farms, then eighteen farms, then twenty-five farms by the middle of July 2020. The Netherlands contained a lot of mink: roughly 900,000 animals at 130 farms. These were American mink (*Neovison vison*), like virtually all farmed mink, preferred for the richness of their fur; they belonged to the mustelid family, which includes also the pine marten, the European polecat, and the Eurasian badger. Dutch exports of mink fur earned about 90 million euros annually in recent years, according to the Dutch Federation of Pelt Farmers. The industry was controversial—fur farming of all sorts is controversial in much of Europe, on grounds of animal welfare—and the Netherlands had already moved toward ending it by 2024. Now that happened more quickly, under government orders to cull all the animals on affected farms, in advance of the usual November doomsday for farmed mink, and not to restock. By the end of June 2020, almost 600,000 Netherlands mink had been slaughtered. The virus wasn't innocuous in mink; it caused respiratory symptoms and some mortality, which was what triggered testing and detection of the virus on those first two farms. But it didn't kill mink as quickly as the culling did.

A team of Dutch scientists investigated the outbreaks, between April and June, and eventually published a paper in *Science*. The senior author on that study was Marion Koopmans, head of virology at the Erasmus Medical Centre in Rotterdam, whom I've already mentioned. "In February, because of the dog infection in Hong Kong," she told me, "we had a meeting." Koopmans, an expert on zoonotic viruses, interacts regularly with the National Public Health Institute, the Veterinary Health Institute, and an independent organization, farmer-supported, called the Animal Health Service. By late spring, everyone was aware that SARS-CoV-2 had appeared not just in one or two Hong Kong dogs but also in domestic cats, tigers, and lions. Among humans, it was raging in Italy, and the Netherlands was suffering its first wave, with almost forty thousand cases by the end of April and a gruesomely high case fatality rate. "We were ramping up human diagnostics," Koopmans said—the labs at her center, as well as veterinary labs in the system. Then came a couple of dead mink, submitted for necropsy. "And I said, 'Hey, well, what the heck. Let's also test these mink.'" It was done at the same veterinary lab that had jumped in to do human diagnostics. Bingo.

As the mink outbreaks turned up on one farm after another, and the human pandemic intensified, Koopmans and her colleagues found time and resources to study the animal phenomenon, which would have implications for public health as well as for the fur industry. They sampled both mink and people on sixteen farms, finding not just lots of infected mink but also eighteen infected people among farm employees and their close contacts. The team sequenced samples and saw that the viral genomes in people generally matched the genomes in that farm's mink. This and other evidence suggested not just human-to-mink transmission starting each outbreak, and mink-to-mink transmission keeping the outbreaks aflame, but also possibly mink-to-human transmission. That last point was ominous and I'll return to it.

In mid-June, it was Denmark's turn. "A herd of mink is being slaughtered at a farm in North Jutland after several of the animals and one employee tested positive for coronavirus," according to a report in *The Local*, an English-language online media service. That farm was quarantined, and all eleven thousand animals would be killed. The news fell heavily because Denmark, with roughly fourteen million mink on more than a thousand farms, produced a large portion of the world's pelts, and the quality of Danish pelts was considered supreme. The virus spread quickly that summer. By early October, forty-one Danish farms had recorded outbreaks and authorities spoke of culling a million mink. This was optimistic. By mid-October: sixty-three farms and plans for culling 2.5 million mink. But that too was just a beginning.

In the meantime, health officials in Spain ordered the culling of 93,000 mink on one farm, after determining that "most of the animals there had been infected with the coronavirus," according to Reuters. Mink tested positive on a farm in Italy. In Sweden, a veterinary official visited a mink farm on the southern coast, reporting, "We tested a number of animals today and all were positive." Mink at two farms in Utah tested positive, and then came some worse news. Veterinary officials from the U.S. Department of Agriculture revealed that a wild, free-ranging mink in Utah had also tested positive. The sequenced virus from that wild mink matched the virus in mink on a farm nearby, so the wild individual had presumably been infected by an escapee—or by schmoozing with captives nose-to-nose through a fence. This raised a concern well beyond the economics of fur: the prospect of SARS-CoV-2 gone rogue into the American landscape. In the lingo of disease ecologists: a sylvatic cycle.

That term comes from the Latin word *sylva*, meaning forest. A virus with a sylvatic cycle is two-faced, like a traveling salesman with another wife and more kids in another town. Yellow fever virus, for example: transmitted by mosquitoes, it infects humans in cities (the urban cycle) when the right mosquitoes are present, but it's broadly enough adapted to infect monkeys also, and it does that in some tropical forests (the sylvatic cycle), circulating in monkey populations. Yellow fever can be eliminated in cities by vaccination and mosquito control, but whenever an unvaccinated person goes into a forest where the virus circulates, that person can become infected, return to the city, and trigger another urban cycle, if some mosquitoes are still there to help. Yellow fever virus has never been eradicated, and travelers to many tropical countries are still obliged to be vaccinated, because the sylvatic cycle will persist, and threaten another urban cycle, until you kill every mosquito or vaccinate every monkey.

Now transfer the concept to SARS-CoV-2 and consider: if the world's forests or other natural ecosystems contain populations of wild animals in which that virus circulates, either because they are the original reservoir hosts (horseshoe bats in southern China?) or because they have become infected by contact with humans (mink in Utah? deer mice in Westchester County?), then there is no end to COVID-19. (There is probably no end to it regardless, but that's another matter, to which I'll return.) There is no herd immunity where there is a sylvatic cycle. An unvaccinated person has contact with an infected wild animal (a mink, a cougar, a monkey, a deer mouse) during some activity (hunting, cutting timber, picking fruit, sweeping up urine-laced dust in a cabin) and becomes infected with the virus, potentially triggering a new outbreak among people. You could vaccinate every person on Earth (that's not gonna happen) and the virus would still be present around us, circulating, replicating, mutating, evolving, generating new variants, ready for its next opportunity.

The chance of a sylvatic cycle in Europe, possibly also derived from mink, is elevated by the fact that many mink escape from farms—a few thousand every year in Denmark alone. Although not native to the European continent, these American mink have established themselves as an invasive population in the wild, their presence reflected in the numbers taken by hunters and trappers. About 5 percent of the farmed Danish mink that escaped in 2020, by one expert's estimate, were infected with SARS-CoV-2. Mink tend to be solitary in the wild, but obviously they meet to mate, and as both predators and prey within the food chain, they come in contact with other animals. Atop the list of other creatures that might be susceptible to a mink-borne virus are their wild mustelid relatives, the pine marten, the European polecat, and the Eurasian badger.

On November 5, 2020, another bit of disquieting news came out of Denmark. The government announced severe restrictions on travel and public gatherings for residents of North Jutland—that low and tapering island curled like a claw toward southwestern Sweden—after discovery that a mink-associated variant of the virus, containing multiple mutations of unknown significance, had spilled back into humans. Twelve people had it. This variant became known as Cluster 5, because it was fifth in a series of mink variants; but it was the first to be detected in humans. It carried four changed amino acids in the spike protein, raising concern that it might evade vaccine protections when vaccines became available. That's it, we're done, said the government statement: all remaining mink would be culled. The mink industry in Denmark was over.

But the rigorous shutdown, the tracing of cases, and the other control measures pinched that variant to a dead end. Within two weeks, a Danish research institute announced that the Cluster 5 lineage seemed to be extinct, at least among humans. Whether it survived in the wild, among escaped mink or their native relatives on the Danish landscape—pine marten, European polecat, Eurasian badger—is another question.

Through the last months of 2020 and well into 2021, reports of SARS-CoV-2 in nonhuman animals continued, sporadic but notable. A tiger at a zoo in Knoxville, Tennessee, tested positive. Four lions of the beleaguered Asiatic population, at a zoo in Singapore, started coughing and sneezing after contact with infected zookeepers. Two gorillas, also coughing, at the San Diego Zoo Safari Park. The two gorillas recovered within weeks, although not before one animal, a forty-eight-year-old silverback with heart disease named Winston, had been treated with monoclonal antibodies. Winston also got cardiac medication and, as a precaution against secondary infection with bacteria, some antibiotics. If he had been a wild gorilla in an African forest, without a Cadillac health plan, he might well be dead. Then again, if he had been a wild gorilla, free of zookeepers, he probably wouldn't have caught this virus.

In October 2021, SARS-CoV-2 reached the Lincoln Children's Zoo, in Lincoln, Nebraska, infecting two Sumatran tigers and three snow leopards. This zoo proclaims a mission to enrich lives, especially children's lives, through "firsthand interaction" with wild creatures, under controlled and educational circumstances. It's a meritorious goal, but as we've all learned, close encounters in the time of Covid carry risks. These snow leopards were less lucky than the three in Louisville a year earlier. In November, despite treatment with steroids, and antibiotics against secondary infection, all three died.

Meanwhile, of course, people were dying too. By the 31st of October 2021—the second Halloween of the pandemic—the state of Nebraska had recorded 2,975 Covid fatalities. For the United States on that date, the cumulative toll was 773,976 dead. Throughout the world, SARS-CoV-2 had killed more than five million humans. In the small nation of Belgium, with a total population less than twelve million, one person in ten had been infected with the virus, the curve was rising steeply, and 26,119 people had died.

In December, also in Belgium, two hippopotamuses at the Antwerp Zoo tested positive. They were luckier than the Nebraska snow leopards or the 26,119 dead Belgians, showing no symptoms beyond runny noses (more runny than usual for hippos), but were put into quarantine.

Other news in late 2021 brought the prospect of a sylvatic cycle from possibility to reality. Scientists at Penn State University, working with colleagues at the Iowa Wildlife Bureau and elsewhere, reported evidence of widespread SARS-CoV-2 infection among white-tailed deer in Iowa. Experimental studies had already shown that captive fawns, inoculated with the virus, could transmit it to other deer. This new work went much further, revealing that wild deer had become infected, somehow, from humans—and not just a few deer. SARS-CoV-2 was rampant throughout the Iowa deer population. That trend began slowly, after the beginning of the pandemic, but by the final months of 2020 it was overwhelming.

The team's trained field staff collected lymph nodes from the throats of almost three hundred deer, mostly free-living animals on the Iowa landscape, a lesser portion contained within nature preserves or game preserves—none of them artificially infected by experiment. The sampled deer had been killed by hunters or in road accidents by vehicles. The field staff dissected out the lymph nodes, in connection with an ongoing surveillance program for another communicable illness, chronic wasting disease. The deer sampled early in the study, during spring and summer 2020, were clean of SARS-CoV-2. (Iowa's initial wave among humans rose in April.) The first positive animal didn't turn up until September 28, 2020. After that, it was like popcorn in a hot pan. Over a seven-week period during hunting season, in late 2020 and early January 2021, the team sampled ninety-seven deer, among whom the positivity rate was 82.5 percent. The research continues, with a second phase of sampling, and if that percentage holds anywhere near steady (confidential updates suggest it will), it's startling evidence of sylvatic SARS-CoV-2 in Iowa.

Iowa is not alone. A different study, done by federal wildlife officials from APHIS, looked for the virus among white-tailed deer in four other states, using blood serum samples rather than lymph nodes. These samples dated from early 2021. Illinois's deer were the most Covid-free, with only a 7 percent rate of infection. If you had announced that statistic alone, at the time, it would have seemed shocking. *Seven* percent of Illinois deer have Covid? But among whitetails sampled in New York, the rate was 31 percent infected; in Pennsylvania it was 44 percent; in Michigan, it was 67 percent.

The United States presently contains an estimated 25 million white-tailed deer, and no one has informed them that SARS-CoV-2 is uniquely, peculiarly well adapted for infecting humans.

66

Two other topics invoked in critiques against the natural origins hypothesis are gain-of-function experiments and the Mojiang mineworkers. These deserve to be considered independently, though they are often muddled together. The Mojiang story has echoed widely, in part for its asserted importance—to which I alluded earlier, in discussing Zhengli Shi's work—and in part because it's a vivid and creepy narrative.

The underground site now famed as "the Mojiang mine" was, as of 2012, an abandoned copper mine within Tongguan Township, Mojiang County, in China's Yunnan province, about two hundred miles southwest of the city of Kunming, Yunnan's capital. It's a hilly and partly forested area, not far from the northern borders of Laos and Vietnam. In April 2012, evidently because someone decided to reactivate the mine operation, a group of workers were sent down to clear large amounts of accumulated bat guano from its tunnels, deposited over decades by bats of several species that roosted in the mine. Six of the laborers, after working for periods between four and fourteen days, fell sick with an unidentified form of pneumonia, their symptoms including cough, fever, chest pain, labored breathing, and (in one case) a secondary condition of chronic hepatitis. They were treated at a hospital attached to the medical university in Kunming. Three of them died, including the hepatic patient. The other three recovered, but only after long stints of hospitalization. These facts come mainly from a 2013 master's thesis by one Li Xu, for a degree in Clinical and Emergency Medicine at Kunming Medical University. Several people have translated sizable portions of the thesis into English, including Alina Chan, alerted to its existence in May 2020 by an anonymous source on Twitter, and my friend Wufei Yu, a journalist from Beijing now based in the U.S., who reviewed and additionally translated a version for me. I've seen three versions, counting Wufei's, and in each "it is inferred" that the six cases "may be caused by viral infections." The thesis concludes—based on a consultation with Nanshan Zhong, a leader of China's response to the 2003 SARS crisis, along with some ambiguous antibody evidence—that the infectious agent was a SARS-like coronavirus from a bat. It mentions the Chinese rufous horseshoe bat, though the author seems unaware that at least five other kinds of bat roosted in that cave. This conclusion, about a killer virus, could be correct—or not. Last on a list of clinical and research "deficiencies" that should be fixed, for the sake of future such situations, Li Xu noted that "it is of great significance to sample bat feces and live bats in the mine."

By the time Xu's thesis was written, Zhengli Shi had already begun doing that. In late summer of 2012, three months after the first of the mineworkers died, she shifted some of her field efforts from caves elsewhere in Yunnan to the Mojiang mine. Her team returned to Mojiang again in April and July 2013, and during those 2012–2013 expeditions they took 276 fecal samples from among bats of six different species. (They made further visits to Mojiang in 2014–2015, but the 276 samples were analyzed as a bundle.) About half of those samples tested positive for some sort of coronavirus, and in a few cases, more than one virus per bat. Shi's group did partial sequencing, targeted to pull out a certain short stretch—about four hundred letters of RNA code for a crucial gene—from each sample. The crucial gene was *RdRp*, coding for RNA-dependent RNA polymerase, which is an enzyme that allows the virus to replicate its RNA within a host cell. A sequence of *RdRp* can signal the identity of its bearer as reliably as a fingerprint. Most of the *RdRp* sequences retrieved by Shi's team indicated the presence of alphacoronaviruses, a group that includes two of the common-cold coronaviruses, but no viruses known to be serious in humans. They also found two sequences indicating betacoronaviruses, more interesting, because that group includes SARS-CoV and MERS-CoV. Betacoronaviruses suggested more likelihood of danger to humans, so Shi's team focused particular attention on those two.

The researchers labeled one of them sample 4991, as I've mentioned. It came from an intermediate horseshoe bat (*Rhinolophus affinis*) so its full tag was RaBtCoV/4991. (I wouldn't trouble you with this labeling business if it hadn't become a matter of some perplexity and heated disputation amid the arcana of argument about SARS-CoV-2.) The *RdRp* sequence from sample 4991 was 440 letters long, representing less than 2 percent of a complete coronavirus genome. It wasn't as close to the human SARS virus as some others among the SARS-like coronaviruses, just similar enough to be notable. And the SARS virus at this time, 2012–2013, was the standard for what a menacing coronavirus might look like, therefore similarity to that virus influenced the weighting of any new find. The 4991 sequence seemed relatively insignificant. Shi, who by this time had coauthored papers in *Nature*, *Science*, and other leading international journals, published this study in *Virologica Sinica*, the house journal of the Wuhan Institute of Virology. The main takeaway of the paper was that some bats in this derelict mine carried "co-infections," more than one coronavirus at a time, "a phenomenon that fosters recombination and promotes the emergence of novel virus strains."

Sample 4991 would attract far more attention later, after Shi's group took it out of the freezer, retrieved almost a full genome sequence from it, and labeled that sequence RaTG13. This label conveyed information, again as I've mentioned, that a four-digit number didn't: Ra for *Rhinolophus affinis*, the bat species from which the sample came; TG for Tongguan, the township within Mojiang County in which the mine was located; 13 for 2013, the year of collection. RaTG13 was the sequence that became famous in January 2020, when Shi and her colleagues announced that they had evidence of a coronavirus, from a bat, that was a 96.2 percent match to the novel virus causing the strange and alarming pneumonias. Critics of Shi and her work, and of the hypothesis that SARS-CoV-2 emerged naturally from an animal, have construed this matter of labels as evidence of guilty concealment.

Zhengli Shi explained it differently during my two-hour Zoom conversation with her. "After we got the RNA-dependent RNA polymerase, we compared it with SARS-CoV-1 and we found that these virus are distantly related to SARS-CoV-1." She meant the two *RdRp* sequences from the betacoronaviruses in the Mojiang mine samples, including the one that has gotten such attention. "It has a simple ID number. It's 4991," she said. "The virus naming is complex," she added, and the more virus samples she and her group collected, the greater the need for orderly, illuminating names. "You know, at the beginning we only have a hundred sample. But then later on we have ten thousand sample." They devised an improved convention. "We decided to name some sequence—we think, some important sequence—based on the bat species, the sampling location, and the sampling year." Ergo, the sample number 4991 gave way to the name of a full sequence, RaTG13. "That's something a little bit confusing," she conceded. "But you know"—and she paused to chuckle, I think in frustration—"we didn't consider it to *make* it confusing."

One other point of confusion needs clarifying. Neither the sample 4991 nor the sequence RaTG13 is a virus. A sample from a bat is a tiny smudge of feces, in which fragments of DNA and RNA may be contained: DNA from the bat itself, DNA from bacteria, DNA or RNA from any number of viruses the bat may carry. A viral sequence obtained from such a sample is the genomic representation of one or more of those virus fragments—either a short sequence, such as the 440 letters of *RdRp*, or a long sequence, patched together from overlapping fragments, representing the entire (or nearly entire) genome of a virus, as RaTG13 does. I'll repeat: *representing* an entire virus. RaTG13 is not a virus, just as the text of *Hamlet, Prince of Denmark* is not a performed play. Laurence Olivier is missing. There's no greasepaint, no costumes, no special effects for the ghost, no rapiers. The text is just words on a page—dramatic words, timeless words, but still only a script, not a performance. Likewise, RaTG13 is the script of a virus. To capture a virus in its wholeness, a *live* virus, requires entirely different techniques. You have to grow it within cells in a culture. That's not easy. The poop from a bat is no ideal environment for the survival of intact, viable viruses. Most attempts at culturing live virus from guano samples fail.

With 4991, Shi's efforts failed. "We couldn't culture any of the sample from this cave at Mojiang," she told me. From that mine, she repeated, "We never culture any coronavirus."

Here's why it matters. Here's why so much attention has been paid to the Mojiang mine, and to the three workers who died in 2012. Some commenters argue that the three men were killed by a virulent coronavirus (which is possible); they propose that Zhengli Shi brought a sample containing that virus or one very much like it back to her Wuhan lab and grew it in cell culture (which she has denied, to me and others); or that, perhaps, she reverse-engineered that (hypothetical) Mojiang killer virus from a complete genome, by expressing the genome through a cell (which she has denied); and that she then allowed it to leak from her lab (which she has denied). Apart from Shi's denials, which can be credited or not, there's a problem with this whole set of scenarios. RaTG13 is not SARS-CoV-2.

Being 96.2 percent similar at the level of nucleotides, as it is, RaTG13 differs by 3.8 percent. Given the pace at which coronaviruses generally mutate and evolve, that reflects about fifty years of evolutionary divergence. RaTG13 differs from the baseline virus first detected in Wuhan (known as Wuhan-Hu-1, the sequence released by Zhang and Holmes) at about 1,150 nucleotide positions, and those positions are scattered throughout the genomes. Some of the world's best evolutionary virologists (professionals in that field, not amateurs visiting it) and coronavirus experts, including Susan Weiss, Stanley Perlman, David Robertson, Robert Garry, and Kristian Andersen, assure us that RaTG13, with or without lab manipulation, is not the answer to the origin question for SARS-CoV-2. So the story of the Mojiang mine and the three dead workers in 2012, though it contains some vivid narrative elements, and has great appeal to certain minds, is probably irrelevant.

67

Gain-of-function research is the second topic aswirl amid arguments for the lab leak hypothesis. In case your television has been off for the past three years, and your computer has been locked on Netflix, here's a basic definition: gain-of-function (GOF) work is any sort of laboratory experiment that increases some biological capacity of an organism. More specifically, "gain-of-function research of concern" (GOFROC) is work on a pathogen (viral or whatever) with pandemic-causing potential that might make it more able to infect humans, to transmit among them, or to cause them greater harm.

The rationale for such work is that it can help scientists anticipate, understand, and prepare for bad luck. It can portend what a dangerous pathogen might look like, and how it might behave, if evolution in the wild happens to make the pathogen more dangerous. That rationale is controversial. Skeptical scientists, including some very sensible and moderate ones, oppose gain-of-function work, or at least certain forms of it, on grounds that giving a pathogen increased capability of any sort is always a lousy idea, because the thing might leak out of a laboratory or be used as a biological weapon. But the phrase itself, "gain-of-function," encompasses some ambiguity. "What we mean by the term," according to Gerald Keusch, associate director of the National Emerging Infectious Diseases Laboratories (NEIDL) at Boston University, speaking to a reporter from *Nature*, "depends on who's using the term."

Keusch is a longtime professor of medicine and infectious disease expert, a former director of the Fogarty International Center at the NIH, and coauthor (with Nicole Lurie) of a 2020 report on public health emergencies to the Global Preparedness Monitoring Board, an arm of the WHO and the World Bank. He has always been a physician-scientist. In the huge lab complex he helps run, NEIDL, researchers study Ebola virus and other menacing microbes within a bio safety level 4 lab. When I asked Keusch about gain-of-function work, he took a breath and answered with a boomerang trajectory. "My mother wanted me to be a GP in the Bronx," he said. Graduating from Columbia University and then from the Harvard Medical School, he chose a different path. "If I stayed in medicine, I wanted to know how things work, so I could then rationally address it. And I wanted to address it both in my research and in the way I practice medicine." That meant doing science as well as seeing patients. "We need to understand how these viruses work." Evolutionary genetics, by experiment as well as observation, helps illuminate them. "The more we understand," Keusch said, the better we can be "at anticipating what might happen. The better we can do at looking for the signs of a troublesome evolutionary path of these viruses. The more we can do in advance." It's a dimension of preparedness, leading toward therapeutics, vaccines, and other forms of response to public health emergencies.

"That's an innocent view of the world," Keusch added. "There is a darker side to the world and that darker side is what gets amplified in the conspiracy theories." The phrase "conspiracy theory" has been a sore point to proponents of the lab leak hypothesis—who insist that they're talking about accident and concealment, not premeditated scheming to do harm—so I'll hasten to add that what Jerry Keusch meant here was the latter: intentional bioterrorism. "There will always be those few who would use science for domestic terror, international terror, for whatever dark reasons," he said. You can't let that hamstring scientific research, basic or applied. "You can't operate in a world where your entire focus then stops you from doing all of the good because of this possibility of the bad."

The core issue over gain-of-function research is whether it creates what policy people call potential pandemic pathogens (PPP, another baleful string of letters for your mental toolbox). A potential pandemic pathogen is a bug that's highly transmissible, capable of uncontrollable spread among humans, and could cause widespread illness and death. By that definition, the first milestone of PPP work in the twenty-first century came in 2005, when a team of researchers from the U.S. CDC and elsewhere reconstructed the 1918 influenza virus. They did that by assembling its genome from old autopsy specimens and some frozen lung tissue of a victim buried in the Alaska permafrost, then activating that into a live virus by reverse genetics, expressing the viral genome in cultured cells. This was controversial. One scientist called it "a recipe for disaster." The researchers defended their work on grounds that reviving the virus and studying it in a secure lab not only illuminated why that virus had been so deadly but also led toward important insights about influenza viruses generally, with relevance to vaccine development, antiviral treatments, and prediction of virulence in other viruses.

The second milestone also involved influenza. In 2011, a tall Dutch virologist named Ron Fouchier announced at a conference in Malta that he and his colleagues had created a version of the highly virulent H5N1 avian influenza that was transmissible, not just among birds (like most avian influenza) but among mammals, and not just by direct contact but through the air. They did it by generating mutations in the virus and then passaging it through a series of ferrets. In their first experiments, they placed infected ferrets together with clean ferrets in the same cage. The clean ferrets became infected. Later, after their virus accumulated mutations, they put a clean ferret in a cage near, but separated from, another cage with an infected ferret. The ferrets couldn't touch each other but there was airflow between the cages. The clean ferret became infected. That happened three out of the four times they tried it. What Fouchier's group had learned was that H5N1 bird flu could become transmissible among ferrets, in airborne or respiratory droplet form and therefore very possibly also that way among humans, by way of mutations that changed as few as five amino acids in the protein construction of the virus. Four of those changes were in the hemagglutinin protein (represented by the H in H5N1), which is the attachment and fusion protein, equivalent to the spike protein in a coronavirus. It's important to note that those mutations weren't the only possible way by which H5N1 might become airborne- or droplet-transmissible. They were exemplary, not exhaustive. So the Fouchier team, by creating such a virus, had identified one version of what to watch for, one version of what might occur in the genome and yield a human-transmissible H5N1 bird flu virus. Some scientists considered this valuable work such as should be done with the utmost caution, and some scientists considered it wildly reckless. Critics also argued that publishing the methodology of this experimental work was like offering a blueprint to bioterrorists.

The result was a serious international discussion of gain-of-function research and, in the United States, a partial moratorium on GOF research, which lasted from 2014 until 2017. The National Institutes of Health suspended its funding of such work. The moratorium was precipitated not just by the Fouchier work, and by similar work on H5N1 at the University of Wisconsin, but also by recent laboratory blunders that involved mishandling of dangerous pathogens (anthrax bacteria that were not properly inactivated, frozen smallpox virus that was supposed to have been destroyed), legitimately concerning but not consequent from gain-of-function research. The blunders reminded everyone that laboratory mistakes and accidents do happen. During the moratorium, two symposia convened scientists from around the world to discuss the risk/benefit balance of GOF work and how to gauge that balance and regulate it. The second symposium made recommendations to the U.S. government, new oversight policies were developed, and in late 2017 the NIH lifted the moratorium. Francis Collins, the NIH director, announced the decision, saying: "GOF research is important in helping us identify, understand, and develop strategies and effective countermeasures against rapidly evolving pathogens that pose a threat to public health."

The new policy framework still has its critics, among whom is David Relman, a microbiologist at Stanford University and a former member of the U.S. National Science Advisory Board for Biosecurity. Relman attended the first GOF symposium, convened in December 2014 by the National Research Council and the Institute of Medicine; he remains dissatisfied with the scope of the framework and the way GOF grants are reviewed. Like Jerry Keusch, he came out of the Harvard Medical School and became a physician-scientist, his specialty being the human microbiome. When I visited Relman in 2015 for an interview toward a different book, he was a youngish sixty-year-old with a mop of brown hair going gray, a genial manner, and a mountain bike parked in his office. Having fought the Wednesday traffic into Palo Alto myself, coming up from Gilroy (the famous Garlic Capital of the World) fifty miles away, where I had found the nearest available motel room, I well understood why Relman would commute to work on two wheels. Getting there by Zoom, more recently, was much easier.

Toward the end of our latest conversation, which ranged across his scientific background, his reactions to early news about a novel virus in January 2020, his views of the Mojiang mineworkers' story and the WHO mission to Wuhan, we discussed gain-of-function research. Some scientists oppose it categorically, I noted—they say it's a terrible idea, in any form. To what extent did he agree?

That's an extreme position along a spectrum on which he lay "someplace closer to the middle," he said. "It's not just because I'm feeling wishy-washy." It was because, Relman said, there are some very important nuances that get lost amid the easy language of the discussion.

That bit in particular. "'Gain-of-function' is a bad term in a way," Relman told me, agreeing with Keusch on this point, "because it glosses over a bunch of stuff that to me is quite distinct." Yes, we have a moral obligation to understand the world around us, he said. The park ranger can't manage her park without an inventory of what's there. That's her park manual. Without it, she can't be a good steward. Where research becomes problematic, he continued, "is when you start to say, 'I'm not just going to understand, or at least recognize, what's there, but I'm going to tinker with it in ways that are predictably more risky than other ways.'" Relman well knew, of course, that experimental science, by definition, is tinkering. Genetic manipulation is a special sort of tinkering (especially presumptuous, you might argue) but it's done every day in laboratories all over the world, and it yields vast benefits to human health—even benefits to other creatures and to ecosystems, in some cases. What's up for debate are the proper boundaries for the tinkering, and the potential value of results versus the possibility of unintended harm. Do you create the very thing that you most fear, with hopes that you might learn something useful? Is that judicious or foolish? In other words, back to risk/benefit analysis. Measuring each, by the powers of foresight, is difficult and contentious.

Just judging what *is* and what *is not* GOF work, under a standard definition, applied to the intricacies of molecular virology, is no trivial task. That's why Tony Fauci, the immunologist from Brooklyn, said to Rand Paul, the ophthalmologist from Kentucky, while testifying under oath on July 20, 2021, "Senator Paul, you do not know what you are talking about, quite frankly. And I want to say that officially. You do not know what you are talking about."

Relman called my attention to a journal paper, published in 2017 by Zhengli Shi and many coauthors, including Linfa Wang of Duke-NUS in Singapore and Peter Daszak of EcoHealth Alliance in New York. I had already read it, but it bore reading again. This was the same paper over which Fauci and Paul had argued—Fauci had even held up a copy of it, as he rebutted Paul's accusations. The long title, "Discovery of a Rich Gene Pool of Bat SARS-Related Coronaviruses Provides New Insights into the Origin of SARS Coronavirus," reflects the broad range of work it describes, and the fact that in 2017 the origin of the SARS virus of 2003 was still a matter of scientific uncertainty. The paper's first author was a young scientist named Ben Hu, so in scientific shorthand it will forever be known as the "Hu et al. (2017)" paper. Critics of the work have mainly targeted Shi, as senior author, as well as Daszak, because some funding support came through EcoHealth Alliance from an NIH grant, and Fauci, because his institute within NIH, the NIAID, made the grant. Shi and Hu and their colleagues reported five years of field sampling among multiple species of bat in a cave (unnamed in the paper, but it was Shitou Cave, as I mentioned earlier) in Yunnan. One salient point in this paper is the detection, by sequencing from samples, of eleven new strains of SARS-related coronavirus circulating among four different kinds of bat in the cave. None of those eleven seemed to be the direct, sole progenitor of the original SARS virus. But the progenitor could have arisen by recombination among several SARS-like viruses found in that cave or another.

That alone was big news: the reservoir host of the 2003 SARS virus had probably been identified, after fourteen years. It was a horseshoe bat in a virus-rich cave in Yunnan, either Shitou itself or one like it. But the paper contained another key point, destined to be more contentious four years later, in the era of SARS-CoV-2. This involved three of the newly detected viruses: they would likely be capable of infecting humans, Hu and his colleagues reported, judging from the capacity of their spikes to attach to human ACE2 receptors in the lab. "Thus, the risk of spillover into people and emergence of a disease similar to SARS is possible," the authors warned.

The experimental work on those three new viruses became the main point of contention, in the context of the pandemic, construed by critics as dangerous gain-of-function research. "That's the experiment I wouldn't do," Relman told me. It was *not* considered GOF work by Fauci or other scientists—including "qualified staff up and down the chain" at NIAID who reviewed the grant, as Fauci told Rand Paul. This disagreement reflects what both Relman and Keusch said about the ambiguity of the "gain-of-function" term, and it merits explaining.

What the Wuhan team had sought to learn was whether those three novel viruses could use the now famous receptor, ACE2, as a point of attachment for infecting human cells and then replicate within such cells. To understand what the team did, it's necessary to remember that they did not possess two of those three viruses. What they possessed were the genome sequences. They had the scripts, not the performances. They had *Hamlet* and *Titus Andronicus* but only on paper. They called those two scripts Rs4231 and Rs7327, the Rs in each case standing for *Rhinolophus sinicus*, the Chinese rufous horseshoe bat. From their sample containing the third sequence, Rs4874, they did manage to grow live virus. Their attempts to grow virus from the other two sequences, Rs4231 and Rs7327, failed.

So they devised a work-around. Using the spike protein sequence from each of those two viruses, along with the backbone sequence from a coronavirus they had earlier succeeded in culturing, called WIV1, they generated hybrid viruses. Previous work, by other scientists, had established that WIV1 could enter human airway cells by way of the ACE2 receptor and, once inside, could replicate efficiently. The question was whether the Rs4231 and Rs7327 viruses, out there in the wild, were capable of the same feats. Shi's team tried those viruses against monkey cells in culture and got more replication of live virus. Then they tried the hybrid viruses against cultured human cells, both with and without the ACE2 receptor. In human cells without ACE2, nothing. In human cells with ACE2, the viruses entered and replicated efficiently. What this told Shi and her colleagues was: beware. Out there in Shitou Cave, and possibly elsewhere, lurked two wild coronaviruses, corresponding to the Rs4231 and Rs7327 genomes, each bearing a spike protein that gave it potential to infect humans. Here were two more coronaviruses that could spill over and cause outbreaks, or worse. Their finding "highlights the necessity of preparedness for future emergence of SARS-like diseases," they wrote.

Had they created a dangerous virus that didn't exist in nature? This is the crux of the argument, but a reasonable answer is: no. They had assembled hybrids, combining elements of two potentially dangerous viruses that already *did* exist in nature with the backbone of another virus, WIV1, that also existed in nature. By instantiating these hybrids, Shi's group sought to test and confirm that both Rs4231 and Rs7327 represent threats to humans. And now I'll leave Shakespeare to the peace of his grave and give you another analogy: these viruses are like the leopards of Mumbai.

68

The leopards of Mumbai are a few dozen large, powerful felids that live within the seventh-largest city in the world. Mumbai contains twelve million people in the city proper, with a population density of roughly 73,000 people per square mile, among the highest anywhere on Earth. Within that vast aggregation of humans and buildings and roads and vehicles live, by one recent count, forty-seven leopards. The leopard numbers fluctuate slightly with births and deaths, of course, but their presence has long been a given, sometimes uncomfortable, to the people of Mumbai. These cats inhabit Sanjay Gandhi National Park (SGNP), a protected forest enclave of roughly forty square miles, graced with flowing water and two lakes and a great diversity of flora and fauna, including chital deer, sambar deer, crocodiles, and cobras, as well as leopards. Some of those leopards are captive within a rescue center at the park, translocated there after becoming unwelcome elsewhere, but most of them roam free. The park lies bounded on three sides by Mumbai neighborhoods such as Aarey Colony and Bhandup West. People visit the place to picnic and walk its trails, to boat on the lakes, to ride a narrow-gauge railway, to board a bus that takes them through a fenced area where a few lions and tigers loll.

And sometimes free-roaming leopards venture out of the park and visit the neighborhoods, seeking territory or food. They eat what they can kill, preying on chital and sambar and whatever else entices them, including stray dogs. Leopards are generally shy of humans, and very secretive, but in SGNP they are crowded, enduring one of the highest leopard densities in the world, and that sometimes pressures an individual leopard to become a "problem animal," bolder and more desperate, crossing boundaries (geographic and behavioral), taking risks. So a leopard grabs a child or attacks a human adult. Back in 2004, a time with too many unsettled leopards, fourteen people were killed. More recently, in autumn 2021, five people were attacked, probably by a two-year-old female. The leopards of Mumbai are innocent of malice, but they are wild animals, hungry predators, and for citizens of the city, especially the adjacent neighborhoods, they constitute a danger.

Here's what I'm getting at. Imagine that you run an animal reproduction laboratory in Mumbai. You decide to clone a Mumbai leopard. This is entirely possible with technology and methods currently available. Scientists have cloned cats, dogs, deer, and other animals. Using microsurgical equipment, you extract the nucleus from one cell of a leopard in captivity at SGNP. This is the leopard you seek to clone—to replicate as closely as possible in a new animal, its genetic twin. It might be a male or a female. Call it Donor 1. You also remove the nucleus from an egg cell, an oocyte, taken from another leopard, necessarily a female (because only females produce eggs). Call her Donor 2. You insert your chosen nucleus into that egg cell, then you activate the thing to begin dividing. When it has developed to the point of comprising a couple hundred cells, those cells will be contained inside a spheroid mass, with a protective layer, called a blastocyst. This is your preimplantation embryo. The nuclear genome in each of those cells is identical to the genome of your clone target, Donor 1. And the cytoplasm of each cell, the gelatinous fluid surrounding the nucleus, contains mitochondrial DNA, a subsidiary genome, inherited only from that female who gave you her egg cell, Donor 2. You surgically implant the embryo in the uterus of a foster mother. For convenience, you might choose a smaller and less fearsome cat for this motherhood role—for instance, a Bornean clouded leopard, a gracefully spotted feline about the size of a border collie. If you are lucky and adept, your Bornean clouded leopard carries the embryo to term and gives birth to a true leopard, the genetic twin—in its nuclear DNA—of Donor 1, the animal from whose cell nucleus you started. Your newborn is actually a hybrid, because it carries the mitochondrial DNA of Donor 2. Anyway, congratulations. You have concocted a leopard in your lab. You coddle and rear this extraordinary little creature.

Within three years, properly fed, your laboratory leopard becomes a full-size adult. It has teeth. It has claws. It has muscles rippling beneath its spotted skin. It might weight 190 pounds, if it's a hefty male, or 130 pounds if it's a female. Anyway, it's far bigger and more powerful than its foster mother, the little clouded leopard. You put it (poor animal) on display, and you say: "Here's what's lurking within your city. Be cautious. Be respectful. Don't let your dogs or your children roam carelessly near the park."

Is this gain-of-function research? No, because leopards just such as your leopard already exist. They are out there, spilling intermittently from Sanjay Gandhi National Park into the streets and alleys of Mumbai. Your leopard has the same functional capacities as those other leopards, except your leopard was conceived in a dish and raised in a laboratory cage.

That's what Zhengli Shi and her colleagues did with two viral genomes, Rs4231 and Rs7327. They created laboratory approximations, in hybrid form, of what exists in the wild. They alerted us, if we will listen, that these potentially ferocious coronaviruses are out there.

69

Some months ago, the veteran virologist Robert Swanepoel, at the University of Pretoria in South Africa, sent me an advance proof of a journal paper on SARS-CoV-2 that he had noticed online. It was by three Frenchmen of whom I'd never heard. Thought you might find this interesting, Swanepoel said. He wasn't endorsing the paper but—because he likes "lateral thinking," and he knew of my interest in spillover—he suggested it was worth a look.

Bob Swanepoel, as I mentioned earlier, is a venerated elder in the field of emerging viruses. He led the team that went looking for Marburg virus during the Durba outbreak in the Democratic Republic of the Congo in 1999, when workers at a gold mine were dying. Before that, he had trained in veterinary medicine as well as virology, worked in Malawi and Zimbabwe, and been chosen to lead the Special Pathogens Unit at the National Institute of Communicable Diseases, in Johannesburg, when it was established in 1980. He's a South African counterpart to Karl Johnson and a few other crusty legends of the American CDC, known for his direct manner and his decades of penetrating, perilous work on many of the more nasty hemorrhagic fever viruses. Such is Swanepoel's reputation that, when I was researching a story on the (still inconclusive) search for Ebola's reservoir in 2015, I knocked politely by email and then flew to South Africa to spend three days at his elbow before a computer screen, looking and listening. So when he called my attention to this new SARS-CoV-2 paper from the Frenchmen in 2021, of course I read it.

The first author was Roger Frutos, a molecular biologist in Montpellier, France. It was a review article, an overview of the origin question, presenting a fresh and somewhat diagonal viewpoint. Yes, SARS-CoV-2 is a naturally occurring virus, Frutos and his coauthors asserted. No, it's not a product of laboratory manipulation. No, it's not RaTG13 from the Mojiang mine, with a furin cleavage site added by design. Nor does a lab leak seem likely, they wrote, adding that, "although a laboratory accident can never be definitively excluded, there is currently no evidence to support it." Then again, the common understanding of how a novel virus might become a human pathogen and cause a pandemic—the spillover model, they called it—was not satisfactory either. They proposed an alternative. They called it the circulation model.

Their logic was this. Viruses with relatively high mutation rates and evolutionary flexibility do not generally abide in just one reservoir host. If they are animal viruses, they aren't limited to one kind of animal. They generate a welter of genetic diversity among their viral populations, by mutation, so that swarms of loosely related viral strains can surge this way and that, exploring various niches and strategies. Such viruses circulate broadly within the animal kingdom, crossing species boundaries, infecting a range of hosts, coming to a dead end in one animal, succeeding temporarily in another animal and its contacts, probing possibilities, evolving, ready for opportunities. They are multi-host viruses. In areas of the world where humans live in close contact with wild animals—rural areas, edge areas, places where people are making incursions into natural landscapes, causing outsized disturbance to ecosystems—humans will be among the variety of hosts within which such a virus circulates. What leads to an epidemic or a pandemic, the Frutos group wrote, is no single incident of an animal virus spilling over into a human, finding itself well adapted, and roaring off to infect millions more humans, but rather "the occurrence of a double accident." They meant something far different from a laboratory accident: a genetic mutation or cluster of mutations or recombination event that yields potential advantage, followed by a societal circumstance within which the advantage is well rewarded. Once that double accident happens, the chains of infection do not come to dead ends—not all of them, anyway. The prevalence of the virus rises to a critical threshold within some aggregation of people. The circulation of a restless virus within multiple hosts, including humans, plus one accident, plus a second accident, yields a new human virus—a new disease emergency. Maybe it's an outbreak. Maybe it's an epidemic. Maybe it becomes a pandemic.

"Explain that to me," I asked Roger Frutos when we Zoomed. "How the double accident can occur, and what the double accident was in the case of COVID-19."

"Okay. It's a double accident of two different origins," he said. First comes the genetic accident. A virus, mutating abundantly as it passes from host to host, this kind of animal to that kind, throws up a mutation that happens to improve its success in one kind of host. Or maybe it's circulating between multiple vectors—mosquitoes or ticks of several kinds—and it improves its success in just one of those. For instance, he said, the case of chikungunya, a vector-borne virus carried by the yellow fever mosquito (*Aedes aegypti*) and therefore limited in its range to the range of that mosquito. The disease was first identified, in the 1950s, amid the borderlands of Mozambique and Tanganyika (now Tanzania). Its mosquito vector also originated in Africa. Then a single mutation gave the virus a vastly improved capacity to dwell in another kind of mosquito, the tiger mosquito (*Aedes albopictus*). The tiger mosquito originated in Southeast Asia. The crucial mutation, known as A226V, seems to have occurred around 2005.

Then came a second accident, the societal one. "In that case, the international trade," Frutos said. The tiger mosquito does well in human surroundings and travels well aboard ships, especially cargo ships carrying containers and bulk goods such as used tires, which trap rainwater and offer excellent egg-laying habitat for mosquitoes. The cargo shipping industry also involves a lot of loading and unloading in places such as Hong Kong, San Francisco, Marseille, Genoa, and seaports around the Indian Ocean, including Mombasa and Colombo. In the process, tiger mosquitoes get themselves loaded and unloaded too. "The mosquitoes, they spread like that," Frutos said. Having expanded their range and established themselves in new places—not just in the tropics but in temperate zones and in cities—tiger mosquitoes are now considered one of the world's most successful and problematic invasive species. They made it to East Africa, Frutos said, a fateful move. "There, the mosquitoes got in touch with this mutated virus." Together, thanks to a mutation accident (in the chikungunya virus) and a societal accident (affecting the distribution of tiger mosquitoes), those mosquitoes and that virus have caused outbreaks in Italy, India, South America, and the island of Réunion in the Indian Ocean. Chikungunya now affects hundreds of thousands of people yearly, especially in Brazil and India, but with intermittent cases and clusters in the U.S., the Caribbean, Europe, and elsewhere.

Very interesting. So how does this model apply to COVID-19? I asked.

The first accident, the genetic one, Frutos posited, was the acquisition of the furin cleavage site. That made the virus more transmissible in humans. "And the second accident?" he asked himself, saving me breath. "Why in Wuhan? Why at that time? Because at that time you had the conjunction of different things in Wuhan. You had different celebrations at the very same time, which brought a *lot* of people—plenty, plenty of people."

He meant the Spring Festival migration, *Chunyun*, occurring every January, with people all over China traveling to visit relatives and celebrate the Chinese New Year, sharing food and other goods transported for these special observances. In Wuhan, the logistics included thousands of passengers daily—to a peak of 100,000—moving through the Hankou railway station, and a potluck banquet involving roughly forty thousand families. The Hankou station is half a mile from the Huanan market. The mayor of Wuhan ignored the warnings of a highly contagious virus spreading through the city, and on January 19, 2020, the potluck went ahead as planned. Four days later, the city government suspended all public transportation, the provincial government cut the highway links, and Wuhan was locked in quarantine. But that was too late. With all those people traveling, convening, eating and drinking, celebrating, "you have the amplification," Frutos said. The chains of infection did not come to dead ends, not all of them. "The outbreak threshold was reached, and then it started." The virus that had been circulating undetected in humans and other animals for months or for years, according to Frutos's model, had gotten a double-accident opportunity and embarked on a whole new phase of its career.

Frutos and his colleagues made their case for this circulation model in two papers published during 2021, the first of which appeared in March—the one Bob Swanepoel sent me. That review article, plus my Zoom conversation with Frutos, primed my interest in a second paper on the origin question when it was published by the same group in October. This time, because we had talked, Roger Frutos sent me that one himself. It carried a provocative title: "There Is No 'Origin' to SARS-CoV-2."

Again, they described the elements of their model: the circulation phase, when the virus is spilling from one kind of animal host to another, remaining broadly adapted, capable of infecting them all; the first accident, when a genetic change happens to give the virus added advantage in humans; and the second accident, a societal circumstance that allows the virus to amplify and spread, approaching a critical point. (This is like what Malcolm Gladwell described in his book *The Tipping Point*.) But now Frutos and his colleagues expanded on their earlier explanation. "An epidemic never starts with only one infected individual. It is a probabilistic process," they wrote. It involves chance and odds. As the virus passes from one human to another, better able to transmit, increasing the number of infections, beginning chains of infection, the odds rise that at least one of those chains might continue indefinitely, rather than coming to a dead end, as all such chains have in the past. It might infect a person in the countryside who then travels to a city, where the population is more dense and further transmission is more likely. It might get into a hospital or an airport, giving it greater opportunity for transmission and dispersal. It might be carried into a crowded market. At a certain point, after the second accident, something happens. The virus gets lucky, humans unlucky. "This is referred to as the epidemic threshold," the Frutos team wrote. The novel virus, once a curious ecological phenomenon, becomes a public health crisis.

So there is no single, deterministic "origin" of SARS-CoV-2 and other emergent viruses, they added, but instead "a permanent process of evolution, adaptation and selection shaped by chance and environment," and that process "gives rise to novel lineages."

High on the list of factors that drive this permanent process is the growth of the human population. The more numerous we become, the more crowded, the more interconnected, the more demanding of resources, the more invasive of wild places, the more disruptive of richly diverse ecosystems—the closer we stand to the epidemic threshold for any new virus that probes us as a possible route to greater evolutionary success.

This circulation model from Roger Frutos and his colleagues is fresh and intriguing but not entirely unique. It roughly resembles what Jonathan Pekar and his colleagues, including Michael Worobey, suggested in their paper about timing the SARS-CoV-2 index case in Hubei (described earlier), when they wrote that "spillover of SARS-CoV-2-like viruses may be frequent, even if pandemics are rare." It was also anticipated, back in

2005, in a paper by Don Burke and several coauthors, including the virologist Nathan Wolfe and Peter Daszak of EcoHealth Alliance. That paper offered a concept they called "viral chatter," meaning repeated transmission of a given virus from its nonhuman animal host to individual humans, dead-end spillovers, with no onward human-to-human transmission, until at a certain point the virus takes hold among humans and an outbreak begins.

The second Frutos paper contained one other notable comment, which caught my attention because the same thought had occurred to me. The story of the Mojiang mineworkers and the suppositions about a laboratory leak have often been bundled together by critics of the natural origins hypothesis, as though they mutually reinforce the suspicion that something sinister, irresponsible, and secretive was done by scientists based in Wuhan. That narrative, according to Frutos and his coauthors, is a composite, comprising several mutually contradictory elements: 1) that SARS-CoV-2 came from the Mojiang mine, and the three mineworkers died of it; 2) that SARS-CoV-2 accidentally escaped from a lab at the Wuhan Institute of Virology; 3) that SARS-CoV-2 was engineered to infect humans. But if it was a dangerous virus that naturally abided in the mine, then it wasn't engineered and it wasn't created by gain-of-function research in the lab of Zhengli Shi. If it infected researchers when they entered the cave, then it didn't leak from a lab. If it was engineered in a lab for nefarious purposes, or produced by reckless gain-of-function research and then leaked from a lab, then the Mojiang mine, as I've said already, is irrelevant. RaTG13 came from the Mojiang mine, yes, but it's a sequence, not a virus, and eminent molecular virologists agree that RaTG13 is not SARS-CoV-2, nor could it be made into SARS-CoV-2 by any imaginable, rational laboratory procedure. You may embrace one of these three hypotheses—lab leak, engineered virus, creature from the Mojiang mine—and argue that evidence supports it, the Frutos group wrote, but you can't claim that they represent a single coherent hypothesis, their details offering mutual confirmation. That's illogical. They don't support one another. They exclude one another.

70

The dark stories about SARS-CoV-2 and its origins washed back and forth throughout much of 2021, variously finding favor, provoking rebuttal, inciting alarm or fascination or sarcasm on the op-ed pages, in certain magazines, and on social media, especially Twitter. Most of that noise was made by amateurs and pundits who had swotted up a bit of virology for the occasion. (I'm an amateur in this field too, but early in 2021 I decided to keep quiet for a while and listen.) The professionals, meanwhile, observed this fracas with varying degrees of dismay and frustration, and stayed at their work. Some of them, smart and honest scientists such as David Relman, called for further investigation of the lab leak hypothesis, and many people, both expert and otherwise, agreed that it would be useful. Further investigation of the natural spillover hypothesis, and of the Frutos circulation model, and of the viruses carried by bats in southern and central China, and of the trade in raccoon dogs and other wild animals throughout Hubei province and beyond, and of viral infections in those animals, would also be useful. But interactions between Chinese scientists and the international scientific community have been chilled and impaired by the political dimension of this pandemic—less delicately, we might say that those relations are in a state of catatonic seizure—so for now, we can draw conclusions only from the data we have.

That effort continues. On September 16, 2021, the journal *Cell*, another of the world's leading outlets for biological research and thought, published a paper titled "The Origins of SARS-CoV-2: A Critical Review." This was more compelling than most of the narratives and ripostes, because of what it said and who said it. The first author was Eddie Holmes, the final author was Andrew Rambaut, and in between, the list of coauthors included Kristian Andersen, Michael Worobey, Susan Weiss, David Robertson, Robert Garry, Angela Rasmussen, Stuart Neil, Wendy Barclay, Maciej Boni, and Jeremy Farrar. These names represent a high pile of credibility and expertise, especially in evolutionary virology and the biology of coronaviruses. You can't fake what they know and do, and you can't swot it up on short notice.

"Coronaviruses have long been known to present a high pandemic risk," these authors began. Consider some facts. SARS-CoV-2 is the latest of seven known coronaviruses that infect humans, and the fifth recognized in the past twenty years. All the previous human coronaviruses, like most human viruses, have animal origins. The first emergence of the original SARS virus, in late 2002, and the reemergence of that virus, in autumn 2003, were associated with markets selling wild animals, notably civets and raccoon dogs. Among animal traders working in such markets in 2003, 13 percent tested positive for SARS antibodies, and of those specializing in civets, more than 50 percent tested positive. (Implication: we need parallel data from animal traders in 2019.)

Another human coronavirus, HKU1, one of the relatively innocuous ones, emerged in a large Chinese city, Shenzhen, in 2004. It contains a furin cleavage site in its spike protein. It was first identified in a case of human pneumonia. (Implication: the Wuhan locale, and the furin cleavage site in SARS-CoV-2, are not unusual.)

Two of the three earliest known cases of COVID-19, and 28 percent of all cases reported in December 2019, were directly linked to the Huanan Seafood Wholesale Market. And 55 percent of the December cases traced either to that market or to another Wuhan market. Most of these market-related cases turned up in the first half of the month, nearer the time of emergence, however it occurred. The other cases, those unconnected to markets, are easily explained by asymptomatic spread. The markets of Wuhan—the Huanan market and several others—traded many thousands of wild animals, alive, during 2019, including known coronavirus carriers such as civets and raccoon dogs. After the Huanan market closed, traces of SARS-CoV-2 showed in environmental samples, especially in the western section of the market, where wild and domestic animals were sold. Some animal carcasses tested negative for the virus, but that testing did not include raccoon dogs or civets. (Inference: yes, Chinese authorities should have done a more thorough job of sampling animals from the Huanan market before they were removed or destroyed.)

The earliest split in the SARS-CoV-2 family tree, as drawn from genome comparisons, occurred very early indeed: probably sometime in mid-December or before. That's reflected in the fact that two distinct lineages, labeled A and B, circulated independently at the time. Lineage B appeared in the cases linked to the Huanan market, and in environmental samples taken there. That lineage spread fast and far, becoming dominant around the world—which is why the Alpha variant (B.1.1.7), first recognized in the U.K., and the Beta variant (B.1.351), first detected in South Africa, and the Delta variant (B.1.617.2) that came roaring out of India, as well as the Omicron variant (B.1.1.529), first recognized again by scientists in South Africa (and about which, more below), all carry the letter B. The only root of that tree located so far is buried in the Huanan market. But not all lineages (at least so it seemed, when this paper was published) trace to that market.

Lineage A, which became more prominent in Wuhan and other parts of China, contains an early case linked to a different market in the city. That pattern, Holmes and his coauthors wrote, is consistent with SARS-CoV-2 having emerged from infected wild animals, or from traders who delivered them to one market and another. (Inference: it is harder to see this pattern emerging from a fateful lab leak at the Wuhan Institute of Virology, ten miles away across the Yangtze River.)

Furthermore, coronaviruses closely related to SARS-CoV-2 have been found in bats (and pangolins too) at multiple sites across southern China, Cambodia, Thailand, Laos, and Japan. Holmes himself had coauthored the publication on discovering RmYN02, the closest match to SARS-CoV-2 throughout most of its genome, sampled from a Malayan horseshoe bat in Mengla County, Yunnan. In Thailand, a scientist named Supaporn Wacharapluesadee, working with a large group of Thai and international colleagues, found another coronavirus that was 91.5 percent similar to SARS-CoV-2 and shared part of its furin cleavage site. They detected it in fecal samples from horseshoe bats roosting in a wildlife sanctuary. More recently, in Laos, a team of Laotian and French researchers detected three coronaviruses, also in horseshoe bats, that are more like SARS-CoV-2 than anything else seen so far, including RaTG13. This match is especially close in the receptor-binding domain. All three kinds of bats in the Laotian study, as well as the kind in the Thailand study, are distributed widely across Southeast Asia, and two of them occur also in southern China. Bats don't respect national boundaries, so it's certain that some individuals among these broad-ranging species are carrying their viruses from one country to another. That factor, along with the propensity of horseshoe bats for gathering in multispecies roosts, and their susceptibility to being infected with more than one virus, offer abundant opportunities for coronaviruses to recombine, swapping portions of genome from one virus to another. (Implication: the pieces from which SARS-CoV-2 assembled itself, and the circumstances in which that assembly occurred, are all available in the skies and the caves of southern China and Southeast Asia.)

Having established this much, Holmes and his coauthors considered the lab leak hypothesis. Yes, there have been laboratory accidents and misguided vaccine trials in which dangerous viruses have gotten into lab workers or the public. But those were known viruses, such as the H1N1 influenza, or small and contained events, such as the 1967 infections with Marburg. "No epidemic has been caused by the escape of a novel virus," the authors noted, and there was no evidence that the Wuhan Institute of Virology, or any other institution, was working on SARS-CoV-2 or any close progenitor of it before this pandemic began. All staff working in Zhengli Shi's laboratory tested negative for SARS-CoV-2 antibodies in March 2020, unless this was a lie to the WHO mission. The WIV has cultured three SARS-like coronaviruses from bats, yes, but none of those closely resembles SARS-CoV-2. And those viruses were grown by amplifying them successively in monkey cells, a process that, when applied to SARS-CoV-2, causes the furin cleavage site to fade away because that isn't necessary in cell cultures. As for gain-of-function research—any work fitting that label seemed very improbable as a source of this virus, the Holmes group wrote, because there was "no rational experimental reason why a new genetic system would be developed using an unknown and un-published virus," given that the logic of gain-of-function research involved knowing a great deal about the virus to which a function was being added. Useful experimentation requires limiting the variables. Under any lab leak scenario, SARS-CoV-2 had to be present in a Wuhan laboratory before the pandemic, but there was no evidence that condition existed, and no reason for concealment if it had. Zhengli Shi was in the business of finding new coronaviruses and announcing them to the world.

There was more. The authors devoted one quite technical section (as though this much isn't technical enough) to their own field of specialty, the genomic structure and ongoing evolution of RNA viruses, notably SARS-CoV-2. There was a reminder that SARS-CoV-2 is a generalist virus, not suspiciously well adapted to humans but, on the contrary, capable of infecting all those mink and tigers and gorillas and other mammals. To this the authors added that, if the virus *had* been unusually well adapted to humans in December 2019, it has certainly done a lot more adapting since. Cases in point, the Doug (D614G) mutation, followed by Nelly (N501Y) and Eek (E484K) and Karen (K417N) and then all the variants. The authors noted, in response to dark claims about the furin cleavage site, that furin cleavage sites are common in spike proteins of other coronaviruses. There was mention of William Gallaher's palindrome, suggesting that a furin cleavage site might have entered this virus by recombination at such a copy-choice breakpoint. And there were brisk refutations of several other lab leak or engineered virus claims, on matters just far enough into the weeds that I haven't asked you to go whacking in there with me.

That brought Holmes and his coauthors to the principle of parsimony: the simplest explanation for a phenomenon is likely the best. A more elaborate explanation brings increased chance of incorporating false facts, pure coincidences, or wrong assumptions. As is true for most human viruses, the group concluded, "the most parsimonious explanation for the origin of SARS-CoV-2 is a zoonotic event." They meant spillover from a nonhuman animal. "Zoonotic spillover by definition selects for viruses able to infect humans." Suspicion that the virus might have a lab origin, they wrote, "stems from the coincidence that it was first detected in a city that houses a major virological laboratory that studies coronaviruses." But the link to Wuhan, they added, more likely reflects the facts that it's the largest city in central China, a hub for travel and commerce, densely populated by eleven million people, with multiple animal markets in its midst.

Yes, Holmes and his coauthors agreed, the possibility of a lab accident can't be entirely dismissed. Furthermore, that hypothesis may be nearly impossible to disprove. But it's "highly unlikely," they judged, "relative to the numerous and repeated human-animal contacts that occur routinely in the wildlife trade." Failure to investigate that zoonotic dimension, with collaborative studies, crossing borders between countries and boundaries between species, would leave this pandemic festering and the world still very vulnerable to the next one.

71

We don't yet have a final answer to the question of origins, as I write this sentence, and we may not have a final answer when you read it. We may never have a final answer, and that would be a grave misfortune, as Eddie Holmes and his colleagues noted. In the meantime, I continue thinking about those words "coincidence" and "unlikely," as germane in their own ways as the fancier word "parsimony." What's a coincidence and what's a revealing pattern? What's likely and what's unlikely? This takes me back to the first few dozen cases and their links, or lack of links, to the Huanan Seafood Wholesale Market.

Daniel Lucey performed a useful service when—as I described many pages ago—he alerted readers of his blog to the fact that early accounts attaching the outbreak to the Huanan market overlooked something important: the exceptions. Of the forty-one earliest cases, twenty-seven had known, direct links to the market. One, a woman, was linked through her husband. Lucey asked: What about those other thirteen, where did *they* get infected? This helped to cast the market-origin premise into doubt.

Michael Worobey returns to the tale again here. Worobey has been a bellwether in the ongoing discussion about origins, because his reputation is sterling and his mind is open as well as keen. When he was researching the origin of HIV-1, as I've described briefly in this book and more expansively elsewhere, he traveled to the Democratic Republic of the Congo and gathered chimpanzee poop, along with the British biologist William Hamilton, when Hamilton wanted to test the Oral Polio Vaccine hypothesis. Worobey was no adherent to that hypothesis, but it seemed worth investigating. One scientific colleague has said, evidently with the Congo adventure in mind, that Worobey is known to have "a soft spot for wild theories." Worobey also initiated a letter to *Science* in May 2021, eventually signed by seventeen other scientists including David Relman and Alina Chan, that called for a deeper investigation of the origins question, something more thorough than the one by the WHO mission, which had been severely constrained by limited time, limited access to records, and limited cooperation by Chinese officialdom. The presence of Worobey among that group of signatories, as well as among the coauthors of the Holmes "Critical Review" on origins, reflects his independence and flexibility. His views tend to be driven by curiosity and data. Curiosity drove him back to the Huanan market.

On November 18, 2021, Worobey published an essay in *Science* titled "Dissecting the Early COVID-19 Cases in Wuhan." His dissection and the conclusion he drew from it were sufficiently newsworthy, especially in light of his having signed the "investigate further" letter, that *The New York Times*, *The Washington Post*, and *The Wall Street Journal* all carried major pieces on it the same day.

Worobey started from the most widely reported data: according to an early study by Huang and colleagues, which I mentioned near the start of this book and which the Holmes group also cited, the first forty-one reported cases in Wuhan included twenty-seven with links to the Huanan market. That's 66 percent. Among just the first nineteen cases, ten were linked to the market. That's 53 percent. The WHO mission on origins, in their report, noted 168 known cases during the entire month of December, of which 33 percent were market-connected—still high, and that number obscures the fact that market linkage was higher during the early weeks of the month. Some people, Worobey wrote, might ask why only a third, or maybe two thirds, of the early cases were linked to the Huanan market if it were the source of the outbreak. But that overlooks the high transmissibility and asymptomatic spread of this virus. A better question would be this: Why would you *expect* to see all cases linked to the market, if the virus was spreading so quickly and subtly? And if the Huanan market *wasn't* the source, what accounted for all those cases?

There was a possible answer to that last question, and it seems to be what piqued Worobey's interest: *ascertainment bias*, a scientific pitfall. In plain words: if you think that a certain factor is what you're looking for, in a scientific study, you will be more likely to find it, because you will look in places where that factor lives. Applied to Wuhan in December 2019: if doctors and public health officials thought that the Huanan market was the source of this strange new pneumonia, they would have looked harder for cases among people connected with that market, and therefore they would have found more, even while the virus was spreading throughout the city. Their expectations, and their search effort, would have biased their results. Worobey examined that possibility by getting ahold of medical records and other documents, and he found that it hadn't occurred. Doctors at the hospitals receiving the earliest cases hadn't known of any suspected link to Huanan. That knowledge hadn't come until December 29. They diagnosed these earlier patients on the basis of clinical symptoms, not epidemiological information such as whether a person had worked in or visited a market. No such bias in the data, Worobey concluded.

Worobey scrutinized an important premise, which had been established back in January 2020 by the report in *The Lancet* and highlighted in good faith by Daniel Lucey: that one or two of the earliest cases confirmed in Wuhan had had no connection to the market. That turned out to be wrong. The very first case, as indicated by that paper, a non-market-connected person who supposedly sickened on December 1, had in fact not fallen ill until late December. That correction was made in the WHO mission report.

Worobey also teased out another insight that was obscured in the WHO report. The first known case, purportedly, was the forty-one-year-old accountant, Mr. Chen, whose symptoms were thought to have begun on December 8. The fact that Mr. Chen never visited the Huanan market— he didn't shop for wild food amid funky stalls and alleys, he got his groceries from a supermarket—seemed, at least to some people, to argue against the market as source. (Worobey: No, you would expect one or two early transmission events to occur away from the epicenter of the outbreak.) Anyway, first case, no market link? But maybe Chen wasn't even the first case. Worobey noted that, in the scientific paper describing Chen's case and in his hospital records, his symptom onset time was shown as December 16, not December 8. In an interview, Mr. Chen himself said his fever started on the 16th. The WHO report mentions none of this. The mission scientists were simply informed by Chinese officials that Chen's symptoms started on December 8.

This left another person, Worobey wrote, as the earliest known case: a female seafood vendor who worked in the Huanan market. Her name was Guixian Wei. She sold shrimp. She took sick with Covid on December 10. She was the first confirmed Covid-positive person of many at her job site. Altogether, including that woman, more than half of the very earliest hospital-treated cases were linked to the Huanan market. "It becomes almost impossible to explain that pattern," Worobey told *The Washington Post*, "if that epidemic didn't start there."

Almost impossible to explain. But we can try. A laboratory worker at the Wuhan Institute of Virology is performing an experiment on a coronavirus. That virus is unknown to the world, and its capacity to infect humans is undetermined, because it has never circulated among people. It's an interesting virus but not terribly interesting, because its genomic similarity to SARS-CoV, the notorious human coronavirus of 2003, is only 79 percent. The laboratory worker makes a mistake, or someone else does, with the consequence that this Unfortunate Worker becomes infected. Maybe a spilled vial, maybe a defective negative-pressure hood, maybe whatever. No one is aware of the infection, not even the Unfortunate Worker, but five days later the UW begins to feel feverish. Starts to cough. Maybe it's a cold or, worse, the flu. After one day of coughing at work, as a precaution and a courtesy to others, UW calls in sick. This event is not recorded in laboratory records, or, if it is, those records will later be concealed or destroyed. Fortunately, no one else at the Wuhan Institute is infected, despite UW's day of coughing. UW stays home for two or three days, feeling increasingly bad, coughing more severely, finding difficulty in even taking a breath; but UW doesn't go to a hospital.

Instead, with a sudden hankering to buy some fresh fish for dinner, or a snake, or a bamboo rat, or maybe a raccoon dog for a family feast, UW travels ten miles, across the Yangtze River, to the Huanan Seafood Wholesale Market. Perhaps UW takes public transportation but better still if UW is affluent enough to own a car. In any case, no one is infected during that transit. Once at Huanan, inside the market, UW suffers an especially severe spasm of coughing and infects another person or possibly several. One of those persons infects another, then another, then two more. A shrimp vendor is infected. The Unfortunate Worker returns across the river, goes home, reports back to work at the Wuhan Institute of Virology, and is never heard of again. The novel virus from the lab has been seeded into the market.

It's not impossible. But it seems unlikely.

VIII

NOBODY KNOWS
EVERYTHING

72

This is a book about the science of SARS-CoV-2. The medical crisis of COVID-19, the heroism of health care workers and other people performing essential services, the unjustly distributed human suffering, and the egregious political malfeasance that made it all worse—those are topics for other books. But science is a human activity too; scientists are people who strive and sacrifice and make blunders and suffer misfortunes and respond to career incentives and personal pressures, much like the rest of us. They are fallible. They know things the rest of us don't know, but they don't have answers to every urgent question about SARS-CoV-2. One thing they do know, the wisest of them, is that their knowledge is fragmentary and provisional. Science is always provisional.

The scientific discussion of this virus has been a firehose of preprints and published studies, of data and analysis and speculation, of honest mistakes and hasty claims and retractions, of correcting this month something said tentatively last month, and of careful inferences from carefully gathered facts, yielding insights that seem likely to stand the test of time. The discourse has also been punctuated by a small number of summary reports, special investigations, sonorous declarations, and testimonial statements, each of which has met mixed reactions. The first of those was a letter published in *The Lancet* on March 7, 2020, cosigned by twenty-seven distinguished scientists from around the world, voicing support for their counterparts in China near the start of what could already be seen as a menacing pandemic. That expression of solidarity wasn't the controversial part. This was: "We stand together to strongly condemn conspiracy theories suggesting that COVID-19 does not have a natural origin." Such theories, the coauthors added, "do nothing but create fear, rumours, and prejudice that jeopardise our global collaboration in the fight against this virus." The letter attracted considerable support when first published but, partly because of the phrase "conspiracy theories," provocative to proponents of the lab leak idea as that hypothesis gained attention, its effect became more inflammatory than emollient.

The list of authors was arranged alphabetically. That put Charles Calisher, a distinguished and very independent-minded virologist with almost three decades of experience at the CDC, then an academic career, and by 2010 an emeritus professor at Colorado State University, in first position. The group also included Dennis Carroll (Texas A&M and the Global Virome Project), Rita Colwell (University of Maryland), Peter Daszak (EcoHealth Alliance), Gerald Keusch (Boston University), Larry Madoff (University of Massachusetts Medical School and ProMED), Jeremy Farrar (Wellcome Trust), and Jonna Mazet (UC-Davis and head of the PREDICT project), some of whom you have met in this book.

Daszak initiated the letter and drafted it, evidently, for which he was later criticized as having a conflict of interest. He certainly had an interest in voicing support for Chinese scientists, having worked closely with Zhengli Shi for fifteen years. Whether that constituted a conflict is another question. I'll be accused of bias on this, I expect, because I've known Peter Daszak a long time, having been fascinated with the mission of his organization and spent arduous field time with some of his EcoHealth people; and he is a friend of mine. Then again, as I mentioned earlier, I've known Matt Ridley—Alina Chan's coauthor—a long time and he is a friend of mine. What does it prove? I've known Charlie Calisher a long time and he is a friend of mine. I've known Karl Johnson, the founder of Ebola studies, a long time and he is a friend of mine. I know quite a few scientists who study emerging viruses, and have shared wild experiences with some of them, because I've covered this subject for twenty years. I'm aware of the proposition that journalists are not supposed to have friends. But authors, working a wider arc of history, character, and narrative, are allowed.

Calisher is a bench virologist of the old school, with a lifetime's experience culturing dangerous viruses and studying what they can do. It was he, more than anyone, who drilled into me the distinction that a viral genome sequence is not a virus. A virus is an organism; a genome sequence is information. Calisher is a New Yorker, grew up in Jamaica and Bayside, Queens, and still speaks like a tough little kid who rode the subway to Stuyvesant High School in Manhattan every day. Tough enough, eventually, with a PhD in microbiology, to work on some nasty pathogens. He's a lifelong Yankees fan who, even in semiretirement in Fort Collins, Colorado, hates to miss a televised game. Months after the *Lancet* letter he told several questioners, including me, that he thought a lab leak origin of SARS-CoV-2 might be possible, or even likely, but that the hypothesis was not grounded in any data. Wait a second, Charlie, I said, what about that statement of support for Chinese scientists? You're first author. Is that because of alphabetical ordering?

"Right. I'm thrilled about it," he said dryly. "I get calls all the time. 'Did you write this paper?' No. I *agreed* with it. I still agree with it." One thing the letter says, he reminded me, is "we don't have data! You don't start blaming somebody without any data." The ruling principle, he added, should not be "guilty until proven innocent."

People kept asking. His name atop the list of authors made Calisher a reluctant spokesman. He had heard from organizations within the U.S. government, never mind which. Strangers sent him reports, messages, in-house communications that had gotten out-of-house, soliciting his comment on the lab leak hypothesis.

"There are no data!" he would answer. Charlie speaks in dialogue, taking all parts, when he tells a story.

"Well, the Chinese government is not cooperating fully."

"That's the *reason* there are no data. But there are no data! I'm not buyin' this until I see data!"

Others called for more data as well, and in May 2020 the World Health Organization took an institutional step toward excavating some. At the 73rd World Health Assembly, in Geneva, the WHO in collaboration with two other international bodies, the World Organisation for Animal Health (OIE) and the Food and Agriculture Organization (FAO) of the United Nations, resolved to "trace the animal origin of the virus," as well as its route into humans, and identify any possible intermediate host. That led to negotiations with China toward "terms of reference" for an international mission to Wuhan. Those terms were laid out in a document at the end of July, setting a course for what became the WHO mission. There would be two phases: phase 1, comprising a group of short-term studies to identify crucial gaps in what was known and to formulate working hypotheses; and phase 2, involving long-term epidemiological, virological, and serological investigations. The serological investigations—that is, examination of archived blood samples for evidence of exposure to the virus beyond the time window and geographical range recognized so far—would screen animal populations as well as humans. The epidemiological investigations would entail interviewing people, far and wide, for clues about where SARS-CoV-2 came from, when, and how. The fact of greatest interest about these phase 2 investigations is that two years into the pandemic, they have not happened and there is no foreseeable prospect that they will. The Chinese government has rejected plans for phase 2 after politicization of the origins question, rising attention to the lab leak hypothesis, and negative reactions to phase 1.

Phase 1 was the WHO-convened mission of international experts to Wuhan for a month of collaborative work. The team consisted of seventeen Chinese members and seventeen foreign members from the Netherlands, Russia, Vietnam, the United Kingdom, Sudan, the United States, and other countries. The foreign members included Marion Koopmans from Erasmus Medical Centre in Rotterdam, Peter Ben Embarek from the WHO, Thea Fischer from Nordsjaellands University Hospital in Denmark, and Peter Daszak from EcoHealth Alliance. The mission extended from January 14 to February 10, 2021, a short period, and for the first two weeks of that time the foreign members were in quarantine, confined to their rooms at a place called the Jade Boutique Hotel, in a software park amid the lakes on the southeast edge of Wuhan. For the second two weeks they met with their Chinese counterparts, divided into three working groups by expertise and topic (epidemiology, molecular epidemiology, animals-and-environment), reviewed data, interviewed witnesses, made site visits (including one to the Huanan market), discussed, agreed on conclusions insofar as agreement was possible, and eventually drafted a report. That was released by the WHO on March 30.

The report discussed, as the joint team had, four scenarios for how SARS-CoV-2 might have gotten into humans: direct spillover from a nonhuman animal host; transmission from a reservoir host through an intermediate animal to humans; introduction to humans on frozen food transported to Wuhan from elsewhere; and a lab leak accident. Frozen food was known as the cold chain hypothesis, reflecting the idea that frozen virus, stuck to the outside or wrapped inside a package of fish or meat, could remain viable for reawakening at the end of a long supply chain, so long as the package remained frozen. The joint team didn't consider an engineered-virus hypothesis, on grounds that other scientists had already persuasively refuted that notion. Their consensual assessment of likelihood for each scenario was this: natural spillover, possible to likely; transmission through an intermediate host, likely to very likely; the cold chain pathway (favored mainly by Chinese members), possible; and a lab leak, "extremely unlikely." Immediately, critics cried foul. The lab leak hypothesis had been given too little credence, they argued, and frozen food imports (such as pig heads and salmon) too much.

Daszak's involvement again became an issue. His group had been pulled into the political limelight back on April 11, 2020, when the *Daily Mail*, a London tabloid, ran a story with a quote from an unnamed government source saying that, while the balance of evidence pointed to a natural spillover of the virus in a Wuhan market, "an accident at the laboratory in the Chinese city was 'no longer being discounted.'" The story didn't mention EcoHealth Alliance but described some coronavirus research at the Wuhan Institute of Virology and suggested that it was "funded by a $3.7 million grant from the U.S. government." This was confusedly based on the reality that, since 2014, EcoHealth had received annual increments of a $3.7 million multiyear grant from the NIH, through the NIAID, for a wide range of its work on infectious diseases, and a fraction of that money had helped support the collaborative work with Zhengli Shi. Six days later in Washington, a reporter for the conservative news website Newsmax amplified the story and the confusion by asking Donald Trump at a press briefing about the grant, in a way suggesting that $3.7 million might have gone to the WIV. "We will end that grant very quickly," Trump said. And within another week, over the misgivings of both Francis Collins and Tony Fauci, the NIH did. At this point, the lab leak scenario seemed to be a political cudgel wielded mainly by sinophobic conservatives, but that began changing around the time Alina Chan and her two coauthors posted their preprint. Still, by late summer and autumn, as the WHO began choosing its members for the study of COVID-19 origins, selecting Peter Daszak was evidently not foreseen to be controversial—or at least, not prohibitively so.

Daszak himself hadn't sought the role. He told me, during a Zoom call soon after his return from Wuhan, that he had been recruited to it by Peter

Ben Embarek, the team leader, and another official at the WHO. An email arrived from Ben Embarek, suggesting that Daszak would be appropriate, and probably acceptable to the Chinese, because of his long-term work there.

"'You have great expertise,' blah blah blah," Daszak quoted loosely, recollecting the email. "I wasn't going to go." Daszak conferred with colleagues at EcoHealth; he talked with his wife about being away for a month. He was disinclined. But he felt he should talk with Ben Embarek. "So I called him up," Daszak told me. "He was just a straight-up, honest guy. And I said to him, 'If I get involved, I'll bring a bunch of politics and conspiracy stuff that might ruin it. Why would— . . . It will bring a lot of problems for WHO.' And he said, 'Well, what's new there? We get criticized every single day.'" Daszak had relevant experience and knowledge, and he was wanted, not just by Geneva. "To be honest," he recalled Ben Embarek telling him, "the Chinese named you as a person who would be good on this trip."

He saw the list of other team members and recognized it was a solid group. They got a "powerful little pep talk" from Mike Ryan, the forthright Irish epidemiologist who led the WHO's Health Emergencies Programme. Daszak's wife said, "You've got to do it." So he went. It seemed like a good idea at the time. But as he emerged as a favored target—of criticisms, accusations, and threats—his scientific collaborations with Zhengli Shi, and the research activities of EcoHealth Alliance generally, as well as his role in the WHO mission, became targeted too. That commotion did more to distract from the question of SARS-CoV-2 origins than to illuminate it.

After the two weeks in quarantine for the foreign team at the Jade Boutique Hotel, during which they could communicate with one another online and study available data, and then the two weeks of joint briefings and interviews and outings, the two teams together wrote their joint report. Orchestrating this task fell to the leaders of the three working groups, each group drafting its respective section. For molecular epidemiology, the leaders were Marion Koopmans and Yang Yungui, deputy director of China's National Center for Bioinformation. For the animals and environment group, the leaders were Daszak and Tong Yigang, a microbiologist at Beijing University of Chemical Technology. Have you ever tried to write a report by committee? This was worse. It involved a lot of high-stakes haggling, voting, and translating.

"The animal report took from nine in the morning until four a.m.," Daszak told me. The discussion was arduous. They sat at a long table, with rows of people on each side, experts and group leaders in the first row, support people behind. On the China side of the table, it went four rows back, with additional scientists and Foreign Ministry personnel and others there to make sure, Daszak said, "that whatever came out of this was not going to damage China's reputation. Fair enough." And that was just the preliminary deliberations.

"By the time it came to actually writing the report text," he said, "suddenly it's there in black and white. And it was a *battle*." After nineteen hours, at 4:00 a.m., "the other side was flaking out, you know, they were falling asleep, they'd gone, people were on long breaks, they're smoking cigarettes outside." Daszak just stood there, he told me, telling the others, "Look, I'll stay here until six a.m., until nine a.m. We need to get this language written."

There were some sticking points. "They wanted less on the wildlife origin, more on the cold chain," Daszak said. The Chinese members argued against the premise that live, wild mammals were even on sale in the Huanan market. No, they said, by Daszak's recollection, the only wild mammals were dead and frozen. Daszak himself doesn't believe that claim—being alive at point of sale was part of what gave wild animals their high value to gourmands—and the photos taken by Eddie Holmes in 2014 stood against it. (So did the wide-eyed but inconspicuous surveys of the market done by Xiao Xiao as recently as November 2019, published after the WHO mission.) Among the Chinese points of counterevidence, Daszak recalled, was a law constraining the sale of live, wild mammals. If it was illegal, it must not have happened. "We ended up with a consensus report that has both in"—both the wild mammals and the cold chain products—"which is important, and is explicit about the data, and is explicit about the further recommendations." High among the recommendations: more data, further study. Daszak's group advised doing surveys of wild animals known to be susceptible to coronavirus infection (such as raccoon dogs and civets and mink) and tracing the supply chain of wild animals farmed for food, as well as the network of cold chain suppliers. A global group of experts should do such tracing jointly. The report's final words called for continuing research toward the origins. This was only supposed to be phase 1, a start, and phase 2 would need to go deeper.

If there is, ever, a phase 2. As we await it, as the WHO remains in deadlock with the Chinese government about further investigation, Daszak's account of that struggle over two scenarios, wildlife origin versus cold chain origin, puts me in mind of another idea. Critics in the West suggest that China has been covering up a laboratory accident. Maybe that's not the point. Maybe that's not the explanation for why Chinese officials have resisted a phase 2 study, and why they favor the cold chain hypothesis, which is plausible but for which there seems to be no evidence. Maybe the point is that China has been embarrassed by, and is covering up, not a lab leak but an animal leak.

This notion isn't unique to me. I've seen it expressed by a few scientists—most cogently by Gigi Kwik Gronvall, an immunologist at Johns Hopkins University and a scholar of global biosecurity, writing in *Survival*, a journal of international strategic studies. Gronvall's essay appeared, at the end of 2021, under the title "The Contested Origin of SARS-CoV-2." She ranged across all the major contentions, all the hypotheses, and focused briefly on the illegal trade in wild animals for food, which had continued in China until the end of 2019, despite laws put in place after the original SARS scare of 2002 to 2004. That trade was worth billions of dollars—by varying estimates, somewhere between $18 billion and $75 billion annually. It was a boon to some local economies. Gronvall cited the same study I mentioned earlier, by Xiao Xiao and colleagues, who surveyed seventeen separate shops at four wet markets in Wuhan between May 2017 and November 2019 while studying a different disease associated with wildlife. It was only fortuitous that the Xiao study, done before the pandemic, yielded data of great relevance to the pandemic in retrospect. Xiao's group found wildlife of thirty-eight species on sale in the markets, including raccoon dogs, hog badgers, civets, bamboo rats, porcupines, and mink. They explained in their paper that, although some wild animals could be legally farmed, such as raccoon dogs for fur, it was illegal to trap them from the wild and to sell them for food. Vendors of protected wildlife were required to display licenses confirming their permission to breed those animals in captivity, as well as certificates indicating the animals' origins and that they had been quarantined to avoid diseases. Xiao, who did the data gathering in the markets, saw that many of the supposedly captive-farmed mammals had (as I mentioned earlier) gunshot or trap wounds, suggesting illegal wild harvesting. None of the seventeen shops that Xiao visited had origin or quarantine certificates posted on their walls, "so all wildlife trade was fundamentally illegal." The wild animals of China are considered state property; penalties for capturing and trading animals of a protected species (such as raccoon dog or bamboo rat or civet) include a three-year prison term and fines. The takeaway from Xiao's study is that, as of November 2019, none of this seemed to be enforced.

Gronvall noted such negligence as a possible reason why the Huanan market was quickly closed, emptied, and sterilized on January 1, 2020, immediately after city officials learned of its connection to the outbreak of pneumonias. The wildlife traders disappeared. "The swift clear-out of the market may have been intended to protect them as well as the law-enforcement officers and local politicians who had looked the other way," Gronvall wrote. "Once it looked as if a disease that came from the market was spreading, sweeping illegal activity under the rug would be a priority to avoid blame and hold onto profits." For the pandemic to have originated in China was "embarrassing" enough, she wrote. Worse still if it reflected negligence in law enforcement and possible corruption. "Although the wildlife trade was a known disease risk, and there were laws restricting it on the books, it wasn't stopped." If the world could be persuaded that the virus reached Wuhan in a package of frozen halibut imported from Greenland, on the other hand, no one was to blame. Except perhaps Greenland fishmongers.

73

The release of the WHO-China joint report, on March 30, occurred with a formal briefing in Geneva, at which Peter Ben Embarek and his Chinese counterpart, Liang Wannian, presented the key findings. When they finished, Director-General Tedros reclaimed the microphone and offered some provocative closing remarks. What made the headlines was: "I do not believe that this assessment was extensive enough . . ." and "Although the team has concluded that a laboratory leak is the least likely hypothesis, this requires further investigation . . ." and "Let me say clearly that as far as WHO is concerned all hypotheses remain on the table." The context of those sentences did not make the headlines, and so I was among many, I'm sure, struck by Tedros's bluntly expressed lack of satisfaction with how his chosen team had executed their impossible task. Or, to put it another way, because I was aware of the constraining terms of reference, I thought: *Interesting. He's throwing them under the bus.* Then I read his complete statement and saw: "This report is a very important beginning, but it is not the end. We have not yet found the source of the virus, and we must continue to follow the science and leave no stone unturned as we do." Who could argue with Director-General Tedros? He was asking for phase 2, as planned from the start. Well, some people could argue, and he hasn't gotten his phase 2.

"As scientists with relevant expertise, we agree with the WHO director-general," wrote a group of eighteen scientists in a letter to *Science* six weeks later. These scientists noted also that the United States, the European Union, and thirteen other countries had called for further investigation of both the lab leak and the natural spillover hypotheses. Among the signatories (not quite in alphabetical order, as for the Calisher letter, but nearly so) were Alina Chan, Ralph Baric, and Marc Lipsitch. Baric, of the University of North Carolina, is well known for his bold laboratory research on coronavirus, and (amid the current controversy) for having collaborated with Zhengli Shi on a study a half dozen years earlier. Lipsitch, of the Harvard T.H. Chan School of Public Health, is a leading critic of gain-of-function research. The final signer, by nonalphabetical placement and therefore suggesting a principal role, was David Relman, of Stanford, the gain-of-function critic I mentioned earlier. Michael Worobey's name came second to last, although it was he who had proposed such a letter to some of the others; this was six months before Worobey decided, based on his own further study of the December 2019 cases and the Huanan market, that those data provide "strong evidence of a live-animal market origin of the pandemic." First author on the new letter was Jesse Bloom, an evolutionary virologist at the Fred Hutchinson Cancer Research Center in

Seattle. The principle on which the group agreed with Tedros of the WHO was that "greater clarity about the origins of this pandemic is necessary and feasible to achieve." Public health agencies and laboratories should open their records to the public. The investigation should be transparent, objective, data-driven, and as free as possible from conflicts of interest.

Some weeks earlier, I spoke with Jesse Bloom. I wasn't aware of his views until then, but he gave me a sort of preview of the *Science* letter. "The origins of SARS coronavirus 2 remain poorly understood," he said. Some scenarios, he said, could be clearly ruled out—among which, I gathered, might be the engineered-virus-for-bioterrorism notion. "But I think at this point we can't rule out that it was some sort of accidental lab escape. We can't rule out that it, you know, directly came from a bat to a human. We can't rule out that it went from a bat to an intermediate host to a human. All those remain possibilities." We don't know much about any viruses that might be closely related to SARS-CoV-2, he added. This was just after the announcements of further SARS-like bat coronaviruses in Thailand and Laos, but I hadn't caught wind of them yet, and I suspect those wouldn't have seemed close enough, not yet, to alter his view.

"There's not a lot of evidence in detail about what actually happened," Bloom told me.

"What kind of evidence do we need?"

He answered my question by comparison to the other two coronaviruses that have shown their lethality in humans, both with known origins in animals. The MERS virus, or something nearly identical, has been found in various animals, such as camels, Bloom noted. The original SARS virus—that was more complicated, that took longer, but eventually it was found in civets as intermediaries and then, all its pieces, in bats. "At this point, we don't have that sort of information for SARS-CoV-2."

This was candid and thoughtful, but it left my question still half unanswered: What sort of evidence do we need for the *other* alternative—to prove or disprove the lab leak hypothesis? Can that one *ever* be disproven? Even if field researchers turn up a bat coronavirus that's 99.5 percent similar to SARS-CoV-2, rather than just 96.2 percent similar (like RaTG13) or 96.8 percent similar (like one of the viruses found in Laotian bats), would that settle the question in all minds?

"Maybe it will never be found," Bloom said. He seemed to be talking now about a wild virus that is the irrefutable match. "We know of only a very, very small fraction of all the viruses out there in the world, right?" Until a very close match is found, he said, "I think what you have is a lot of people strongly defaulting to their prior beliefs." He laughed. "For some people, that prior belief is strongly centered on the lab work. And for some people, that prior belief is strongly centered on a natural zoonosis." I didn't disagree. He was reminding me of a very important reality about science: it's a rational process leading toward ever-clearer understanding of the material world, but it's also an activity performed by humans.

On May 26, 2021, almost as though he had read the Bloom-Relman letter and mulled it over for two weeks, President Joe Biden announced that he had asked the Intelligence Community of the United States to investigate the question of the pandemic's origins and report their judgment to him in ninety days. Yes, the Intelligence Community (IC) of the U.S. is a thing, established in 1981 as a formal assemblage of agencies that now number seventeen, including of course the CIA and the Office of Naval Intelligence (ONI), the Defense Intelligence Agency (DIA), the National Security Agency (NSA), and the Intelligence Branch (IB) of the FBI, but also Marine Corps Intelligence (MCI), Space Delta 7 (within the United States Space Force, whatever that is), as well as branches of the Coast Guard, the Department of the Treasury, the State Department, the Department of Homeland Security, and the Drug Enforcement Administration, among others. A bodacious aggregation of intelligences. The IC members furrowed their brows, they presumably read some scientific papers, presumably interviewed some scientists, presumably reviewed all their human and signals intel from China during late 2019 and early 2020, maybe they tortured some people (it has happened before), who knows what they did, and then delivered their findings to their boss, the Director of National Intelligence (DNI), or to one of her key underlings, who was obliged to make sense of it all. The Office of the DNI then produced a confidential report to the president, as ordered, within ninety days, of which an unclassified version was made public in October, sharing the IC's considered wisdom on the origins of SARS-CoV-2. It said: Well, we're not sure, and we don't agree with one another.

74

As the pandemic entered its third year, so did the questioning about its origins. Certitude hasn't arrived—it may never—but confidence about one crucial point seemed to increase further, on February 26, 2022, when Michael Worobey and a long list of coauthors posted a new preprint, with a title declaring plainly: "The Huanan Market Was the Epicenter of SARS-CoV-2 Emergence." The signatories included Kristian Andersen, Eddie Holmes, Andrew Rambaut, Marion Koopmans, David Robertson, Angela Rasmussen, Robert Garry, and others—a formidable assemblage of expertise, easily matching for formidability the assemblage who signed the Bloom-Relman-Chan letter. But this wasn't a matter of *argumentum ab auctoritate*, argument from authority, rightly scorned by logicians. This was new primary work from experts. Worobey, now with these many coauthors, pressed further into the matter of *just when* during December 2019, and *just where* within the city of Wuhan, did the earliest confirmed cases of COVID-19 appear. They found a very clear signal that December's cases had turned up either in or centered around the Huanan Seafood Wholesale Market, indicating that the virus first spread from the market into the surrounding community. Among those patients who had visited the market or worked there, "the overwhelming majority were specifically linked to the western section of the Huanan market, where most of the live-mammal vendors were located." The earliest two supposed cases, which had seemed to lack such connection back at the time when Daniel Lucey called attention to the paper by Huang and colleagues, both failed to withstand closer scrutiny of the case data.

Among the live mammals on sale for food in that western part of the market were raccoon dogs, known to be susceptible to SARS-CoV-2. One of the vendor stalls there, which had been selling live mammals, was especially hot with signs of the virus. It yielded five environmental samples positive for SARS-CoV-2. Those samples all came from objects associated with animal sales: a metal cage, carts for moving animals, a hair remover. And this was the same stall in which Eddie Holmes, on his visit in 2014, had photographed live raccoon dogs held in cages for sale.

Another signal revelation in this preprint involved the two-origins scenario, which was based on the fact that two distinct lineages of the virus, labeled Lineage A and Lineage B, were detected among the earliest patients. B was the lineage seen in almost all the December 2019 human cases, including all those directly linked to the Huanan market. Lineage A had appeared first in a sample taken on December 30, 2019, then in another on January 5, 2020. Neither of those Lineage A cases was directly linked to the market. For that reason, it seemed natural to suppose that Lineage A had gotten its start among humans somewhere else. But in the new preprint, with its fine-grain geographical analysis, Worobey and his colleagues reported that—to their surprise—the two early Lineage A cases, though not sited within the market, were centered around it in a very nonrandom way. Wuhan is a big city. Population density had been accounted for. Still, here were two dots on the data map, representing Lineage A, plotted within a mile of the market. Why? "These findings suggest," Worobey's team wrote, "that both lineages may have spilled over at the Huanan market during the early stages of the COVID-19 pandemic in Wuhan."

The same day that this preprint went up, another combination of scientists (also including Worobey, Holmes, Andersen, Jonathan Pekar as first author, and others from the "Epicenter" paper, plus additional authors) posted a companion preprint (again, not yet peer-reviewed), with new data and analysis, further supporting the two-origins hypothesis. These authors offered a different form of evidence that both spillovers likely occurred in the Huanan market. That evidence was genomic, not geographical, and included the fact that no verified intermediate genomes between lineages A and B had turned up in samples. There was no evidence that the two had diverged from a common line during evolution within humans. They seemed to have diverged *before* getting into humans. That implied two spillovers. Another inference from their evidence, these authors noted, was that SARS-CoV-2 didn't need further adaptation, and didn't need laboratory manipulation, to make it capable of infecting humans. It was ready to go, and it went—twice.

Both preprints were bolstered by a third, authored by a totally different team, which had been posted just a day earlier. This paper came from a group of mostly Chinese scientists led by George Gao, director-general of the China CDC. It offered a wealth of new detail on the environmental sampling done in and around the market—the swabbing of walls and door handles and left-behind animal carcasses and stray cats and waste bins and gutters—by field staff from Gao's CCDC and two other health agencies, between January 1 and early March 2020. The Gao paper reported, amid other data, 828 environmental samples collected inside the market. Of those, sixty-four samples tested positive for the virus, in most cases by the detection of genomic fragments. Complete genomes of SARS-CoV-2 were recovered from just three samples. One of the genomes belonged to Lineage A. It came from a sample taken the day the market closed, January 1. That affirmed the likelihood of two independent spillovers, from animals, having occurred in the market.

There's an irony: here was an evidentiary synergy between the Gao group and the Zhang-Holmes group, a bit of positive overlap to heal away the implicit competition for priority with which this story began. Although the point will still be argued, at least by holdouts, until more evidence emerges, these new findings further supported the proposition that SARS-CoV-2 came to humans by way of some direct, catastrophically unfortunate interaction with wild animals.

75

Nobody knows everything about this virus, and our efforts to comprehend it have only begun. As lengthy as the dreary months and years of the COVID-19 pandemic—the pandemic *so far*—may have felt to us, the time is early. We've scarcely started the effort of adapting ourselves and our societies for its next challenges and later stages. This virus is going to be with us forever. It will be in humans—always somewhere—and it will be in some of the animals that surround us. The rule "Never say never" is a sensible one, but no expert can tell us right now how SARS-CoV-2 might ever be eradicated. We haven't eradicated polio, despite decades of effort. We haven't eradicated measles. And those viruses have nowhere to hide except within humans. This virus has many more options. We might clear it from every human on Earth (not likely) and it will still be there in the white-tailed deer of Iowa, the stray mink on the landscape of Denmark.

It will continue to change. It will adapt to our adaptations. The latest variant as I write this, Omicron, seems a dramatic example of that.

Omicron lurched into view internationally in late November 2021, when scientists in South Africa reported its existence to the WHO in Geneva. The leader of the South African team was Tulio de Oliveira, now director of the Centre for Epidemic Response and Innovation at Stellenbosch University; he is also head of the country's Network for Genomic Surveillance, and still a professor at the University of KwaZulu-Natal. In mid-November, de Oliveira and his colleagues noticed a peculiar uptick of case numbers in Gauteng, the small but densely populated province that includes Johannesburg and Pretoria. The network scientists increased their genomic surveillance and, from one lab, de Oliveira's group received six genomes all sharing a high incidence of mutations. That was November 23, a Tuesday. Concerned, the team looked at other data and found evidence of the same strain, increasing in prevalence among Gauteng cases. By the following morning, November 24, as de Oliveira later told *The New Yorker*, "we began to see it might be a very suddenly emerging variant." He alerted the WHO. One day later, de Oliveira's team got shorthand results from a hundred more samples, randomly picked from around Gauteng, signaling that these infections were all the same variant. That morning, he briefed the health minister and then the president of South Africa, Cyril Ramaphosa. On Friday, with unusual quickness, the WHO declared this new strain a Variant of Concern. It was another variant within the B lineage, but very different from all the others, and by the PANGOLIN system it was classified as B.1.1.529. The WHO, jumping forward in the Greek alphabet, named it Omicron.

What made Omicron alarming was the presence of fifty-three mutations, fifty-three differences from the baseline Wuhan genome, most of them in the spike protein, causing more than thirty amino acid changes to the spike, half of those within the receptor-binding domain. There were also two mutations near the furin cleavage site. No one could discern from just a glance what those mutations might have been doing as the variant spread through Gauteng, or predict what they might do elsewhere, but the news got around fast—even faster than Omicron itself. On the morning of

November 24 in La Jolla, afternoon in Edinburgh, Kristian Andersen got a message via Slack from Andrew Rambaut: "This variant is completely insane." Andersen wrote back within minutes: "Just had a look at the list of mutations—so nuts."

Two major questions loomed immediately regarding Omicron, and they are the same two questions that everyone still wants answered, on a broader scale, with SARS-CoV-2 itself: Where did it come from and what will it do?

Omicron's panoply of mutations reflects a period of active, extensive evolution—because the mutations not only occurred but they were preserved, within the lineage, suggesting they offered adaptive value (or maybe some of them, but not all, had just been lucky). It had happened in some context so far unidentified. Suddenly the changes were simply there, bundled within a single virus strain that was thriving, as reflected in the genomes that de Oliveira's group saw. There were no traces in the available data of intermediate forms, containing just half these mutations, say. What could account for the absence of intermediates? A large group of scientists, including Tulio de Oliveira himself, soon addressed that mystery and others in a long, two-part post on the Virological website. First author among this group was Darren P. Martin, a computational biologist at the University of Cape Town.

Martin and his coauthors posited three possible explanations for the missing intermediates along the trail to Omicron. Maybe the sampling and sequencing in South Africa had been too sparse to detect what was happening among the population of patients. By this scenario, the intermediate stages were there, spread among the crowd, but science didn't see them. Or maybe the Omicron variant evolved in a chronically infected patient, with the intermediate stages occurring all within that patient, rather than from one patient to another. By this scenario, the intermediates had arisen and existed as a swarm within one sick person (or maybe several), but science didn't see them because that patient wasn't sampled repeatedly (or those several patients weren't), and the variant emerged fully fledged. Immunocompromised patients are more likely to suffer prolonged infections with SARS-CoV-2, and as Penny Moore reminded me, South Africa has a high number of people immunocompromised because they are living with HIV.

The third possibility harked back to sylvatic cycles and wildlife. Maybe Omicron derived from a reverse zoonotic event—human transmission to a nonhuman animal, followed by a period of evolution in the population of animals, followed then by a new spillover into humans. One paradigm for this scenario was the Cluster 5 variant that came out of mink, into people, in Denmark a year earlier. The difference would be that Omicron, unlike Cluster 5, happened to combine mutations making it supremely transmissible among humans. The intermediate stages might be there, out in the woods, but science didn't see them because science wasn't sampling wild animals (lions, leopards, striped polecats?) in South Africa. "At present there is no direct evidence to support or reject any of these hypotheses on the origin of Omicron," Martin's group wrote, "but as new data are collected, its origin may be more precisely defined." Or not.

The many innovations embodied in Omicron—and, particularly, thirteen changes to amino acids in the spike protein that haven't appeared in other SARS-CoV-2 variants—suggested to Martin and his coauthors that Darwinian natural selection favored the mutations either individually or collectively, because they increased the capacity of the virus to replicate, or to transmit, or to evade immune defenses. (Important to remember: some mutations, with neutral or even negative effect individually, can get passed along by sheer chance—but not likely fifty-three of them in one lineage.) How much better did the variant transmit? How effectively would it evade immune defenses, including the sort mustered by vaccines and boosters? Those questions too will be better answered "as new data are collected." The situation is fluid. You'll be able to see Omicron more clearly, by the time you read this, than Darren Martin and Tulio de Oliveira (let alone I) can right now.

The coauthors discussed one other intriguing uncertainty, and I alluded to it just above: Did evolutionary selection favor all of Omicron's significant changes individually, amino acid by amino acid, or did it favor them as a bundle—for their combined effect, their intricate interaction, their collective bottom line? The second of those propositions, as Martin and his coauthors noted, has a fancy name in genetics: positive epistasis. This concept is simple (only the details are complex). Epistasis refers to the interactive effects of genes in different parts of the genome, playing off one another or harmonizing, like instruments in different parts of an orchestra. A mutation inclined to have neutral or even negative impact individually may perform a beneficial function when that gene interacts with others. Further, the effect of one mutated gene may depend on the presence or absence of mutations in other genes. *Positive* epistasis in the context of Omicron means that multiple mutations are enhancing the adaptive value of one another. This variant, with its many mutations, may be a creature in which great intricacies of epistasis are making it more formidable.

Kristian Andersen has a postdoc in his lab, Edyth Parker, who put it more vividly when she said that "the epistatic fucking cirque du soleil of this virus is discombobulating."

76

Meanwhile, two years into the pandemic, people were still dying. On the day when Darren Martin and his coauthors posted their analysis of positive epistasis among the mutations in Omicron, December 5, 2021, South Africa recorded "only" a few dozen deaths, but the country's total Covid fatalities reached 90,466. The United Kingdom had a total of 146,622 deaths since the pandemic began. Italy was amid its fifth wave, with case numbers high again but, thanks no doubt partly to vaccinations, deaths not rising so steeply as during earlier waves: "just" forty-eight dead Italians that day. Germany was now getting its turn, with new cases, total deaths, and daily deaths all sharply up. Cases and deaths in South Korea too were sharply up, after all the long months during which that nation had seemed an exemplar in controlling the virus. Singapore, another exemplar, had also been hit with its worst wave in autumn 2021, from which cases and deaths had just begun to trend down.

What explained these irregular geographical patterns throughout the pandemic? What explained why one country was pummeled while another breezed by, to be pummeled later, the waves of misery and death rising and falling here and there like chop on a windswept ocean? There may be many partial answers, but there isn't a single, comprehensive one.

In earlier centuries and millennia, prophets and preachers might have as-cribed such variously allocated miseries to the whims and judgments of a capricious, punishing God. Nowadays we have science, but science hasn't supplied the one answer either. Not yet.

Among the last words that need saying here are the most obvious: COVID-19 has been a horrific sorrow to humanity, and particularly to those who by disadvantage, misfortune, age, or their own courageous choices have been most vulnerably exposed to it. Our societies should have done better to protect them. This book is no amelioration to the big sorrow, obviously, just an attempt to understand the biology and history of the responsible virus, seen over the shoulders of scientists. It's only a book.

The scientists can tell us a lot, about where a virus came from and where it might be going, but they can't tell us everything. And they know that. Molecular evolutionary virology, of the sort practiced by Eddie Holmes and Kristian Andersen and Susan Weiss and Michael Worobey and Áine O'Toole and Edyth Parker and others, is an extraordinarily pow-erful set of methods and foundational principles and tools, but it has limits and constraints. It gives us pieces of the whole. Those pieces are often in exquisitely fine focus, but still just pieces. The evolutionary virologists can only work with what they get, or what comes to them, from the wider world: samples of bat guano, samples of human saliva, living viruses that can be grown in a cell culture, images of viral particles as made visible by electron microscopy, and molecular sequences of genomic RNA or DNA—most of all, those molecular sequences. Those sequences are the Code of Hammurabi as unearthed on an eight-foot monolith, the three versions of a decree on the Rosetta Stone, the Gnostic Gospels in the orig-inal Coptic. The genomic sequences of viruses from the wild are quite often assembled from fragments put together using the clues of overlap-ping sections. If you assembled a jigsaw puzzle of the *Mona Lisa*, pull-ing pieces from five different boxes each containing the same puzzle, you would begin to appreciate the challenge. The viral sequence RmYN02, for instance, which is 29,671 letters long, was assembled from thousands of genomic fragments read from a pool of eleven fecal samples from eleven Malayan horseshoe bats captured and swabbed in a cave in southern Yun-nan. The paper announcing that discovery has thirteen authors, including

Holmes. Nobody knows everything, not even him. This is a poor man's variant of the uncertainty principle: even a scientist who has acquired great certitude on some aspects of a question will remain ignorant, or at least deeply uncertain, of other aspects.

A final personal note: I vastly admire the work of molecular evolutionary virologists, but I do that through a long, blurry, turned-around telescope of ignorance. I got my own academic education not in science but mainly in literature, and this uncertainty principle reached me not from the physicist Werner Heisenberg but from the novelist William Faulkner. More than fifty years ago, when I first read Faulkner and fell under his spell, the single impression that struck me most, the single bit of wisdom I saw underpinning his tales and the way that he told them, was that the truth of any event or person is fragmented, and those fragments are only available from diverse points of view. Anyone who has read the best of his work, the great books of his middle years—*The Sound and the Fury*, *As I Lay Dying*, *Light in August*, *The Wild Palms*, and *Go Down, Moses*, as well as, and most formidably, *Absalom, Absalom!*—will know what I mean. Others won't have needed novels to teach them the same: that reality in the round can only be grasped by adding up disparate perspectives. The discernment of truth—let's make that "truth," because it's such an imperious and suspect word—comes from listening to many voices. For a case in point, our pandemic. We need to hear many voices, and we need to help one another understand. Maybe that's our human version of positive epistasis.

One thing is nearly certain, I believe, amid the swirl of uncertainties. COVID-19 won't be our last pandemic of the twenty-first century. It probably won't be our worst. There are many more leopards in the environs of Mumbai. There are many more fearsome viruses where SARS-CoV-2 came from, wherever that was.

CREDITS

Many brilliant scientists and a few courageous public health officials gave their time, trust, and patience to help educate me on this subject. Beginning on January 7, 2021, I interviewed ninety-five such people by Zoom, in most cases for at least an hour and a half. Some questions I asked were unique and appropriate to each, pertaining to that person's work and scientific views; other questions were somewhat more personal, and those I asked everyone as part of a uniform protocol. I wanted to hear about their lives and their experiences during COVID-19, as well as about their professional judgments and insights. These ninety-five good people allowed me to record our conversations, and their voices, when quoted, come from transcriptions done by the redoubtable Gloria Thiede, my faithful and trusted transcriber for thirty years. If a person gasped, if a person chuckled, if a person hesitated, made a grammatical mistake, started over—Gloria put that in the transcript. There is no "reconstructed dialogue" in this book. A statement is not in quotation marks unless I had a direct, verbatim source.

Some of these ninety-five people—but only a subset dictated by narrative structure—appear in the book. A ninety-sixth source, Ali Khan, also appears; I interviewed him repeatedly between 2009 and 2020. (During 2020, before I began work on this book, I also benefited from interviews with other experts on topics related to SARS-CoV-2 and the pandemic—such as viral evolution in general, vaccine development, bats, and pangolins—toward journalistic pieces I was writing for *The New Yorker*, *National Geographic*, and the op-ed pages of *The New York Times*. Those people are thanked toward the end of this appendix; and any quotes from those interviews are cited by date in my Notes.) Others among the ninety-five provided illumination and shared personal experiences related to the pandemic that contributed to my account vastly but invisibly. I came to think of them all, collectively, as my Greek Chorus, although unlike the Chorus in classical Greek tragedy they did not speak in unison, and each said something different from the others. I am very grateful to them all. Quotes from these people are not sourced in my Notes but can be understood to come from the Zoom interviews, which occurred on the dates listed with these thumbnail bits of biography. My ninety-five plus Ali Khan, in alphabetical order:

JESSIE ABBATE
Interviewed: February 18, 2021

Jesse Abbate is an infectious disease ecologist focused on geospatial patterns of pathogen outbreaks. Based in Montpellier, France, she does epidemiology and translational data science for Geomatys, a computing development company that provides services in geospatial information processing and analysis. She also consults for WHO-AFRO on disease events, including COVID-19, in Francophone Africa. In the early weeks of January 2020, she was contacted by a U.S.-based international company that provided distance-learning services in China and elsewhere. They wanted her to write a report on how the schools in China might be affected by this new virus. You might want to consider that question globally, she told them. "Because this is not staying in China."

KRISTIAN G. ANDERSEN
Interviewed: January 7, 2021

Kristian Andersen is an infectious disease researcher, trained as an immunologist and now active along the boundaries of evolutionary biology, genomics, and virology. He's a professor in the Department of Immunology and Microbiology at the Scripps Research Institute, La Jolla, California. Working on Ebola virus and Lassa virus in West Africa, beginning in 2009, he and colleagues did much to develop the discipline of genomic epidemiology. He expects that vaccines can bring us to a point where COVID-19, in future years, is merely a recurrent tribulation on the scale of tuberculosis or measles (which still kill tens of thousands of people annually), but that it will likely never recede to the point, as some claim, of being no more serious than a common cold. Near the end of our two-hour talk, I asked him whether he feels the COVID-19 pandemic will have been bad enough to change human understanding and behavior such that we will be much better prepared next time. "I'm going to be a 'no' on that," he said.

DANIELLE ANDERSON
Interviewed: July 6, 2021

Danielle Anderson, a virologist, is a senior research fellow at the Peter Doherty Institute for Infection and Immunity at the University of Melbourne. Previously she has served as an assistant professor and scientific director of the ABSL-3 lab at Duke-National University of Singapore Medical School in Singapore. She has also been a visiting scientist at the Wuhan Institute of Virology, training in the BSL-4 lab. Anderson was at the WIV during October and November 2019, the last foreigner working there before the pandemic struck. Some proponents of the lab leak hypothesis have cited an undisclosed "intelligence report" that three employees of the WIV sought hospital care for respiratory symptoms in November 2019. "I was not aware of anything going on," Anderson told me. She was careful to note that it wasn't *impossible* such an event escaped her notice. "Someone could have been sick, and I could have not known about it," she said. "But three people being hospitalized? That would have been some discussion, I would have thought." She added: "I didn't hear anything about that."

SIMON ANTHONY
Interviewed: June 9, 2021

Simon Anthony is an associate professor in the Department of Pathology, Microbiology and Immunology at the University of California-Davis. He works on the genetics and ecology of coronaviruses, among other emerging viruses, and has done extensive field and laboratory studies of their relationships with bats.

RALPH S. BARIC
Interviewed: March 23, 2021

Ralph Baric is the William R. Kenan Jr. Distinguished Professor in the Department of Epidemiology at the University of North Carolina at Chapel Hill, and a professor in the Department of Microbiology and Immunology. He's considered one of the world's leading experts on the genetics of coronaviruses. He went through North Carolina State on a swimming scholarship in the mid-1970s, and stayed around to do a PhD in microbiology. In 2015 he was senior author on a paper, "A SARS-like Cluster of Circulating Bat Coronaviruses Shows Potential for Human Emergence," coauthored with thirteen other scientists, including Zhengli Shi. The work it described, criticized as gain-of-function research by some, valued as highly revelatory by others, was performed in Chapel Hill, North Carolina, not in Wuhan.

JESSE BLOOM
Interviewed: February 16, 2021

Jesse Bloom is an evolutionary biologist and a professor at the Fred Hutchinson Cancer Research Center in Seattle. He has a long-standing interest in how molecular properties of an organism relate to its more abstract evolutionary properties, such as evolvability and epistasis. Those questions are especially well illuminated by viruses, with their high rates of evolution.

BRANDON J. BONIN
Interviewed: April 14, 2021

Brandon Bonin is now director of the Santa Clara County Public Health Laboratory, San Jose, California. He has a master's degree in Forensic DNA and Serology and is finishing a doctorate in Public Health. He served four years in the United States Navy.

DONALD S. BURKE
Interviewed: July 8, 2021

Donald Burke is dean emeritus of the University of Pittsburgh School of Public Health, and a professor of Epidemiology and of Medicine at the University of Pittsburgh. He served twenty-three years in the United States Army, including a period as director of the U.S. Military HIV/AIDS Research Program, and a period as associate director for Emerging Threats and Biotechnology at the Walter Reed Army Institute of Research.

CHARLES H. CALISHER
Interviewed: April 9, 2021

Charlie Calisher is a professor emeritus of microbiology at Colorado State University in the College of Veterinary Medicine and Biomedical Sciences. He was chief of the Arbovirus Reference Branch at the CDC for sixteen years. He is also an expert on the taxonomy of viruses—the crucial business of delineating, sorting, and naming that allows scientists to organize and communicate knowledge of viral identities, characteristics, and diversity. He has a keen and critical eye for grammatical infelicities and taxonomic mistakes (and for bullshit) and has served as an editor of many books. His own book, *Lifting the Impenetrable Veil: From Yellow Fever to Ebola Hemorrhagic Fever and SARS*, was published in 2013.

ILARIA CAPUA
Interviewed: March 17, 2021

Ilaria Capua is a professor at the University of Florida and director of the One Health Center of Excellence for Research and Training there. She is also a former member of the Italian Parliament. She has described herself as a veterinarian by training and a virologist by passion, reflecting her fascination with the capacities of viruses. "It takes you a while to even get your head around what they're doing and how they're doing things," she told me. Capua did some of her early work on infectious bronchitis in chickens, a coronavirus disease. She has been an active advocate of the One Health perspective, which views animal health and human health as two inseparable and interactive concerns.

COLIN J. CARLSON
Interviewed: June 21, 2021

Colin Carlson is an assistant research professor at the Center for Global Health Science and Security of Georgetown University. He studies the interrelatedness of global climate change, loss of biological diversity, and emerging infectious diseases. He approaches infectious disease with the tools and perspective of a mathematical modeler, using quantitative data to attempt provisional predictions of what has occurred and will occur. Some of his work, with colleagues, has yielded the estimate that over two hundred bat species may harbor betacoronaviruses—viruses of the genus that includes the original SARS virus, SARS-CoV-2, and the MERS virus.

DENNIS CARROLL
Interviewed: February 9, 2021

Dennis Carroll, who trained as a molecular biochemist, served for fifteen years as director of the Emerging Threats Division of the U.S. Agency for International Development. He designed and oversaw the Emerging Pandemic Threats Program, which included the PREDICT project, making grants totaling $200 million over five years for research to identify pathogens, especially viruses, that seem poised to spill from their animal hosts into humans. Carroll is now a senior advisor on global health security at University Research Co (URC). He lives on a boat in Washington, D.C.

ALINA CHAN
Interviewed: June 7, 2021

Alina Chan, formerly a postdoctoral researcher at the Broad Institute of Harvard and MIT, is now a scientific advisor there. Her research, in the lab of Ben Deverman, involves the study and engineering of nonpathogenic viral vectors for use in human gene therapy. Chan is coauthor, with Matt Ridley, of *Viral: The Search for the Origin of COVID-19.*

SARA H. CODY
Interviewed: April 7, 2021

Sara Cody is a physician and epidemiologist who serves as health officer and public health director for Santa Clara County, California. Following medical school and internship, she did a two-year fellowship with the CDC's famed Epidemic Intelligence Service, investigating disease outbreaks. Early in the COVID-19 pandemic, she was the first official on the U.S. mainland to issue and implement a stay-at-home order. She could take that bold step, she told me, because Santa Clara County not only had a strong team of public health professionals; they also had a group of attorneys, in the Office of County Counsel, that "really, really, really understand public health law, and what we can and can't do."

PETER DASZAK
Interviewed: February 15, 2021

Peter Daszak is president of EcoHealth Alliance. He was educated in the U.K. as a parasite ecologist, and his early work included identifying the fungal disease chytridiomycosis as the cause of catastrophic global amphibian declines. That led to broader concerns with wildlife diseases, and with the dynamics between wildlife diseases and emerging infections in humans. Daszak became executive director of the Consortium for Conservation Medicine, and then head of the organization into which it morphed, EcoHealth Alliance. I met him in 2006, when I was asked by *National Geographic* to write a story about zoonotic diseases.

JESSICA DAVIS
Interviewed: March 22, 2021

Jessica Davis is now a postdoctoral researcher in the Network Science program at Northeastern University in Boston, working with Alessandro Vespignani (see below), a physicist who studies networks and spreading phenomena, including networks of spreading disease. When I interviewed Vespignani, he told me how the early news of a novel virus spreading out of Wuhan affected his young grad students. Soon their modeling told them it could become a pandemic. "I will always remember the eyes of the junior people," Vespignani told me. "Because they were like, okay, so this is real . . . what do we do? So, I think that evening I went home with such a burden." One of those grad students was Davis. Vespignani came back to the lab and asked her if she had ever seen the movie *Contagion.* When she said no, he suggested she watch it to prepare herself. "I think that was the moment where I thought, oh, this is going to be a problem," Davis told me.

ANDREW DOBSON
Interviewed: May 11, 2021

Andy Dobson is a disease ecologist and a professor in the Department of Ecology and Evolutionary Biology at Princeton. He has written much and influentially on the ecological dynamics of wildlife diseases, on human actions resulting in losses of biological diversity, and on the points of intersection between those two fields. Among the things Dobson understands keenly is the evolution of virulence. Will this virus, SARS-CoV-2, evolve toward being innocuous, I asked him, like one of the coronaviruses that causes a common cold? No, not necessarily, he said. Why? "Transmission is occurring before virulence is expressed." In other words, the virus is succeeding regardless of the number of deaths it leaves behind. It doesn't "care" whether it kills many people or few, so long as it can seize every chance to increase its proliferation.

PAUL DUPREX

Interviewed: February 17, 2021; March 4, 2021; March 12, 2021

Paul Duprex is a molecular virologist, a professor in the Department of Microbiology and Molecular Genetics at the University of Pittsburgh, and director of the Center for Vaccine Research there. He's a proud Ulsterman, born in County Armagh, educated at Queen's University of Belfast, and so exuberantly engaging that my interview with him stretched across three sessions. He studies the molecular basis of pathogenesis and attenuation of respiratory RNA viruses. He and a group of colleagues detected a mechanism by which SARS-CoV-2 transcends its relatively slow (for an RNA virus) mutation rate to acquire variance in its spike protein that gives resistance to neutralizing antibodies: by outright deletions of, rather than changes to, certain amino acids.

ISABELLA ECKERLE

Interviewed: March 12, 2021

Isabella Eckerle is a German virologist and physician, an associate professor and head of the Centre for Emerging Viral Diseases at the University of Geneva. Earlier in her career, during fieldwork in Africa, she devised a way to fast-freeze organ samples from bats in order to culture bat-borne cell lines in the laboratory. Her recent work includes study of human immune responses in adults and children to SARS-CoV-2.

JONATHAN H. EPSTEIN

Interviewed: May 17, 2021; June 23, 2021

Jon Epstein, a veterinarian and disease ecologist, is vice president for science and outreach at EcoHealth Alliance. He has done extensive fieldwork with colleagues in China, Australia, Saudi Arabia, and elsewhere on the ecology of bat-borne viruses, including Nipah, Hendra, Ebola, MERS-CoV, and SARS- CoV. He was part of the team, along with Zhengli Shi and Linfa Wang and others, who in 2005 showed that certain bats are reservoir hosts of coronaviruses similar to the original SARS virus. I have followed him up a bad ladder and across the rooftop of a derelict warehouse in Bangladesh, in the middle of the night, in order to watch him and his team capture and sample bats. When holding a giant fruit bat that might harbor a lethal virus, Epstein once told me, raise your arm above your head, because the bat wants to ascend and, if your arm is lowered, the bat will claw its way up your sleeve to your face—a piece of valuable advice that I have not yet needed to apply.

ANTHONY S. FAUCI
Interviewed: February 1, 2021

Tony Fauci is director of the National Institute of Allergy and Infectious Diseases (NIAID) and has been since 1984. He was born in Brooklyn and attended a Jesuit high school in Manhattan, at which he captained the basketball team as a quick five-foot-seven guard with a good shot. It wasn't the last time he would play above his size. He took a medical degree, did laboratory work in immunology, and led the NIAID, as a research scientist as well as a public health official, through the excruciating early decades of the AIDS pandemic. At the end of a very serious Zoom interview, I changed the tone for a moment and asked him who was the better Tony Fauci impersonator, Brad Pitt or Kate McKinnon. "I thought they were both terrific," he said. To see Pitt get an Emmy nomination for his portrayal on *Saturday Night Live* was great, Fauci added, but Kate McKinnon is the most hysterically funny actress he has ever seen. "She is *really* talented."

HUME FIELD
Interviewed: June 21, 2021

Hume Field is a veterinarian, environmental scientist, and emerging diseases epidemilogist based in Brisbane. He played a key role in identifying which kinds of bats serve as the natural reservoirs of Hendra virus (in Australia), Nipah virus (in Malaysia), SARS-CoV (in China), and Reston virus (an ebolavirus, in the Philippines). Field is an adjunct professor in the School of

Veterinary Science at the University of Queensland, a science and policy advisor to EcoHealth Alliance on China and Southeast Asia, and heads a private consultancy on wildlife-associated emerging diseases.

ROGER FRUTOS
Interviewed: March 25, 2021

Roger Frutos is a molecular biologist who studies the dynamics of emerging infectious diseases. He is a professor and research director at CIRAD, the French Agricultural Research Centre for International Development, in Montpellier. He devised his circulation model for the origin of SARS-CoV-2 after becoming dissatisfied with how other explanations, including the simplest spillover model, failed to fit the available data. "There is something wrong," he told me. "It doesn't match. To me there was something wrong. The pieces of the puzzle were not going well together." Among the implications he saw, for that lack of fit, would be failure of preparedness for the next pandemic caused by a virus of animal origin. "If we do what we do today, it's too late. Okay? And if we keep using the software, the medical software, then for the next disease to happen we will still be in the same situation. We are going to react instead of doing preventive action. And my question is, what will happen? Is the next one to come—because there will be a next one to come, okay—what happens if the next one to come is both highly virulent and highly transmissible? Something like the Spanish flu. Wow! We're going to be in real trouble." Late in our conversation, I drew him back to that, asking if this pandemic will have been horrible enough that we humans will learn what's necessary to avoid another. "I'm afraid no," he said. "What I see is that people are not going to change the way they do things. So we will not be ready for the next one." After a pause, he repeated: "And there will be a next one."

GEORGE FU GAO
Interviewed: June 7, 2021

George Gao is director-general of the Chinese Center for Disease Control and Prevention (CCDC). He grew up in Ying Xian, a county in the northwestern boondocks of Shanxi province, one of six siblings, his father a carpenter, his mother a homemaker who couldn't read. Qualifying for university, he was assigned to a veterinary program at Shanxi Agricultural University. "But I didn't want to be a vet," Gao told me. He spent half his time studying English, and then realized that he could link veterinarian science with human medical science. "So I decided to spend more time on microbiology." He got to Beijing for a master's degree, studying duck hepatitis virus, and then to Oxford University for a doctorate, working on another virus. He stayed another four years at Oxford on a postdoc fellowship, followed by three years at the Harvard Medical School, then back to Oxford for a lectureship, eventually returning to a professorship in China in 2004. It was a long journey for a carpenter's son from Ying Xian. His research before the pandemic included a study of how MERS-CoV binds to and enters human cells, using a different receptor protein from the one used by SARS-CoV (and, later, SARS-CoV-2). That paper suggested obliquely that variation in the receptor-binding domain could make betacoronaviruses versatile in their use of hosts. Gao's group developed, with Eli-Lilly and Junshi, the first monoclonal antibody (etesevimab) for use in COVID-19 patients under age twelve, and, with Zhifei Longcom, the protein subunit vaccine ZF2001, for use against the virus.

ROBERT F. GARRY
Interviewed: January 13, 2021

Robert Garry is a professor of microbiology and immunology at the Tulane University School of Medicine in New Orleans. Much of his career's work has focused on mechanisms of pathogenesis in retroviruses, most notably HIV. He has also worked on Ebola, Marburg, and other menacing RNA viruses, and established a laboratory in Sierra Leone for the long-term study of Lassa virus. Together with William Gallaher, now an emeritus professor at the Louisiana State University Medical School, he first illuminated the function of the spike proteins of coronaviruses such as SARS-CoV in binding to and entering cells. SARS-CoV-2 presented similarities but also important differences. "You can look at the spike protein," Garry told me, "and you can see from the sequence what this thing is likely to be like." *He* can look and see that, anyway. "There aren't too many of us left," he said, modest but confident, "that can take a look at a protein sequence and start to discern, you know, what that protein might be doing."

MARINO GATTO

Interviewed: February 22, 2021

Marino Gatto trained as an engineer but gravitated into ecology. He is now professor emeritus of ecology at Politecnico di Milano. Gatto and colleagues have charted the geographic spread of COVID-19 in Italy and modeled the potential effects of varied containment and control measures for bringing the epidemic curve down.

THOMAS R. GILLESPIE

Interviewed: February 22, 2021

Tom Gillespie is a professor in the Department of Environmental Sciences at Emory University in Atlanta. He trained as a disease ecologist and did a postdoc in molecular epidemiology. His research includes the study of zoonotic pathogens spilling between humans and other primates, and how ecological disturbance of wild landscapes by humans affects that process. Among his concerns during the pandemic has been the possibility of SARS-CoV-2 infecting wild chimpanzees, perhaps to the extent of establishing a sylvatic cycle.

BARNEY S. GRAHAM

Interviewed: June 1, 2021

Barney Graham recently retired as deputy director of the Virus Research Center of the NIAID and head of its Viral Pathogenesis Laboratory. His work on respiratory syncytial virus (RSV), which led to his ideas for an mRNA vaccine, stretches back thirty years. After his retirement, Graham and his wife moved to Atlanta to be near their children and grandchildren.

LISA GRALINSKI

Interviewed: June 29, 2021

Lisa Gralinski is an assistant professor in the Department of Epidemiology at the University of North Carolina. She spent five years as a postdoc in the lab of Ralph Baric. Gralinski studies the interactions between coronaviruses and the human immune system.

BARBARA A. HAN

Interviewed: March 9, 2021

Barbara Han is a disease ecologist at the Cary Institute of Ecosystem Studies in Millbrook, New York. She uses computer algorithms and machine learning (by which algorithms can improve) to analyze patterns and processes involved in zoonotic spillovers of pathogens and to try to forecast coming outbreaks. Her first information about a novel virus in Wuhan came, as for others, in late 2019. "As soon as I heard about it," she told me, "I thought, 'Here we go.'"

VERITY HILL

Interviewed: February 2, 2021

Verity Hill is now a postdoc in the lab of Nathan Grubaugh at the Yale School of Public Health. As a graduate student in molecular evolution, phylogenetics, and epidemiology at the University of Edinburgh, she worked in the lab of Andrew Rambaut. She was halfway through the third year of her PhD studies, using genomics to study how Ebola spread through West Africa during the 2014 epidemic, when the news from Wuhan took hold. Rambaut had warned her, when he approved her Ebola idea, "If there's an epidemic, you'll probably have to change." She did change her focus, like the rest of Rambaut's lab, to SARS-CoV-2. She wasn't surprised; she had realized that, in the four years it might take her to finish a dissertation, a dangerous new virus was likely to emerge.

EMMA HODCROFT

Interviewed: February 9, 2021

Emma Hodcroft is a molecular phylogeneticist, currently a postdoc in the lab of Christian Althaus at the University of Bern. She is part of the Nextstrain team, an international group of collaborators who track the evolution and relatedness of strains of pathogens, including SARS-CoV-2, using the latest available genome data. In mid-January 2020, at a point when only about ten sequences of SARS-CoV-2 samples had become available online, Hodcroft took part in a virtual meeting with Nextstrain colleagues at which they decided to build a phylogenetic tree of the virus's sequences. "Because we think this will be useful for people to be able to see how the sequences are related to each other, what the mutations are," she told me. "Of course, with Nextstrain we can put it on a little map, and we can draw the little lines, and we thought this will help people to understand the information that's coming out."

EDWARD C. HOLMES
Interviewed: February 8, 2021

Eddie Holmes is an ARC Australian Laureate Fellow and Professor at the University of Sydney, and a Fellow of the Royal Society, London. He wrote the book on the evolution of RNA viruses.

PETER J. HOTEZ
Interviewed: March 18, 2021

Peter Hotez is a physician-scientist, a professor in two departments at the Baylor College of Medicine in Houston, dean of the National School of Tropical Medicine there, co-director of the Texas Children's Hospital Center for Vaccine, the author of *Forgotten People, Forgotten Diseases* and other books, a coauthor on roughly six hundred articles, and a frequently seen commentator on national television. He sleeps fewer hours than you or I do. A text search of his 300-page curriculum vitae does not yield the word "hobbies," for which there can't possibly be time. But he is nonetheless a genial, conversable man, generous with his efforts to communicate science as well as to do it. His team has helped to develop low-cost COVID-19 vaccines, using recombinant-protein methodology and in partnership with vaccine manufacturers in developing countries; one of those vaccines, developed with the company Biological E, has been released for emergency use authorization in India and may soon be available globally. The point is to create thermally stable vaccines, cheap and widely accessible, that can be taken orally or as a nasal squirt. "I think it's doable," he told me. "It's a matter of time and additional money." Hotez is also an ardent public defender of vaccines, against a hostile anti-science movement that has grown in volume nationally, his closeness to the subject reflected in his 2018 book about his youngest daughter, *Vaccines Did Not Cause Rachel's Autism*.

PETER J. HUDSON
Interviewed: April 12, 2021; May 3, 2021

Pete Hudson, a wildlife disease ecologist, is Willaman Professor of Biology and former director of the Huck Institutes at Penn State University. His studies include, but are not limited to, those wildlife diseases that become human diseases as well. It was his birthday around which a group of like-minded souls, small at first but eventually numbering hundreds, coalesced to create the Ecology and Evolution of Infectious Diseases (EEID) annual meeting. He and his wife live on and manage a ninety-acre woodland nature reserve. He takes photos of wildlife and builds furniture.

WILLIAM B. KARESH
Interviewed: April 23, 2021

Billy Karesh is a wildlife veterinarian, formerly head of the International Field Veterinary Program and vice president of the Wildlife Conservation Society and now executive vice president for health and policy at EcoHealth Alliance. As nearly as I can trace, he coined the term "One Health" for the enterprise that views animal health, human health, and ecosystem health as inseparable. Karesh travels widely to study and treat sick wildlife—as described in his 1999 book, *Appointment at the Ends of the World*—and to investigate zoonotic diseases, and he sees a lot of wildlife viruses. When the novel coronavirus first emerged among humans in Wuhan, before its capacity for asymptomatic transmission was recognized, Karesh thought it could be controlled, like SARS-CoV. By early February 2020, though, "I was in the camp that, you know—that we were going to live with this forever," he told me. "And I think we will."

MATT KELLEY
Interviewed: April 22, 2021

Matt Kelley was, for eleven years, including the first year and a half of the pandemic, health officer of Gallatin County, Montana. That happened to give me (because I live in Gallatin County) a close view of how hard he worked, how sorely his efforts were impeded, and how badly he was abused by certain elements of the population he tried to serve during COVID-19. (Angry, menacing people lurked outside his house. We picketed on the streets of downtown, in countering support.) Kelley grew up in Wisconsin, a Packers fan, and signed on as a business reporter for the *Omaha World-Herald* after college. The *World-Herald* sent him to Washington, D.C., and he became a political reporter for a stretch of years. Then, wanting a big change, he did a stint in the Peace Corps with his wife, two years in a little village in Mali, West

Africa, as a water and sanitation extension agent. Back in the U.S., Kelley got a master's degree in public health and worked on public health and mental health systems for the mayor's office in Washington, D.C., then was offered a job in this place called Bozeman, Montana. When he got the decision call from Bozeman after having interviewed, he was frankly informed that he had been second choice, but the first candidate had declined. "I always note that Vince Lombardi was second choice too," Kelley told me. "So I felt I could rationalize it." He has left the county job, but he still works on public health in Montana.

GERALD T. KEUSCH
Interviewed: March 19, 2021

Jerry Keusch is a professor of medicine and international health, and associate director of the National Emerging Infectious Diseases Laboratory, at Boston University. He is also a former director of the Fogarty International Center of the National Institutes of Health, which supports medical research and the training of researchers internationally. He has thought deeply and carefully about the matter of preparedness against pandemic threats, as reflected in his coauthorship (with Nicole Lurie) of a 2020 report to the Global Preparedness Monitoring Board (an arm of the WHO and the World Bank), *The R&D Preparedness Ecosystem: Preparedness for Health Emergencies.* Keusch's "favorite" lesson from a lifetime's work in public health, he told me with gentle mordancy, is that "when public health works, nothing happens. When nothing happens, the politicians say, 'What, are we paying for nothing? Let's put the money somewhere else.' So they steal the money away from public health until something happens, and then they say, 'Where was public health when we needed them?' Well, you didn't fund them. And it's this cycle over and over and over again. And that's something that needs to be absolutely reversed."

ALI S. KHAN
Interviewed: August 11, 2009; March 17, 2020; March 19, 2020; March 23, 2020

Ali Khan is dean of the College of Public Health at the University of Nebraska Medical Center and professor of epidemiology there. When I first met him, in 2006, he was deputy director of NCZVED (the National Centre for Zoonotic,

Vector-Borne, and Enteric Diseases) at the CDC in Atlanta. In 2010, he became director of the CDC's Office of Public Health Preparedness and Response. In 2014, he moved to the role in Nebraska. In 2015, he was part of a WHO response team in Sierra Leone during the West African epidemic of Ebola, a virus with which he was already harrowingly familiar. In 2016, he published his book, *The Next Pandemic.* In late 2021, he volunteered for a hitch working against COVID-19 in the Northern Mariana Islands. During the past thirty years, he has done more than two dozen other field postings for infectious disease response, from Crimean-Congo hemorrhagic fever in the Sultanate of Oman to hantavirus pulmonary syndrome in Brazil to monkeypox in Indiana. Blessed is a man who so loves his work, so generously gives his skills to the aid of suffering people, and so calmly retains a sense of humor amidst it all.

EMER KINIRY
Interviewed: June 13, 2021

Emer Kiniry is a senior administrative assistant at Canuck Place Children's Hospice in Vancouver, British Columbia, the first freestanding hospice in North America to provide comprehensive services for children with complex medical challenges. Kiniry's first name is Irish and so is her background; she was born and raised in Dublin. At Canuck Place, she told me, the children generally can't be vaccinated against COVID-19 because of compromised immune systems. So the hospice has taken drastic measures to enable continued service—staff reductions, staff working remotely where possible, volunteer shifts on hold, regular health screening, physical distancing, mandatory masks, no extended family and friends visiting, virtual counseling and virtual health care meetings where possible. There is one compensation. "I feel like it's almost the safest place you can be during Covid," Kiniry said. The extreme vigilance protects children and parents who have enough to concern them already.

MARION KOOPMANS
Interviewed: March 8, 2021

Marion Koopmans, trained as a veterinarian and veterinary internal medicine specialist, is an endowed professor and head of the Department of Viroscience at Erasmus Medical Center in Rotterdam. She led the research response against an avian influenza outbreak in the Netherlands, in 2003; played a key role in tracing MERS-CoV to dromedary camels as an intermediate host, on the Arabian Peninsula and in Africa, in 2014; and was responsible for deploying mobile diagnostic laboratories from the Netherlands to Sierra Leone and Liberia during the Ebola epidemic of 2013–2016. She also leads a WHO collaborating center and an international research consortium (VEO) focusing on emerging diseases.

JEFFREY P. KOPLAN
Interviewed: February 18, 2021

Jeff Koplan is a physician, a public health professional, and a former director of the CDC. Following his time at the CDC, he became director of the Emory Global Health Institute at Emory University, and later vice president for Global Health at the university. When I asked him about the tenure of Robert Redfield as CDC director during the Trump administration, he said, "The CDC director plays a tricky role, has a tricky task to perform." Your immediate boss is the political appointee serving as secretary of health and human services. That person's boss is in the White House. "It's less tough when the folks above you believe in science." Do you have any wisdom, I asked Koplan, on how we can reverse the science denialism that has become part of the American ethos? "Boy, is that depressing," he said.

BETTE KORBER
Interviewed: June 18, 2021

Bette Korber is a computational biologist and a laboratory fellow in Theoretical Biology and Biophysics at the Los Alamos National Laboratory. She oversees the HIV Database and Analysis Project at Los Alamos and has devoted much of her career to HIV research. Her primary work has been the study of viral evolution under immune pressure; she uses this information to design vaccine strategies against highly variable viruses. In 2000, she and colleagues published results of a study to determine when the pandemic HIV subtype (HIV-1 group M) began to diverge from its chimpanzee virus progenitor; that is, when the fateful spillover occurred that marked the start of that pandemic. They estimated 1930. This was a few years before Michael Worobey and colleagues, with the benefit of some older samples, pushed the estimate back to around 1908. Korber and her collaborators, in their work on COVID-19, have drawn heavily and gratefully from the GISAID database of SARS-CoV-2 genomes, so abundant and so quickly available as the virus continues to mutate and evolve. In the field of HIV studies, she told me, "new" genomic data are often one or two years old before they are published and shared. "In contrast, new data in the SARS-CoV-2 field was sampled last week; this change is because of GISAID." She and her colleagues receive a daily data feed that enables their work, and GISAID provides the same service to bioinformatics groups all over the world.

JENS H. KUHN

Interviewed: April 15, 2021; May 3, 2021

Jens Kuhn is a virologist, a historian of virology, and an expert on biodefense. He is a principal scientist and director of virology at the Integrated Research Facility at Fort Detrick, Maryland. In 2001, he became the first Western scientist welcomed to do a laboratory rotation at a former Soviet bioweapons facility, Vektor, in Novosibirsk. He is also author of *Filoviruses*, a compendium of forty years of studies on Ebola virus, Marburg virus, and their cousins. I first met Jens at a filovirus conference in Libreville, Gabon, where, thanks to the accident that we were staying at a smaller and less fancy hotel and rode a bus together to the conference hotel each morning, we became friends. He has an MD and two PhDs but he's fun.

MARCUS V. G. dE LACERDA

Interviewed: February 10, 2022

Marcus Lacerda is a practicing physician at the Fundação de Medicina Tropical Doutor Heitor Vieira Dourado, in Manaus; a researcher with the Oswaldo Cruz Foundation (Fiocruz); and a professor of tropical medicine at the University of Amazonas. He also coordinates the Institute of Clinical Research Carlos Borborema in Manaus. Lacerda has worked on malaria, and prescribed chloroquine as a treatment against it, for more than two decades. The malaria pathogen is not a virus. His recent research, with colleagues, has offered strong data against use of chloroquine at high doses against COVID-19. Has anyone in your family had a case of Covid? I asked him. "Oh yeah, every-body!" he said. "Everybody." But they were lucky.

HEIDI J. LARSON
Interviewed: June 11, 2021

Heidi Larson is an anthropologist and the founder of the Vaccine Confidence Project. Her 2020 book is *Stuck: How Vaccine Rumors Start—And Why They Don't Go Away*. Before we discussed antivaccine rumors, I asked her about rumors that SARS-CoV-2 originated from a laboratory leak. "With rumors," she said, "until you have a clear answer, they're going to keep resurfacing." Especially in the context of uncertainty. "And in this case, you know, it's the perfect fertile ground for rumors, because you've got kind of incomplete information."

RAMANAN LAXMINARAYAN
Interviewed: April 23, 2021

Ramanan Laxminarayan is founder and director of the Center for Disease Dynamics, Economics and Policy in Washington, D.C., and a senior research scholar at Princeton University. He has worked much on the problem of antibiotic-resistant bacteria, and on antibiotic effectiveness—seen in terms of policy and justice—as a shared global resource. Some of his recent writings have examined the epidemiology and transmission of COVID-19 in India and estimates of Covid mortality.

PHILIPPE LEMEY
Interviewed: June 18, 2021

Philippe Lemey is an associate professor in the Laboratory for Clinical and Epidemiological Virology, Department of Microbiology, Immunology and Transplantation at KU Leuven University, in Belgium. He works on virus evolution and molecular epidemiology, and has coauthored papers on the evolution and spread of SARS-CoV-2 in Europe, Brazil, the United States, and elsewhere. I asked him whether the virus, from the beginning, seemed suspiciously well adapted to infecting humans, as asserted by proponents of the lab leak hypothesis. No, he said. "What we're seeing is that, you know, this has evolved into a generalist pathogen with a reasonable capability of transmitting to humans already in the bat population." He paused. I waited, letting those last four words sink in; in the bats, before spillover, he meant. "There's no need to kind of invoke the lab theory for that," he added. I wanted to be clear, so I pressed: It became a virus broadly well suited to using the ACE2 receptors in a whole range of mammals? "Exactly. Yeah," he said.

YIZE (HENRY) LI
Interviewed: February 10, 2021

Yize Li is an assistant professor in the Biodesign Institute at Arizona State University. His colleagues and friends in the West call him Henry. He studied bioengineering in Chongqing, virology in Shanghai, and did a postdoc in virology and immunology in the lab of Susan Weiss at the University of Pennsylvania. Like his mentor, Weiss, and not so very many others, he studied coronaviruses before it was fashionable, with a focus on virus-host interactions and innate immune responses. From an encounter at a conference in Shanghai in 2018, and some subsequent WeChat communications, Li knew Yong-Zhen Zhang, Eddie Holmes's collaborator on the first public release of a full genomic sequence of SARS-CoV-2. "He didn't get permission from the Chinese government. That make them angry, very angry," Li told me. Zhang is a brave man and something of a nonconformist, he said. "And then they shut down his lab." Zhang was told, according to Li, "You cannot work on SARS-CoV-2 anymore."

POH LIAN LIM
Interviewed: June 16, 2021

Poh Lian Lim, a physician and public health official, is director of the High Level Isolation Unit at the National Centre for Infectious Diseases in Singapore, and a senior consultant to the Ministry of Health. Back in 2004, she led a study of an apparent lab accident at Singapore General Hospital during August 2003, when a graduate student became infected with the original SARS virus, three months after the SARS outbreak in Singapore had been brought to an end. The grad student was working on West Nile virus, and possibly suffered exposure to the SARS virus because both viruses were growing simultaneously in the monkey kidney cells in which the student was culturing

West Nile. Do you have any thoughts, I asked Lim, about the possibility of a lab leak, somehow related to SARS-CoV-2, at the Wuhan Institute of Virology? "I usually, generally try not to comment on these things," she said. Then she added, "There's a difference between what can happen versus what did happen, right?"

W. IAN LIPKIN
Interviewed: January 9, 2021

Ian Lipkin, a physician and research virologist, is John Snow Professor of Epidemiology at Columbia University, and director of the Center for Infection and Immunity at Columbia's Mailman School of Public Health. He's an expert on the use and development of molecular methods for identifying new pathogens, such as Nipah virus. He served as scientific advisor for the 2011 movie *Contagion*, directed by Steven Soderbergh, in which the pandemic pathogen is based loosely on Nipah. Lipkin was a coauthor on the early 2020 paper by Andersen and colleagues, "The Proximal Origin of SARS-CoV-2," though he told me almost a year later that he was less comfortable than some of his co-authors with dismissing the possibility of a laboratory accident. Maybe some graduate student or trainee in the lab of Zhengli Shi tried to grow a new virus from bat samples, and that person succeeded but got sloppy. Zhengli Shi herself never would have concealed such a virus, he posited. She is conscientious and she is professionally motivated to publish discoveries. "If they found a virus that looked like this," Lipkin said, "and she knew about it, she would have sequenced it and she would have published it." So her cognizance can be ruled out. "But that's not the same thing as saying it can't possibly have come from this lab." He had no reason to believe some-such person had gotten sloppy, he added, "but I can't rule it out."

MARC LIPSITCH
Interviewed: June 30, 2021

Marc Lipsitch is a professor in the Department of Epidemiology, and director of the Center for Communicable Disease Dynamics, at the Harvard T.H. Chan School of Public Health. He has been an outspoken critic of gain-of-function research with potential pandemic pathogens. But he discreetly declined to speak on the record about that topic in our conversation because of a pending change in his role in the scientific community. As now announced, he will be director for science at a new center within the CDC, the Center for Forecasting and Outbreak Analytics. Back in May 2020, Lipsitch and coauthors of a paper in *Science* projected that "prolonged or intermittent social distancing may be necessary into 2022," to prevent critical care facilities being overwhelmed by COVID-19 cases. "Even in the event of apparent elimination" of the virus, they added, "SARS-CoV-2 surveillance should be maintained because a resurgence in contagion could be possible as late as 2024."

DANIEL R. LUCEY
Interviewed: January 11, 2021; January 14, 2021

Daniel Lucey is a physician, public health expert, and professor at the Geisel School of Medicine at Dartmouth. Beginning in January 2020, he posted a long series of influential blogs about COVID-19 and SARS-CoV-2 (the disease and the virus) to the "Science Speaks" page of the website of the Infectious Diseases Society of America, reporting on and challenging various facts and ideas about the pandemic. His first post, in the form of questions and answers about the possible origin and nature of the outbreak, was written on January 6, 2020. Exactly one year later, the day of the mob attack on the U.S. Capitol, Lucey emerged from his apartment, just off Pennsylvania Avenue, and walked through the crowd of people, some with their signs and their flags and their red MAGA hats, some with their costumes and their weapons, who had just left Donald Trump's incendiary speech at the Ellipse and marched toward the Capitol. Lucey himself moved into that flow "like a salmon," he told me. "I walked upstream against them because I wanted to see what they looked like. And I wore this hat." He was modeling it for me on Zoom: a yellow ball cap, custom-made for him, bearing the words: "MENE, TEKEL Daniel 5:25." It referred to the story in the Book of Daniel about Belshazzar's Feast, in which the king of Babylon is warned, through magical handwriting on a wall: "God hath numbered thy kingdom and finished it. Thou art weighed in the balances, and art found wanting." Lucey is a man of intense conscience and intense views. His one-man counterdemonstration was probably caught on FBI footage somewhere, he told me. He had been a little concerned that someone among the angry marchers would recognize the allusion. "But nobody did."

NICOLE LURIE

Interviewed: March 25, 2021

Nicole Lurie, a physician and public health expert, was assistant secretary for preparedness and response (ASPR) in the Department of Health and Human Services during the Obama administration. She is coauthor, with Gerald Keusch, of the 2020 report to the Global Preparedness Monitoring Board (an arm of the WHO and the World Bank), *The R&D Preparedness Ecosystem: Preparedness for Health Emergencies.* She was primary author of the World Bank International Vaccines Task Force Report. Lurie lectures at the Harvard Medical School and serves as strategic advisor to the CEO of the Coalition for Epidemic Preparedness Innovations, among other consultative work.

HOLLY L. LUTZ

Interviewed: May 10, 2021

Holly Lutz is an evolutionary biologist who studies, among other things, the microbiomes of bats. She is a postdoctoral affiliate at the Scripps Research Institute in La Jolla, California, and a research associate within the Negaunee Integrative Research Center of the Field Museum of Natural History, in Chicago. She has done fieldwork on mammals and their pathogens in Kenya, Mozambique, and other parts of Africa. In 2013, while trapping bats within a large hollow tree in Uganda, she and several of her colleagues contracted lung infections, later diagnosed as histoplasmosis, caused by spores of a fungus carried in bat guano. The same tree had been implicated, two years earlier, in another outbreak of histoplasmosis among visiting biology students. Lutz's symptoms included fever, headache, weakness, weight loss, and a dry cough. Unlike the three Mojiang mineworkers, whose infections have been presumed but never proven to have been caused by a virus, not a fungus, Lutz and her colleagues survived.

SPYROS LYTRAS

Interviewed: June 24, 2021

Spyros Lytras is a PhD student in virology at the University of Glasgow, working with David L. Robertson and other supervisors. He studies the molecular evolution of viruses, including SARS-CoV-2 and its relatives among the SARS-like coronaviruses. He is co-lead author, with Oscar A. MacLean, of a paper in *PLOS Biology* titled "Natural Selection in the Evolution of SARS-CoV-2 in Bats Created a Generalist Virus and Highly Capable Human Pathogen."

LAWRENCE C. MADOFF
Interviewed: March 4, 2021

Larry Madoff is a professor of medicine at the University of Massachusetts Chan Medical School. He is an infectious disease doctor specializing in the epidemiology of emerging pathogens and international public health. He has served since 2018 as the medical director for infectious disease for the Massachusetts Department of Public Health; more recently, he has retired as editor of ProMED-Mail.

JONNA A. K. MAZET
Interviewed: May 11, 2021

Jonna Mazet, trained as a wildlife veterinarian and an epidemiologist, is a vice provost at the University of California-Davis, and Chancellor's Leadership Professor of Epidemiology and Disease Ecology in the One Health Institute of the university's School of Veterinary Medicine. For more than a decade she was global director of the PREDICT project of the U.S. Agency for International Development, leading a multinational consortium to collect samples from wildlife and detect novel viruses with potential to become human pathogens. Project teams identified 1,200 animal viruses with apparent potential to cause human illness, including more than 160 coronaviruses. The PREDICT project embodied the discovery side of a complicated discovery-versus-surveillance dichotomy of scientific opinion with regard to pandemic preparedness and response: "discovery" meaning find dangerous viruses before they spill over, "surveillance" meaning watch for outbreaks and control them before they become epidemics. The project was targeted for closure in 2020 by the Trump administration, at the completion of two five-year funding cycles. Then it was partially extended with a modest grant soon after SARS-CoV-2 arrived in America, when even Trumpian officials (some of them) were incapable of denying the gravity of emerging pandemic threats. The preceding sentence represents my language and view, for which Jonna Mazet bears no responsibility.

PLACIDE MBALA-KINGEBENI
Interviewed: April 18, 2021

Placide Mbala-Kingebeni, a physician and microbiologist, is head of the Epidemiology Department and the pathogen sequencing laboratory at the Institut National de Recherche Biomédicale in Kinshasa, Democratic Republic of the Congo. He has worked with the PREDICT project (see above, under Dennis Carroll and Jonna Mazet), studied HIV prevalence among the armed forces of the DRC, and led the Viral Hemorrhagic Fever Unit of the Institut during outbreaks of Ebola virus fever. It has been a difficult time for DRC recently, Mbala-Kingebeni told me. They had an outbreak of Ebola in Bas Uele province in 2017, then another outbreak, this one in Equateur province, in 2018. Having brought that to an end, they detected still another outbreak, in North Kivu province, starting in August 2018, finally ended in June 2020. "At the same time, during the same period, we also face the pandemic of COVID-19." And measles? I asked. "A new outbreak," he concurred. "Measles, a new outbreak, with Ebola again in Equateur in 2020, and then a new outbreak of Ebola in North Kivu in 2021." The medical professionals and disease scientists of DRC, such as Mbala-Kingebeni and Jean-Jacques Muyembe Tamfum, perform at amazing and heroic levels against dangerous viruses despite a dire shortage of resources. They have experience.

JASON S. McLELLAN
Interviewed: August 12, 2021

Jason McLellan is a professor of molecular biosciences at the University of Texas, Austin. As a postdoc under Peter D. Kwong at the Vaccine Research Center of the NIAID, and later while in his academic positions at Dartmouth and then the University of Texas, he worked with Kwong, Barney Graham, and other colleagues at determining the three-dimensional structures and consequent properties of fusion proteins used for cell attachment and entry by various viruses, including respiratory syncytial virus and SARS-CoV-2. This led to his role, with his own lab group and others, in creating a stabilized form of the spike protein of SARS-CoV-2, a crucial element in developing the mRNA vaccines of Pfizer and Moderna.

VINEET DAVID MENACHERY

Interviewed: April 16, 2021

Vineet Menachery studies the dynamics of virus-host interactions that generate disease in the hosts, and the factors that suggest a given animal virus might be capable of spillover into humans. He uses reverse genetic systems (viruses brought to life from genomes), animal experiments, and other methods. He's an assistant professor in the Department of Microbiology and Immunology, University of Texas Medical Branch, in Galveston. Menachery did postdoctoral work for almost seven years in the lab of Ralph Baric, in Chapel Hill. One study published during that period, of which he was lead researcher and Baric was senior author, involved using reverse genetics to produce a chimeric virus, composed of the spike protein of a wild coronavirus from a horseshoe bat in China, mounted in a backbone of the original SARS virus that had been laboratory-adapted to grow in mice. The primary question was whether the bat coronavirus, SHC014, posed a threat of emerging into humans. The chimeric virus grew in human cells, so the answer was yes. This work, done in Chapel Hill, was controversial—praised by some scientists for the warning it offered, condemned by others as dangerous gain-of-function work. "There is some risk with that, I don't disagree," Menachery told me. "But I don't know if we're better off not knowing that those viruses existed"—that is, remaining oblivious to SCH014, versus showing the threat it might present. "And unfortunately, this was the only way we could have shown it."

PENNY L. MOORE

Interviewed: June 15, 2021

Penny Moore holds the South African Research Chair of Virus-Host Dynamics at the University of Witwatersrand and the National Institute for Communicable Diseases, South Africa. She studies HIV and its capacity to evolve, escaping immune defenses by changing its susceptibility to antibodies. This topic is relevant to HIV vaccine design efforts. It also has parallels, in some degree, to the evolution of SARS-CoV-2 and its variants. The Beta variant had recently emerged in South Africa when I spoke with her, and the Omicron variant has emerged since. Like some other scientists, she was concerned that the variants might be most likely to emerge from immunocompromised patients, in whom a lingering infection—and therefore continuous viral mutation and evolution—is possible. South Africa has 7.5 million people living with HIV, but they aren't the only humans at risk of lingering infections. "These variants are clearly not just popping out of HIV positives," Moore told me. "I think there are lots of studies in the U.S. now that have shown other immunosuppressed people who struggle to clear the virus, for whatever reason."

CARLOS MEDICIS MOREL
Interviewed: March 26, 2021; April 28, 2021

Carlos Morel is director of the Center for Technological Development in Health, of the Oswaldo Cruz Foundation (Fiocruz) in Rio de Janeiro. He is an emeritus director of Fiocruz itself. He is also a former director of the Programme for Research and Training in Tropical Diseases (TDR) of the World Health Organization. Besides answering my questions and recounting his own struggle with COVID-19, over the course of two long and genial Zoom conversations, he was kind enough to put me in touch with his friend George Fu Gao.

DAVID M. MORENS
Interviewed: February 26, 2021

David Morens, a physician and epidemiologist, is senior advisor to the director of the NIAID, Anthony Fauci. That means, among other things, that he coauthors scientific papers with Fauci. "Emerging Pandemic Diseases: How We Got to Covid-19," published in the journal *Cell*, was one example. And sometimes Morens publishes papers a little too controversial to carry Fauci's name, such as "The Origin of COVID-19 and Why It Matters," coauthored with Charlie Calisher, Jerry Keusch, and seven other distinguished scientists. At the end of that one, the authors note that it was "highly unlikely that SARS-CoV-2 was released from a laboratory by accident because no laboratory had the virus nor did its genetic sequence exist in any sequence database before its initial GenBank deposition (early January 2020)." As for the notion that the virus was engineered for sinister purposes, "Mother Nature 'knows' how to make bad viruses and human beings know that those viruses are bad once Mother Nature has made them," Morens told me. "But human beings don't have the knowledge to manipulate viruses that will turn something that is starting material into something new that is really bad." Try it in a million experiments and you'll fail 999,999 times, he added. And on the millionth try, you won't even know you've succeeded—unless you experiment on people.

JOHAN NEYTS
Interviewed: June 10, 2021

Johan Neyts is professor of virology in the Faculty of Medicine at KU Leuven (University of Leuven), in Belgium, and a past president of the International Society for Antiviral Research. He works on vaccines and antiviral drug candidates against a range of viruses, including coronaviruses, paramyxoviruses (such as RSV), and flaviviruses (such as dengue viruses). Neyts was on a ski vacation in France with his son on January 20, 2020, when they stopped for a coffee between runs and checked the news. He learned that the novel coronavirus in China had just been revealed to be capable of human-to-human transmission. He immediately called his lab and said, "Okay, now we'll start on the vaccine."

KEVIN J. OLIVAL
Interviewed: February 25, 2021

Kevin Olival is an ecologist and evolutionary biologist who studies bats and the viruses they carry. He is vice president for research at EcoHealth Alliance. He warned, as first author of a 2020 paper with a long list of coauthors, that SARS-CoV-2 could possibly spill back from humans into free-ranging wildlife, including not just mink and other terrestrial mammals but also perhaps bats throughout the world. Once it gets into bats, with their communal and multispecies roosting, the virus could spread fast and far. There might eventually be sylvatic cycles of SARS-CoV-2, he told me, moving intermittently between bat populations and humans around the world. The danger of such cycles, Olival noted, would be not just the circumstances they offer for reinfecting people but also the possibility of new variants or recombinant viruses emerging from the bats.

MICHAEL T. OSTERHOLM
Interviewed: April 28, 2021

Michael Osterholm, an epidemiologist, is Regents Professor at the University of Minnesota and founding director of the Center for Infectious Disease Research and Policy (CIDRAP), which has several functions, including the issuance of daily online updates on emerging infectious diseases. Osterholm has served widely as an advisor on public and occupational health—in contexts ranging from the World Economic Forum to the Council on Foreign Relations, to the Biden-Harris COVID-19 Advisory Board, to the National Football League. "When you see a virus, it goes into humans, but then all of a sudden it quickly goes into cats and dogs and gorillas and lions and tigers," he told me, "that thing is pretty damn well adapted." His sense, he added, "is that this really was something from nature that got into humans, just as SARS and MERS did."

ÁINE O'TOOLE
Interviewed: February 3, 2021

Áine O'Toole is now a postdoctoral researcher in the lab of Andrew Rambaut at the University of Edinburgh, which works on molecular evolution, phylogenetics, and epidemiology. She is chief creator of the software tool PANGOLIN, which classifies genome sequences of SARS-CoV-2, assigning them to their appropriate positions of relatedness on the virus's family tree and giving each lineage a label (for instance B.1.1.7). PANGOLIN has been used all over the world for putting SARS-CoV-2 samples into evolutionary context. O'Toole stayed up late one night, and next morning, there it was.

GABRIELE PAGANI
Interviewed: April 16, 2021

Gabriele Pagani is an infectious disease physician now working at Legnano Hospital, northwest of Milan. He was an ID resident during the early months of 2020 at Luigi Sacco Hospital. When he was putting in twelve, fourteen, sixteen hours a day at the hospital and on the Castiglione d'Adda study (described in my text), he socially distanced from his parents, who were seventyish, but his mother kept him well fed. "Yeah, Italian mother, you know," he told me. She cooked for an extra person each evening and left him a tray. "It's one of the things that made me survive." Otherwise, he would have eaten pizza six days a week and probably nothing on the seventh, he said.

SHARON J. PEACOCK
Interviewed: March 31, 2021

Sharon Peacock, a physician and microbiologist, is professor of public health and microbiology at the University of Cambridge. She is executive director of the COVID-19 Genomics UK Consortium (COG-UK), a collaboration of agencies and university laboratories, created in April 2020 (at Peacock's initiative) to collect, sequence, and analyze genomes of SARS-CoV-2. Her life and career trajectory—the intellectual hunger and pluck that carried her through closed doors, from shop assistant to dental nurse to the highest rungs of public health in the United Kingdom—have a Dickensian grandeur, but she speaks of all that offhandedly and without melodrama. Helen Mirren should play her in the movie.

JOSEPH F. PETROSINO
Interviewed: August 26, 2021

Joseph Petrosino is a professor of virology and microbiology, and founding director of the Center for Metagenomics and Microbiome Research, at the Baylor College of Medicine, in Houston. He began his research career with a focus on biodefense, searching for vaccine targets in potentially weaponized pathogens such as the bacteria that cause anthrax and tularemia. After the NIH launched its Human Microbiome Project in 2007, Petrosino switched his attention "from the bad guys to the good guys," he told me, and began to study the commensal microbes of the human microbiome, using genetics and genomics. Matt Wong came to his lab as a computational specialist to help design tools for mining viral genomic data from microbiomic jumbles.

PETER PIOT
Interviewed: April 1, 2021; April 6, 2021

Peter Piot has recently retired as director of the London School of Hygiene and Tropical Medicine (LSHTM); he remains its Handa Professor of Global Health. He is also the author of *No Time to Lose: A Life in Pursuit of Deadly Viruses.* Piot trained as a physician in Ghent, and was a research associate in microbiology in Antwerp, at work toward a PhD in 1976, when circumstances took him to Zaire (now the Democratic Republic of the Congo), where he was part of the team, led by Karl Johnson, that responded to a disease outbreak centered on a remote mission hospital, isolated the virus that was the causing the chain of disease, and named that virus Ebola. Through the following years, Piot worked often in Africa, as well as holding professorships in his native Belgium, in Singapore, and then in London. He was founding executive director of the United Nations Programme on HIV/

AIDS (UNAIDS) and served as under-secretary-general of the United Nations. In mid-March 2020, just as LSHTM was shifting to remote learning and working from home, Piot became infected with SARS-CoV-2. "And that came really sudden," he told me. "Suddenly a splitting headache. I never coughed. I mean, until later." He had muscle aches, sore throat, diarrhea, and exhaustion, but the lack of a cough meant he didn't fit the case definition, and therefore he couldn't get tested at a public hospital. He went to a private clinic, tested positive, and rode it out at home until his fever spiked to 104 degrees. Then his wife (the anthropologist Heidi Larson, see above) helped him get to a hospital in a taxi, and X-rays of his lungs showed secondary bacterial pneumonia. He was hospitalized for seven days. "One of the things I've personally learned, but also that we know from clinical experience," he told me, is that COVID-19, although transmitted by the respiratory route, is "really a systemic infection that affects the whole." It wasn't a typical viral thing. It was much worse than he expected.

RAINA K. PLOWRIGHT
Interviewed: March 10, 2021

Raina Plowright, trained as a veterinarian and an ecologist, is an associate professor of epidemiology at Montana State University. She has long studied the ecology of zoonotic viruses, in particular Hendra virus, which has its reservoir in flying foxes (a group of fruit bats), in her native Australia, and spills over through horses, an intermediate host. Plowright and colleagues have illuminated how pregnancy status in such bats, the nursing of their pups, and nutritional stress increase the likelihood of their being infected with Hendra virus. She has also written on how land-use change—such as forest habitat destruction—drives the cycle of viral spread among bat populations, viral shedding, and spillover to humans.

MARJORIE P. POLLACK
Interviewed: February 3, 2021

Marjorie Pollack, a physician and epidemiologist, is deputy editor of ProMED-mail. She spent two years in the CDC's Epidemic Intelligence Service (EIS) soon after her medical residency, then an additional year completing a residency in preventive medicine, and has been a consulting medical epidemiologist for forty-plus years. She was on the desk for ProMED-mail on the night of December 30, 2019, when the first alarm bells in Wuhan began to be heard elsewhere.

VINCENT RACANIELLO
Interviewed: March 29, 2021

Vincent Racaniello is Higgins Professor of Microbiology and Immunology at Columbia University. His research specialty is picornaviruses, a family that includes poliovirus, the hepatitis A virus, and some common-cold viruses. His lab identified the receptor, CD155, that poliovirus uses to grasp and infect human cells. Racaniello is also the host of a probing but lively podcast, *This Week in Virology.* I asked for his thoughts about the origins of the virus, and whether the lab leak hypothesis deserved further consideration. "We *are* trying to figure it out. We're trying to do wildlife sampling! That's the way to do it. We don't need to go and look at lab records and find out what they were working with. It's not going to help us." The closest known virus at the time he and I spoke, RaTG13, was only 96 percent similar to SARS-CoV-2. That couldn't have been the origin, he said, either by engineering or by accidental release. "*Nobody* had anything close in the laboratory. And if they did, they would have published it, because that's how science works! You publish *cool* stuff, right? And the Wuhan Institute of Virology didn't have it."

ANDREW RAMBAUT
Interviewed: March 8, 2021

Andrew Rambaut is a professor of molecular evolution at the University of Edinburgh. He is a co-creator of the software platform BEAST (Bayesian Evolutionary Analysis Sampling Trees), an influential tool for assigning molecular sequences to their positions on family trees. "Bayesian" alludes to a form of inference in which the probability of a hypothesis is updated as more data become available. It's useful in science and it would be useful in public discourse too. Rambaut is also the creator of the website Virological.org, on which some of the most interesting and important ruminations of scientists about SARS-CoV-2 have appeared.

ANGELA L. RASMUSSEN
Interviewed: February 2, 2021

Angela Rasmussen, a virologist, is an associate professor at the Vaccine and Infectious Disease Organization (VIDO), University of Saskatchewan. She is also affiliated with the Georgetown Center for Global Health Science and Security. "One of the criticisms of something like the PREDICT program," she told me, "is that it's essentially stamp collecting. Because how do you say, out of all these thousands, potentially millions, of viruses that are circulating in the wild, which one of them is actually a risk?" Which can infect a human? Which can transmit between humans? Which can cause severe harm? "And I think that is where gain-of-function research is useful," she added, referring to very specific GOF studies, such as making chimeras to investigate the function of a specific viral element (a receptor-binding domain or a furin cleavage site, for example) in the context of a known viral pathogen. Such research can be valuable, in her view, for understanding the potential of a given virus as a human pathogen, or to illuminate its mechanisms of virulence.

DAVID A. RELMAN
Interviewed: March 23, 2021

David Relman is the Thomas C. and Joan M. Merigan Professor in Medicine, and a professor of microbiology and immunology, as well as a Senior Fellow at the Center for International Security and Cooperation, at Stanford University. He's also chief of infectious diseases at the Veterans Affairs Health Care System in Palo Alto. He was a pioneer in the study of the human microbiome and has served on multiple national advisory boards and committees concerned with biosecurity. He has been skeptical of gain-of-function research involving potential pandemic pathogens, and critical of the WHO-convened Global Study of the Origins of SARS-CoV-2.

ANNE W. RIMOIN
Interviewed: March 24, 2021

Anne Rimoin holds the Gordon-Levin Endowed Chair in Infectious Diseases and Public Health at the UCLA Fielding School of Public Health, and is director of the Center for Global and Immigrant Health at UCLA. She has worked for two decades in the Democratic Republic of the Congo, focused on infectious diseases such as monkeypox, Ebola, and Marburg, and their outbreaks at the interface where humans and nonhuman animals interact. She founded the UCLA-DRC Health Research and Training program to train U.S. and Congolese epidemiologists for work in challenging circumstances. "An infection anywhere is potentially an infection everywhere," Rimoin told me. "And if this pandemic hasn't driven that home, I don't know what will."

DAVID L. ROBERTSON
Interviewed: February 22, 2021

David Robertson is a research professor, and head of Bioinformatics, at the MRC-University of Glasgow Centre for Virus Research. He uses computational tools to study virus evolution, the dynamics of infection within and among hosts, and host-species specificity. His group—Spyros Lytras among them, with his co-supervisor Joseph Hughes—are all computational people and they have played an active role in the COG-UK Consortium, gathering and analyzing genomic sequences at an unprecedented scale to discern evolutionary trends and the emergence of variants of concern. It was like the early days of HIV/AIDS research, Robertson told me. "Why I got into science. It was the sense that you were trying to do something about something." About something *important*, he added. There was no need to worry about grants, and publishing papers wasn't the first priority. You were trying to understand something deadly and unknown. "That sense of urgency," he said, "was just very compelling and interesting. Especially if you spend, I guess, twenty-five years studying viruses and how they evolve." Then suddenly the urgency is back, with COVID-19, and the importance of evolutionary virology is again severe and global. Robertson paused, searching for adequate words. "And now we're just overwhelmed," he said. So much information, so many preprints and published papers, so much data.

DAVID RODRÍGUEZ-LÁZARO
Interviewed: April 13, 2021

David Rodríguez-Lázaro is an associate professor of microbiology and head of the Microbiology Division at the University of Burgos, Spain. Trained as a veterinary doctor and microbiologist, he has specialized in food science. He and a group of Brazilian and Spanish colleagues did the PCR study of human sewage from the city of Florianópolis, on the Brazilian coast, reporting that they detected evidence of SARS-CoV-2 from as early as November 27, 2019, ninety-one days before the first confirmed COVID-19 case in Brazil. Do you think that this pandemic will have been bad enough that we will learn from it, I asked him, and be better prepared next time? He laughed gently. "No," he said. Then he spoke a proverb in Spanish, which translates as "Humans are the only animal that will touch the stove twice."

FOREST ROHWER
Interviewed: May 4, 2021

Forest Rohwer is a virologist of deep and broad curiosity. He studies marine viruses, and the global role of viruses as evolutionary factors and repositories of information, for the sheer sake of knowing. He also researches cystic fibrosis, a genetic disorder that allows some bacterial infections to thwart the human immune system and grow out of control, especially in the lungs. I trust Forest's sound judgment, foresight, and humanity because I once spent six weeks on a research ship in the Russian Arctic with him, and he had brought along an espresso maker plus coffee to feed it, which, in the very early mornings when others weren't up, he shared. Forest came out of the College of Idaho and is now a professor at San Diego State University. He got his first substantive view of the novel coronavirus at a meeting of virologists in Lake Tahoe, March 2020, at which Eddie Holmes did a presentation. Afterward, Forest stayed up all night reading the literature then available, he told me, and thinking, "'We better figure out what to do.' Because it was clear that the CDC had no idea what they were doing." He wanted to understand why the diagnostic tests weren't working. He wanted to understand the virus's pathology. "Because I was, in particular, worried about the CF population." The *which* population? I asked. "The cystic fibrosis population," he said.

PARDIS C. SABETI
Interviewed: April 29, 2021

Pardis Sabeti is a professor in the Center for Systems Biology at Harvard, and in the Harvard T.H. Chan School of Public Health. Her lab is focused on developing genomic and computational tools to help detect, contain, and treat deadly viral diseases. She co-led an effort, in 2014, to sequence Ebola virus genomes during the outbreak in Sierra Leone, which illuminated patterns of transmission in the early weeks of that epidemic. Sabeti is coauthor (with Lara Salahi) of the 2018 book *Outbreak Culture: The Ebola Crisis and the Next Epidemic.*

PEI-YONG SHI
Interviewed: February 13, 2021

Pei-Yong Shi holds the John Sealy Distinguished Chair in Innovations in Molecular Biology at the University of Texas Medical Branch, Galveston. He has worked in the private sector (Novartis, Bristol Myers Squibb) and the public sector (New York State Department of Health) as well as in university labs. His research focus is RNA viruses, in particular the mechanisms of viral replication, toward the goals of developing antiviral drugs, vaccines, and diagnostic tools. He and colleagues (including Vineet Menachery, see above) developed a reverse-genetic system for rapidly engineering viral variants of SARS-CoV-2, useful for purposes of vaccine evaluation and screening of antiviral drug candidates. It's a six-step system, which sounds almost simple when they say that; but there are 108 sub-steps.

ZHENGLI SHI
Interviewed: July 30, 2021

Zhengli Shi is a senior scientist at the Wuhan Institute of Virology. She did her undergraduate and master's degrees in Wuhan, then a PhD in virology at the University of Montpellier, France. She has coauthored more than seventy scientific articles on coronaviruses, dating back to a landmark 2005 paper in *Science,* "Bats Are Natural Reservoirs of SARS-like Coronaviruses," pointing the way toward the origin of SARS-CoV.

EMMA C. THOMSON
Interviewed: March 5, 2021

Emma Thomson, a physician and virologist, is a professor of infectious diseases in the MRC-Centre for Virus Research (CVR) at the University of Glasgow and at the London School of Hygiene and Tropical Medicine (LSHTM). She continues to do clinical work, seeing patients at Queen Elizabeth University Hospital, while leading laboratory and field research on the detection of viral infections in Uganda and other sub-Saharan African countries, as well as the U.K. In early 2020 her lab started sequencing SARS-CoV-2 genomes. "We made a strategic decision in March," she told me. There was a meeting of the steering committee of the CVR, "and we decided that we were just going to have to turn off everything other than SARS-CoV-2, that it was going to become a really significant problem, and that we couldn't stand by and watch an outbreak in our own country without responding to it." When I spoke with her, she hadn't traveled for a year. "It's annoying," she said, "because I would like to be in Uganda at the moment."

NATALIE J. THORNBURG
Interviewed: May 6, 2021

Natalie Thornburg is a lead research microbiologist at the Centers for Disease Control and Prevention (CDC) in Atlanta. She is a viral immunologist and vaccine researcher who has worked on respiratory syncytial virus, Epstein-Barr virus, cowpox virus, MERS virus, and other viruses, including most of the human coronaviruses. She was co-leader of the group that isolated and characterized SARS-CoV-2 from the first confirmed COVID-19 patient in the United States. She came home from a visit on December 31, 2019, and was putting away dishes when her husband, scanning Twitter, said, "Uh, did you hear that there is an outbreak of pneumonia in China?" She said, "Oh . . . shit. No, I did not hear that." Three weeks later a specimen arrived at the CDC, having been express-shipped from Snohomish, Washington, and it tested positive for the novel virus, constituting America's first known case. "And that was," Thornburg told me, "the second 'Oh shit' moment."

ALESSANDRO VESPIGNANI
Interviewed: March 12, 2021

Alessandro Vespignani is Sternberg Distinguished University Professor, and director of the Network Science Institute, at Northeastern University, in Boston. Trained as a physicist in Rome, he gravitated to computational sciences and the study of how complex social and technological networks evolve. Those fields overlap with epidemiology, and his recent research has included studies of how travel restrictions affected the early spread of SARS-CoV-2 from Wuhan, and of how the Alpha variant was expected (as of February

2021) to spread across Europe. I asked him, as I asked others: What was the most important decision you made in 2020? "I think, for me, the day that I decided to tell people, 'Look, this is going to be very bad. This is going to be a pandemic and we are going to be, like, in a science fiction movie.' That was a big decision," he said. "The feeling that the other people, in February, are looking at you like a completely insane person was tangible." One person to whom he said that was his graduate student Jessica Davis (see above), a coauthor on the SARS-CoV-2 studies. "I remember, you know, her face." Perhaps you should watch the movie *Contagion*, he told her.

SUPAPORN WACHARAPLUESADEE
Interviewed: July 25, 2021

Supaporn Wacharapluesadee is a molecular biologist, based at the Thai Red Cross Emerging Infectious Diseases Clinical Center, King Chulalongkorn Memorial Hospital, in Bangkok. She studies emerging infectious pathogens, especially viruses hosted in bats. She led the team that detected the first MERS case in Thailand, and her group was the first to identify a case of COVID-19 outside of China, in January 2020. Five months later, she and colleagues sampled horseshoe bats roosting in a wildlife sanctuary east of Bangkok and found RNA fragments, from which they assembled a whole-genome sequence, designated RacCS203, which is 91.5 percent similar to SARS-CoV-2. That work was funded in part by King Chulalongkorn Memorial Hospital, in part by the Biological Threat Reduction Program, within the U.S. Department of Defense.

LINFA WANG
Interviewed: March 9, 2021

Linfa Wang is a molecular biologist who studies bat viruses. He is a coauthor on many of the most interesting bat virus papers in recent decades, including the one that first revealed (in 2005) bats as reservoirs for SARS-like coronaviruses, and the one that persuasively established (in 2017) horseshoe bats as reservoirs of SARS-CoV. He was born in Shanghai and hoped to study engineering at East China Normal University, a top school; he qualified to attend, but his math skills didn't get him into the physics-and-engineering program, and he was assigned to biology. He shifted to biochemistry because it involved molecular work, not live animals. "I'm not an animal person," Wang told me. He loves bats for their mystery, their unique biology, their behavior, but wouldn't want one as a pet. Not many of us do. (When I was a boy, sorry to admit, I tried.) He did his doctorate at the University of California-Davis, in biochemistry, and then established his lab at the Australian Animal Health Laboratory (AAHL) in Geelong, Victoria, where I once visited him and got the tour of the BSL-4 facility. Wang is an Australian citizen but now works in Singapore, as a professor in the Program in Emerging and Infectious Diseases at Duke-NUS (National University of Singapore) Medical School. He is a brilliant lab man, happy to leave the cave crawling and bat catching and guano sampling to others.

ROBERT G. WEBSTER
Interviewed: June 3, 2021

Robert Webster can be considered the dean of influenza virologists. He held the Rose Marie Thomas Chair in the Department of Infectious Diseases at St. Jude's Children's Research Hospital, in Memphis, where he has worked since 1968. It was Webster, along with his friend and fellow scientist Graeme Laver, who walked down a beach on the southeast coast of Australia in 1967 and noticed a clue that led to the modern understanding of the origins of the influenza viruses. The clue was a group of dead muttonbirds, washed up on the sand. Webster and Laver mused that these birds might have been killed by an influenza virus, and that started the two men on a chain of investigations by which they eventually established a cardinal fact in the realm of zoonotic diseases: that novel human influenza viruses have their origins in wild aquatic birds. The influenzas are RNA viruses with great capacity for variation and speedy evolution, which is what gives them pandemic potential. It's what makes them, like some coronaviruses, not only dangerous but very unpredictable. Webster himself, as well as influenza experts at the WHO, expected that the next human pandemic would likely be caused by a highly pathogenic avian influenza, such as H5N1, but of a strain evolved for transmissibility among humans. When he first heard of the novel coronavirus in Wuhan, he thought that this virus might not be a major concern, because humans have been exposed to many relatively mild coronaviruses. "I didn't take it seriously, to be quite honest," he told me. Moral: if an RNA virus can surprise Robert Webster, it can surprise anyone.

SUSAN R. WEISS
Interviewed: February 2, 2021

Susan Weiss has studied coronaviruses for more than forty years. For thirty of those years, she has been a professor in the Department of Microbiology at the University of Pennsylvania. She remembers the first international coronavirus conference, held in Würzburg, Germany, in autumn 1980 and bringing together virtually all the world's coronavirus researchers: about sixty people. Her recent work includes a paper coauthored with her former postdoc Yize (Henry) Li and others, describing interactions in the immune response to SARS-CoV-2. Their findings indicate that this virus is less able to antagonize the innate immune system than MERS-CoV, which may in part explain why SARS-CoV-2 is often less pathogenic in a human host.

HEATHER L. WELLS
Interviewed: June 1, 2021

Heather Wells is a PhD student in the Department of Ecology, Evolution, and Environmental Biology at Columbia University, working on the genetic and ecological drivers of recombination in coronaviruses, under the advisorship of Simon Anthony and Maria Diuk-Wasser. Wells is first author on an interesting study of the evolutionary history of ACE2 receptor binding by coronaviruses in the lineage of SARS-like viruses. She and other team members sampled bats in Uganda and Rwanda and found fragments of a SARS-like coronavirus intermediate between SARS-CoV and SARS-CoV-2, but with a receptor-binding domain (RBD) incapable (like those in many known viruses on the SARS-CoV branch of the family) of using ACE2 receptors. Wells and her colleagues constructed a most-likely family tree, putting these three viruses into context among many other bat coronaviruses, which points to the possibility that SARS-CoV got its RBD by a recombination event and SARS-CoV-2, the ancestral form, has carried it for a long time.

MATTHEW WONG
Interviewed: September 9, 2021

Matt Wong is a bioinformatics specialist in the Program for Innovative Microbiome and Translational Research, under the leadership of Jennifer Wargo and Nadim Adjami, at the MD Anderson Cancer Institute, in Houston.

Previously he filled the same role in the lab of Joseph Petrosino (see above) at the Baylor College of Medicine. His sparse but piquant observations online can be found @torptube.

MICHAEL WOROBEY
Interviewed: June 14, 2021

Michael Worobey is the Louise Foucar Marshall Science Research Professor at the University of Arizona. He's a molecular virologist who studies the evolution of infectious diseases. Among the more significant papers Worobey has coauthored, in years preceding this pandemic, are one that placed the spillover into humans of the pandemic HIV strain to around 1908 (Worobey et al. 2008), and one that illuminated both the origin and the pathogenicity of the 1918 influenza virus (Worobey, Han, and Rambaut 2014). The latter paper suggested that the 1918 virus, an H1N1 strain, had inflicted especially high mortality on twenty- to forty-year-olds (a long-persisting mystery) because those individuals, unlike older or younger people, had experienced their first childhood exposure to influenza in the form of a very different virus, an H3N8, which had circulated from roughly 1889 to 1900 and had primed their immune systems for the wrong sort of challenge. That 2014 paper may represent Worobey's most important contribution—at least until the "Epicenter" preprint is published. I interview him whenever I have an excuse, and I read him whenever I see his byline.

KWOK-YUNG YUEN
Interviewed: May 25, 2021

K.Y. Yuen is a physician, a surgeon, and a microbiologist. He is Henry Fok Professor in Infectious Diseases, and chair of the Microbiology Department, at Hong Kong University. He has studied avian influenzas in humans since 1997 and coronaviruses in humans since 2003. In 2005, he led a group that found horseshoe bats within the Hong Kong Special Administrative Region serving as hosts of SARS-like coronaviruses, at the same time as other scientists (including Linfa Wang, Zhengli Shi, Wendong Li, Peter Daszak, and Jon Epstein) reported bats elsewhere in China playing the same role. Yuen was also part of the group that identified palm civets, sold for food in wet markets, as the likely intermediate hosts from which SARS-CoV spilled into humans. Besides discovering the human coronavirus HKU1 (still circulating globally as a common-cold coronavirus), he also found the bat coronavirus

HKU2 (associated with outbreaks of porcine epidemic diarrhea disease) and several other coronaviruses of potential zoonotic significance. He has been vocal on the danger presented by live-animal markets generally for the spill-over of novel viruses—from poultry and mammals—into humans. But people have their culinary customs, their persistent tastes. A frozen chicken at a Hong Kong market goes for half the price of a live chicken butchered fresh, Yuen told me. There is a difference in the meat, the texture. "To me it's not worth it," he said—meaning, I think, the disease risk as well as the price. Do you eat chicken? I asked him. "I do. I eat chicken, yes," he said. But you don't care if it's frozen? "I don't care if it's frozen." He offered many other interesting observations about zoonotic viruses and human behavior, which we agreed would be off the record. Do you think this pandemic will have been bad enough, I asked him, as I had asked others, that people and governments will learn from it? "I'm sorry to say," he answered, "that's unlikely." Except for a short period, while memory is fresh, he added.

In addition, I benefited from exchanges with other scientists and conservationists during the pandemic, by phone or Skype or email, on several topics, including virus evolution, emerging viral pathogens, the international traffic in pangolins, and bats. Ronald Swanstrom, of the University of North Carolina at Chapel Hill, was very generous, during the late stage of my work, in helping me understand some of the complexities and complex histories of certain antiviral drugs. Stephen Goldstein, of the Elde lab in the Eccles Institute of Human Genetics at the University of Utah, gave his time to a close reading of a couple crucial sections on the matter of origins. I owe warm thanks also to: Chantal Abergel, Brenda Ang, Steve Blake, Gustavo Caetano-Anollés, Beth Cameron, Dan Challender, Jean-Michel Claverie, Luc Evouna Embolo, Mike Fay, Amanda Fine, Patrick Forterre, Winifred Frick, Sarah Heinrich, Alice Hughes, Lisa Hywood, Zhou Jinfeng, Karl Johnson, Vivek Kapur, Thomas Ksiazek, Ade Kurniawan, Fabian Leendertz, David Lehman, Sonja Luz, Olajumoke Morenikeji, Paul Offit, Jonathan Pekar, C. J. Peters, Jane Qiu, Pierre Rollin, Chris Shepherd, Jason Shepherd, Brent Stirton, Bob Swanepoel, Eric Kaba Tah, Paul Thomson, Johanna Wysocka, Zhaomin Zhou, and others whose names I apologize for inadvertently omitting.

Thanks also to my editorial partners for some of those Covid-related projects: David Remnick and Willing Davidson of *The New Yorker* (in which small portions of this book first appeared), John Hoeffel and Susan Goldberg of *National Geographic* (in which other small portions appeared), and Stephanie Giry of *The New York Times* (who started me on this virus by requesting an op-ed in January 2020). Christian Frei has been generous in sharing sources and thoughts as he involved me in conversations toward a film related to this subject.

I owe particular thanks to Charlie Calisher, Larry Gold, Jens Kuhn, Kristian Andersen, David Luce, and Mike Gilpin, who read the entire book for scientific accuracy and gave me valuable corrections and other feedback; to Sheli Radoshitzky, who did likewise with a sizable chunk; and to most members of the Greek Chorus (above), who reviewed portions for accuracy and returned them annotated. Gloria Thiede and Emily Krieger assisted me, as they have on past books, in essential ways. It's now thirty-some years during which Gloria has transcribed my recorded interviews, and her ear has only gotten better, her attention to vocal nuance more acute. Emily expertly provides the backstopping that all nonfiction writers should want: she checks the facts. Wudan Yan also gave her keen attention to portions of the fact-checking, pitching in as we ran short of time. Wufei Yu provided special and essential help, with his own journalistic work and his interpreting and translating from Mandarin. Dan Krza and Dan Smith, the two crucial Dans, have again been my go-to guys for computer expertise and website operations, respectively.

The other essential partners in this effort to whom I offer deep thanks are my editor, Bob Bender, CEO Jonathan Karp, Johanna Li, and all the team at Simon & Schuster; Fred Chase, who gave the book a fine, astute copyedit; and my agent, the incomparable Amanda Urban, along with her team at ICM.

My wife, Betsy Gaines Quammen, writes books too, and we both work out of home offices, so the stay-at-home-and-isolate imperatives of COVID-19 have not fallen upon us, as they have upon so many people, as unfamiliar hardships. I thank goodness, and I thank Betsy, that we operate within this turreted wooden house in a space so full of laughter and love and lively conversation and mutual support and dogs. Even the cat and the python seem to appreciate all that.

NOTES

These source notes apply only to quotations from published material. Scientific sources are cited simply by first author, except where more is necessary to disambiguate; full citations of those sources appear in the bibliography. Quotes spoken directly to me come from recorded interviews, transcribed by an expert human (Gloria Thiede); dates of those interviews are given in Credits, above. Sourcing for all other facts is available on request through www.davidquammen.com.

7 *"urgent notice on the treatment"*: Quoted in ProMED-mail, 12/30/19, from a machine translation of a report on Finance Sina. https://scholar.harvard.edu/files/kleeler ner/files/20191230_promed_-_undiagnosed_pneumonia_-_china_hu-_rfi_archive _number-_20191230.6864153.pdf.

8 *"REQUEST FOR INFORMATION"*: ProMED-mail post, 12/30/19.

9 *"Patients with unknown cause"*: ProMED-mail post, 12/31/19.

9 *"They just called us and said"*: Caixin Global, 2/29/20. https://www.caixinglobal .com/2020-02-29/in-depth-how-early-signs-of-a-sars-like-virus-were-spotted-spread -and-throttled-101521745.html.

10 *"many similar patients"*: Caixin 2/29/20, behind paywall. https://www.caixinglobal .com/2020-02-29/in-depth-how-early-signs-of-a-sars-like-virus-were-spotted -spread-and-throttled-101521745.html.

10 *"7 confirmed cases of SARS"*: Jianxing Tan, 1/30/20. *Caixin* (in Chinese). Archived from *the original* on 1/31/20. Retrieved to Wikipedia on 2/6/20.

10 *"Don't circulate this information"*: BBC/*Frontline*, 2/2/21. https://www.pbs.org/wgbh /frontline/article/a-timeline-of-chinas-response-in-the-first-days-of-COVID-19/.

10 *"Other severe pneumonia"*: Reuters, 12/31/21. https://www.reuters.com/article /us-china-health-pneumonia/chinese-officials-investigate-cause-of-pneumo nia-outbreak-in-wuhan-idUSKBN1YZ0GP.

11 *"stop testing and destroy"*: Caixin Global, 2/29/20. https://www.caixinglobal .com/2020-02-29/in-depth-how-early-signs-of-a-sars-like-virus-were-spotted -spread-and-throttled-101521745.html.

12 *"But there's no need to panic"*: South China Morning Post, 12/31/20. https://www .scmp.com/news/china/politics/article/3044050/mystery-illness-hits-chinas-wuhan -city-nearly-30-hospitalised.

13 *"sanitation and renovation"*: South China Morning Post, 1/1/20. https://www.scmp.com/news/china/politics/article/3044207/china-shuts-seafood-market-linked-mystery-viral-pneumonia.

13 *"It took us less than forty hours"*: Charlie Campbell, *Time*, 8/24/20. https://time.com/5882918/zhang-yongzhen-interview-china-coronavirus-genome/.

18 *"I asked Eddie to give me"*: Charlie Campbell, *Time*, 8/24/20. https://time.com/5882918/zhang-vongzhen-interview-china-coronavirus-genome/.

19 *"Please feel free to download"*: Virological.org, 1/10/20. https://virological.org/t/novel-2019-coronavirus-genome/319.

23 *"These cryptic cases"*: Chan et al. (2020).

23 *"novel coronavirus-infected pneumonia"*: China CDC Weekly, Vol. 2, No. 5, 1/21/20. http://weekly.chinacdc.cn/en/article/id/e3c63ca9-dedb-4fb6-9c1c-d057adb77b57.

27 *"Evolvability of Emerging Viruses"*: Burke (1998).

28 *"serves both to hybridize"*: Ibid.

30 *"with near-optimal efficiency"*: Ibid.

31 *"I made a lucky guess"*: First DQ interview with Don Burke, 11/30/11.

37 *"The guy to my left was praying"*: Khan (2016), p. 4.

41 *"All of them will be killed today"*: The New York Times, 1/7/04. https://www.nytimes.com/2004/01/07/world/the-sars-scare-in-china-slaughter-of-the-animals.html.

42 *"That was the defining moment"*: DQ interview with Brenda Ang, Singapore, 1/30/09.

45 *"The concept of 'super-spreader'"*: Khan et al. (1999), S76, S84.

49 *"What were the animals?"*: Pollack et al. (2012), 143–44.

55 *"camel flu"*: https://www.thetimes.co.uk/article/travel-alert-after-eighth-camel-flu-death-2k8j83mzgq2.

57 *"Nobody had any idea"*: https://www.youtube.com/watch?v=AE8G4cVj038; https://www.thebulwark.com/a-timeline-of-trumps-press-briefing-lies/; https://www.yahoo.com/entertainment/trump-claims-nobody-had-any-idea-coronavirus-deadly-despite-saying-otherwise-recording-055843938.html.

64 *"Bats Are Natural Reservoirs"*: Li et al. (2005).

67 *"may become infectious to humans"*: Ren et al. (2008), 1900.

67 *"Our results provide the strongest evidence"*: Ge et al. (2013), 535.

68 *"severe pulmonary infection"*: Xu (2013), 2.

70 *"a phenomenon that fosters recombination"*: Ge et al. (2016), 31.

70 *"smoking gun"*: Cyranoski (2017), 15.

70 *"This work provides new insights"*: Hu et al. (2017), 1.

77 *"Snakes were also sold"*: Ji et al. (2020), 436.

77 *"provides some insights to the question"*: Ibid., 438.

78 *"unique" stretches of amino acids*: Pradhan et al. (2020), 1.

78 *"To avoid further misinterpretation"*: This was posted, at least temporarily, to the Comments page on bioRxiv; hard copy in DQ files. Also see https://www.biorxiv.org/content/10.1101/2020.01.30.927871v2.

83 *"Mining Coronavirus Genomes"*: Cohen (2020a).

83 *"It seems humans can't resist controversy"*: Jon Cohen, *Science*, "Mining Coronavirus Genomes for Clues to the Outbreak's Origins," 1/31/20. https://www.science.org /content/article/mining-coronavirus-genome-clues-outbreak-s-origins.

83 *"This just came out today"*: Fauci email to Andersen, and Andersen reply, 1/31/20. Variously published on the web after a FOIA-request release. Hard copies in DQ files.

85 *"What the email shows"*: Andersen tweet, June 1, 2021; hard copy in DQ files.

87 *"There's a lot of finger-pointing"*: Interview with Sarah Heinrich, 7/6/20.

88 *"Such was the magnitude"*: Challender et al. (2020), 265.

88 *"If you go into a restaurant"*: Interview with Daniel Challender, 5/29/20.

89 *"I know we're serving as a transit point"*: Interview with Olajumoke Morenikeji, 5/28/20.

90 *"lurking ailments in our stomachs"*: Wufei Yu, *The New York Times*, March 5, 2020. https://www.nytimes.com/2020/03/05/opinion/coronavirus-china-pangolins .html.

90 *"It's not a matter of tradition"*: Interview with Zhou Jinfeng, 6/4/20.

97 *"This result indicates"*: torptube on Virological.org. https://virological.org/t/ncov -2019-spike-protein-receptor-binding-domain-shares-high-amino-acid-identity -with-a-coronavirus-recovered-from-a-pangolin-viral-metagenomic-dataset/362.

99 *"The Proximal Origin of SARS-CoV-2"*: Andersen et al. (2020).

99 *"considerable discussion"*: Andersen et al. (2020), 450.

101 *"provides a much stronger"*: Ibid., 452.

101 *"the involvement of an immune system"*: Ibid.

101 *"More scientific data could swing"*: Ibid.

103 *"a temporary public panic"*: Chen et al. (2020), 2.

104 *"to isolate patients and trace"*: Chan et al. (2020), 523.

105 *"If the cruise ship epidemic"*: Hung et al. (2020), 1058.

110 *"I shall defend"*: Lwoff (1957), 240.

111 *"No virus is* known *to do good"*: Medawar and Medawar (1983), 275.

113 *"the surprises expected"*: Philippe et al. (2013), 281.

114 *"ancestral protocell"*: Abergel et al. (2015), 793.

114 *"Our analyses clearly show"*: Andersen et al. (2020), 450.

116 *"might have"* originated by recombination: Xiao et al. (2020), 287.

116 *"gradually showed signs of respiratory disease"*: Ibid., 286.

116 *"were mostly inactive and sobbing"*: Ibid., 7 (in the accelerated preview version; "cry-ing," 290, in the published version).

120 *"suggests that pangolins"*: Lam et al. (2020), 282.

120 *"should be removed from wet markets"*: Ibid.

126 *"the shared history of exposure"*: Huang et al. (2020), 498.

126 *"which has been closed down"*: Sarah Boseley, *The Guardian*, 1/24/20. https://www .theguardian.com/science/2020/jan/24/calls-for-global-ban-wild-animal-markets -amid-coronavirus-outbreak.

126 *"an evidence-based hypothesis"*: Daniel Lucey, *Science Speaks*, 1/25/20. https:// www.idsociety.org/science-speaks-blog/2020/update-wuhan-coronavirus—2019 -ncov-qa-6-an-evidence-based-hypothesis/.

128 *"appreciate the criticism"*: Cao is quoted in Jon Cohen, *Science*, 1/26/20. https://

www.science.org/content/article/wuhan-seafood-market-may-not-be-source
-novel-virus-spreading-globally.

130 *"The number of deaths"*: Huang et al. (2020), 501.

132 *"the best place to live"*: https://www.mundopositivo.com.br/noticias/turismo/2018
1033-veja_o_que_fazer_em_florianopolis_e_se_encante.html.

134 *"cough" and "diarrhea" were trending:* Nsoesie et al. (2020), preprint posted on
DASH, 4. https://dash.harvard.edu/bitstream/handle/1/42669767/Satellite_Images
_Baidu_COVID19_manuscript_DASH.pdf?isAllowed=y&sequence=3.

140 *BavPat1, as in "Bavarian Patient 1":* Worobey et al. (2020), 564.

141 *"The public health response":* Ibid., 569.

141 *"The value of detecting cases early":* Ibid.

145 *acting a little "crazy":* Abutaleb and Paletta (2021), 231.

145 *"my job at the White House":* https://www.foxnews.com/transcript/peter-navarro
-on-how-us-is-fighting-the-spread-of-coronavirus.

146 *"They knew Trump would have a fit":* Abutaleb and Paletta (2021), 97.

146 *"The global novel coronavirus situation":* https://www.cdc.gov/media/releases
/2020/t0225-cdc-telebriefing-covid-19.html.

147 *"People have their televisions on":* Abutaleb and Paletta (2021), 101.

148 *"A Dynamic Nomenclature Proposal":* Rambaut et al. (2020).

149 *"a technician," Zhou told me modestly:* Email from Zhaomin Zhou, 9/27/21.

149 *"an objective observer unconnected":* Xiao et al. (2020), 2.

149 *"the sort of cachet attached":* Ibid., 3.

149 *"a substantial desire to purchase":* Ibid. 5.

150 *"lift all protein boats":* https://research.rabobank.com/far/en/sectors/animal-protein
/rising-african-swine-fever-losses-to-lift-all-protein.html.

150 *"will create challenges":* Ibid.

150 *"may have increased the transmission":* Xia et al. (2021), preprint, 1. https://www
.preprints.org/manuscript/202102.0590/v1.

152 *"It is highly probable that SARS-CoV-2":* Pekar et al. (2021), 414.

153 *"unlikely to be valid":* Ibid., 416.

153 *"spillover of SARS-CoV-2-like viruses":* Ibid., 415.

158 *"remarkably low" genetic diversity:* Rausch et al. (2020), 24614.

158 *"little evidence" of natural selection:* Dearlove et al. (2020), 23652.

158 *"is being transmitted more rapidly":* Ibid.

161 *"Our data show that":* Korber et al. (2020), 819.

161 *"To this end":* Ibid., 823.

162 *"The speed with which":* Ibid.

163 *"red zone":* https://www.politico.eu/article/italy-coronavirus-covid19-lombardy-lodi/.

165 *"Unfortunately, we couldn't have known":* https://www.si.com/soccer/2020/03/25
/atalanta-valencia-coronavirus-champions-league-san-siro-milan-italy.

167 *"This is a lower-than-expected prevalence":* Pagani et al. (2020), 9.

167 *"a large part of the population":* Ibid., 1.

170 *"There were doubters":* Sharon Peacock (2020), December 17.

173 *"snake flu" virus had reached Scotland:* The Scottish Sun, 1/24/20, as reported by the
BBC, https://www.bbc.com/news/uk-scotland-51233161.

173 *"stay at home"*: https://www.instituteforgovernment.org.uk/sites/default/files/time line-lockdown-web.pdf.

175 *"an emergent SARS-CoV-2 lineage"*: https://virological.org/t/preliminary-genomic -characterisation-of-an-emergent-sars-cov-2-lineage-in-the-uk-defined-by-a-novel -set-of-spike- mutations/563.

175 *"enhanced genomic surveillance worldwide"*: Ibid.

178 *"We have enough sequencers"*: Amy Maxmen, *Nature*, 4/7/21. https://www.nature.com /articles/d41586-021-00908-0.

178 *"leading to further surges"*: Washington et al. (2021), preprint posted on medRxiv, 2/7/21, 3.

180 *"So what?" he said with a shrug*: Tom Phillips, *The Guardian*, 4/29/20. https://www .theguardian.com/world/2020/apr/29/so-what-bolsonaro-shrugs-off-brazil-rising -coronavirus-death-toll.

181 *"an explosive epidemic"*: Buss et al. (2021), 288.

181 *"excess mortality"*: Ibid.

181 *"attack rate"*: Ibid.

185 *"spurt" in the number*: Cherian et al. (2021), 4.

186 *"variant of concern"*: Stephanie Nebehay and Emma Farge, Reuters, 5/10/21. https:// www.reuters.com/business/healthcare-pharmaceuticals/who-designates-india-variant -being-global-concern-2021-05-10/.

186 *"to either the epidemiological containment measures"*: Tchesnokova (2021), 15.

187 *"viral loads" more than a thousand*: Li et al. (2021) on Virological.org, 7/7/21. https:// virological.org/t/viral-infection-and-transmission-in-a-large-well-traced-outbreak -caused-by-the-delta-sars-cov-2-variant/724/1.

191 *"We have done an incredible job"*: https://www.cnn.com/interactive/2020/10/politics /covid-disappearing-trump-comment-tracker/; https://www.c-span.org/video/?46 9786-1/president-trump-hosts-african-american-history-month-reception.

191 *"like a miracle—it will disappear"*: https://www.cnn.com/interactive/2020/10/politic s/covid-disappearing-trump-comment-tracker/index.html.

191 *"powers of resistance"*: Robertson (2021), 1474.

191 *"These facts show something"*: Ibid.

192 *"abortion disease," now known as brucellosis*: Eichhorn and Potter (1917), 3.

192 *"Thus a herd immunity seems"*: Ibid., 9.

192 *"Our aim," Vallance said*: https://www.bbc.com/news/uk-politics-54252272.

193 *"a clear plan"*: https://www.gov.uk/government/speeches/pm-statement-on-coro navirus-12-march-2020.

193 *"We want to suppress it"*: https://www.youtube.com/watch?v=2XRc389TvG8.

194 *"A Contribution to the Mathematical Theory of Epidemics"*: Kermack and McKend- rick (1927).

198 *"During the years that the herd"*: Eichhorn and Potter (1917), 9.

199 *"mop-up" clinics*: Bowes (1967), 413. Bowes spelled it "mopup," but I judged that might be confusing.

202 *"suggests a possible prophylactic and therapeutic use"*: Vincent et al. (2005), 9.

203 *"If they work"*: https://www.c-span.org/video/?470503-1/president-trump-corona virus-task-force-hold-briefing-white-house.

203 *"The answer is no":* Abutaleb and Paletta (2021), 223–24.

203 *"anecdotal":* https://abcnews.go.com/Politics/fauci-throws-cold-water-trumps-dec laration-malaria-drug/story?id=69716324.

204 *"I've spent my life being 'against'":* Scott Sayare, *The New York Times Magazine,* 5/12/20. https://www.nytimes.com/2020/05/12/magazine/didier-raoult-hydroxy chloroquine.html.

204 *"I'm a big fan":* Abutaleb and Paletta (2021), 224.

204 *"are unlikely to be effective":* https://www.fda.gov/news-events/press-announce ments/coronavirus-COVID-19-update-fda-revokes-emergency-use-authorization -chloroquine-and.

205 *"circulating zoonotic strains":* Sheahan et al. (2017), 5.

205 *"highly effective" against SARS-CoV-2:* Manli Wang et al. (2020), 271.

206 *"did not significantly improve":* Yeming Wang et al. (2020), 1575.

206 *"an adult drank an injectable ivermectin":* https://emergency.cdc.gov/han/2021/han 00449.asp.

208 *"the reliable evidence does not":* https://www.cochranelibrary.com/cdsr/doi/10 .1002/14651858.CD015017.pub2/epdf/full.

209 *"has potent antiviral activity":* Shuntai Zhou et al. (2021), 415.

209 *"I would probably take molnupiravir":* RS email to DQ, 10/26/21.

210 *"The biochemical pathways and our data":* Ibid.

211 *"molnupiravir reduced the risk":* https://www.merck.com/news/merck-and-ridge backs-investigational-oral-antiviral-molnupiravir-reduced-the-risk-of-hospital ization-or-death-by-approximately-50-percent-compared-to-placebo-for-patients -with-mild-or-mod erat/.

215 *"Clinton summoned Fauci":* Another account of this meeting appears in Gina Kolata and Benjamin Mueller, *The New York Times,* 1/15/22. https://www.nytimes .com/2022/01/15/health/mrna-vaccine.html?searchResultPosition=6.

215 *"vaccine-enhanced disease":* Ibid., 149.

216 *by one account it "haunted" him:* David Heath and Gus Garcia-Roberts, *USA Today,* 1/26/21.

221 *"provide an important step":* Pallesen et al. (2017), E7354.

225 *"Vaccine inequity is the world's biggest obstacle":* https://www.who.int/news/item /22-07-2021-vaccine-inequity-undermining-global-economic-recovery.

227 *"gave up on waiting":* Olivia Goldhill, Rosa Furneaux, and Madlen Davies, *STAT News,* 10/8/21. https://www.statnews.com/2021/10/08/how-covax-failed-on-its -promise-to-vaccinate-the-world/.

227 *"appalled," he told a news conference:* James Keaton, AP, 9/8/21. https://apnews .com/article/business-health-coronavirus-pandemic-united-nations-world-health -organization-6384ff91c399679824311ac26e3c768a.

238 *"More and more epidemiological data":* DQ interview with Jonathan Towner, 8/11/09.

238 *"It was really unnerving":* DQ interview with Brian Amman, 8/11/09.

239 *"The Possible Origins of 2019-nCov Coronavirus":* Xiao and Xiao (2020).

240 *"The probability was very low":* Ibid., 2.

240 *"The speculation about the possible origins":* James T. Areddy, *The Wall Street Journal,*

3/5/20. https://www.wsj.com/articles/coronavirus-epidemic-draws-scrutiny-to -labs-handling-deadly-pathogens-11583349777.

241 *"uncanny similarity":* Pradhan et al. (2020), 1.

241 *"insertions" in the SARS-CoV-2 spike:* Ibid.

241 *"unconventional evolution":* Ibid., 9.

242 *"It was not our intention":* Note by Prashant Pradhan on the Comments page of bioRxiv; hard copy in DQ files. See also Jessica McDonald, 2/7/20, posting on FactCheck.org, "Baseless Conspiracy Theories Claim New Coronavirus Was Bioengineered." https://www.factcheck.org/2020/02/baseless-conspiracy-theories-claim -new-coronavirus-was-bioengineered/.

242 *"We still stand by what we had published":* Abhinandan Mishra and Dibyendu Mondal, *The Sunday Guardian,* 6/5/21. https://www.sundayguardianlive.com /news/fauci-described-indian-research-man-made-covid-outlandish.

243 *"The Indian paper is really outlandish":* Ibid.

244 *"I have been privately dealing":* William R. Gallaher (writing as profbillg1901) on Virological.org, 2/6/20. https://virological.org/t/tackling-rumors-of-a-suspicious -origin-of-ncov2019/384.

245 *"I have found a probable source":* Ibid.

245 *"The only laboratory required":* Ibid.

245 *"This is a really nice finding, Bill":* Andrew Rambaut (posting as arambaut) on Virological.org, 5/3/20. https://virological.org/t/tackling-rumors-of-a-suspicious-origin -of-ncov2019/384/5.

246 *"The sequence analysis all seems very cogent":* Steve Barger (posting as swbarg), ibid.

246 *"The entirely natural origin":* William R. Gallaher (posting as profbillg1901), 5/7/20, ibid.

246 *"an overlooked fragment":* Spyros Lytras (posting as spyroslytras) on Virological.org, 8/8/20. https://virological.org/t/the-sarbecovirus-origin-of-sars-cov-2-s-furin-cleavage -site/536.

247 *"these viruses must have co-circulated":* Ibid.

248 *"collectively, our results support":* MacLean et al. (2021), 1.

249 *"Our observations suggest":* Zhan, Deverman, and Chan (2020), 1.

249 *"I got whacked so many times":* Rowan Jacobsen, *Boston Magazine,* 9/9/20. https:// www.bostonmagazine.com/news/2020/09/09/alina-chan-broad-institute-corona virus/.

251 *"Chan's puzzle detectors pulsed again":* Ibid.

252 *"was already pre-adapted":* Zhan et al. (2020), 1.

252 *"And that, Chan says":* Jacobsen, *Boston Magazine.*

253 *"very likely already well adapted":* Chan and Ridley (2021), 96.

254 *"remained bright and alert":* Sit et al. (2020), 776.

254 *"Our data demonstrated":* Qiang Zhang et al. (2020), 2013.

256 *"The minks showed various symptoms":* ProMED-mail post, 4/26/20.

257 *"a herd of mink is being slaughtered":* ProMED-mail post, 6/17/21.

258 *"most of the animals there had been infected":* Reuters, 7/16/20.

258 *"We tested a number of animals":* ProMED-mail post, 10/24/20.

260 *"firsthand interaction":* https://www.lincolnzoo.org.

262 *"it is inferred":* Xu (2013), anonymous translation, corrected by Wufei Yu, 19.

263 *"deficiencies" that should be fixed:* Ibid., 20.

264 *"a phenomenon that fosters":* Ge et al. (2016), 31.

267 *"What we mean by the term":* Amber Dance, *Nature,* 10/27/21.

268 *"a recipe for disaster":* van Aken (2007), 1.

269 *"GOF research is important":* https://www.nih.gov/about-nih/who-we-are/nih -director/statements/nih-lifts-funding-pause-gain-function-research.

271 *"Senator Paul, you do not know what you are talking about":* https://www.cnbc.com /2021/07/20/if-anybody-is-lying-here-senator-it-is-you-fauci-tells-sen-paul-in -heated-exchange-at-senate-hearing.html. This statement is also available as video on YouTube (with transcript) here: https://www.youtube.com/watch?v=pFoaBV _cTek; and here from *The Guardian:* https://www.theguardian.com/us-news/video /2021/jul/20/fauci-to-rand-paul-you-do-not-know-what-you-are-talking-about -video.

271 *"Discovery of a Rich Gene Pool":* Hu et al. (2017).

272 *"Thus, the risk of spillover":* Ibid., 19.

273 *"highlights the necessity":* Ibid., 1.

276 *"although a laboratory accident":* Frutos, Gavotte, and Devaux (2021), 3.

277 *"the occurrence of a double accident":* Ibid., 5.

279 *"There Is No 'Origin' to SARS-CoV-2":* Frutos, et al. (2021).

280 *"An epidemic never starts":* Ibid., 7.

280 *"This is referred to as the epidemic threshold":* Ibid.

280 *"a permanent process of evolution":* Ibid.

280 *"spillover of SARS-CoV-2-like viruses":* Pekar et al. (2021), 415.

281 *"viral chatter":* Wolfe et al. (2005), 1824.

282 *"The Origins of SARS-CoV-2: A Critical Review":* Holmes et al. (2021).

283 *"Coronaviruses have long been known":* Ibid., 1.

285 *"No epidemic has been caused by":* Ibid., 3.

285 *"no rational experimental reason":* Ibid., 4.

286 *"the most parsimonious explanation":* Ibid., 5.

286 *"stems from the coincidence":* Ibid., 6.

286 *"highly unlikely," they judged:* Ibid.

287 *"a soft spot for wild theories":* David Robertson, quoted in Jane Qiu, *MIT Technology Review,* 11/19/21. https://www.technologyreview.com/2021/11/19/1040390 /covid-wuhan-natural-spillover-wuhan-wet-market-huanan/.

288 *"Dissecting the Early COVID-19 Cases in Wuhan":* Worobey (2021).

290 *"It becomes almost impossible":* Joel Achenbach, *The Washington Post,* 11/18/21. https://www.washingtonpost.com/health/2021/11/18/coronavirus-origins-wuhan -market-animals-science-journal/.

293 *"We stand together to strongly condemn":* Calisher et al. (2020), e42.

295 *"trace the animal origin of the virus":* WHO-convened Global Study of the Origins of SARS-CoV-2: Terms of References for the China Part, 11/5/20, 2. https://www .who.int/publications/m/item/who-convened-global-study-of-the-origins-of-sars -cov-2.

297 *and a lab leak, "extremely unlikely":* WHO-convened Global Study of Origins of SARS-CoV-2: China Part (2021).

297 *"an accident at the laboratory":* Glen Owen, *Daily Mail*, 4/11/20. https://www.daily mail.co.uk/news/article-8211257/Wuhan-lab-performing-experiments-bats-corona virus-caves.html.

297 *"We will end that grant":* Sarah Owermohle, *Politico*, 4/27/20. https://www.politico .com/news/2020/04/27/trump-cuts-research-bat-human-virus-china-213076.

300 *"The Contested Origin of SARS-CoV-2":* Gronvall (2021).

301 *"so all wildlife trade was fundamentally illegal":* Xiao et al. (2021), 3.

301 *"The swift clear-out":* Gronvall (2021), 12.

301 *"Once it looked as if":* Ibid., 21–22.

302 *"I do not believe that this assessment":* https://www.who.int/director-general/speeches /detail/who-director-general-s-remarks-at-the-member-state-briefing-on-the-report -of-the-international-team-studying-the-origins-of-sars-cov-2.

302 *"As scientists with relevant expertise":* Bloom et al. (2021), 694.

302 *"strong evidence of a live-animal market origin":* Worobey (2021), 1204.

303 *"greater clarity about the origins":* Bloom et al. (2021), 694.

305 *"The Huanan Market Was the Epicenter":* Worobey et al. (2022), 1.

305 *"the overwhelming majority were specifically linked":* Ibid., 4.

306 *"These findings suggest":* Ibid., 11.

308 *"we began to see it might be":* Isaac Chotiner, *The New Yorker*, 11/30/21. https://www .newyorker.com/news/q-and-a/how-south-african-researchers-identified-the-omicron -variant-of-covid.

309 *"This variant is completely insane":* Kai Kupferschmidt, *Science*, 11/27/21. https:// www.science.org/content/article/patience-crucial-why-we-won-t-know-weeks -how-dangerous-omicron.

310 *"At present there is no direct evidence":* Martin et al. post on Virological.org, 12/5/21. https://virological.org/t/selection-analysis-identifies-significant-mutational -changes-in-omicron-that-are-likely-to-influence-both-antibody-neutralization -and-spike-function-part-1-of-2/771.

311 *"the epistatic fucking cirque du soleil":* Quoted on Twitter by Kristian Andersen, 12/5/21; used here with permission of Edyth Parker.

BIBLIOGRAPHY

Abbate, Jessie L., et al. 2020. "Pathogen Community Composition and Co-Infection Patterns in a Wild Community of Rodents." Preprint, bioRxiv, posted September 2, 2020.

Abdelnabi, Rana, et al. 2021. "Comparing Infectivity and Virulence of Emerging SARS-CoV-2 Variants in Syrian Hamsters." *EBioMedicine* 68 (103403).

Abergel, Chantal, Matthiew Legendre, and Jean-Michel Claverie. 2015. "The Rapidly Expanding Universe of Giant Viruses: Mimivirus, Pandoravirus, Pithovirus and Mollivirus." *FEMS Microbiology Reviews* 39 (6).

Abraham, Thomas. 2004. *Twenty-First Century Plague: The Story of SARS*. Baltimore: Johns Hopkins University Press.

Abutaleb, Yasmeen, and Damian Paletta. 2021. *Nightmare Scenario: Inside the Trump Administration's Response to the Pandemic That Changed History*. New York: HarperCollins.

Afelt, Aneta, Roger Frutos, and Christian Devaux. 2018. "Bats, Coronaviruses, and Deforestation: Toward the Emergence of Novel Infectious Diseases?" *Frontiers in Microbiology* 9 (702).

Albery, Gregory F., et al. 2020. "Predicting the Global Mammalian Viral Sharing Network Using Phylogeography." *Nature Communications* 11 (1).

Allen, Arthur. 2020. "Government-Funded Scientists Laid the Groundwork for Billion-Dollar Vaccines." *Kaiser Health News*, November 18.

Al-Tawfiq, Jaffar A., 2013. "Middle East Respiratory Syndrome-Coronavirus Infection: An Overview." *Journal of Infection and Public Health* 6 (5).

Alwan, Nisreen A., et al. 2020. "Scientific Consensus on the COVID-19 Pandemic: We Need to Act Now." *The Lancet* 396 (10260).

Aly, Mahmoud, et al. 2017. "Occurrence of the Middle East Respiratory Syndrome Coronavirus (MERS-CoV) Across the Gulf Corporation Council Countries: Four Years Update." *PLOS ONE* 12 (10).

Amendola, Antonella, et al. 2021a. "Evidence of SARS-CoV-2 RNA in an Oropharyngeal Swab Specimen, Milan, Italy, Early December 2019." *Emerging Infectious Diseases* 27 (2).

Amendola, Antonella, et al. 2021b. "Molecular Evidence for SARS-CoV-2 in Samples Collected from Patients with Morbilliform Eruptions Since Late Summer 2019 in Lombardy, Northern Italy." Preprint at *The Lancet*, posted August 6, 2021.

Amman, Brian R., et al. 2014. "Marburgvirus Resurgence in Kitaka Mine Bat Population after Extermination Attempts, Uganda." *Emerging Infectious Diseases* 20 (10).

Andersen, Kristian G., et al. 2015. "Clinical Sequencing Uncovers Origins and Evolution of Lassa Virus." *Cell* 162 (4).

Andersen, Kristian G., et al. 2020. "The Proximal Origin of SARS-CoV-2." *Nature Medicine* 4.

Anderson, Danielle E., et al. 2020. "Lack of Cross-Neutralization by SARS Patient Sera Towards SARS-CoV-2." *Emerging Microbes & Infections* 9 (1).

Anderson, Roy, et al. 2020. "Challenges in Creating Herd Immunity to SARS-CoV-2 Infection by Mass Vaccination." *The Lancet* 396 (10263).

Anthony, Simon J., et al. 2017. "Global Patterns in Coronavirus Diversity." *Virus Evolution* 3 (1).

Apolone, Giovanni, et al. 2020. "Unexpected Detection of SARS-CoV-2 Antibodies in the Prepandemic Period in Italy." *Tumori Journal* 107 (5).

Aschwanden, Christie. 2020. "The False Promise of Herd Immunity for COVID-19." *Nature* 587 (7832).

Assiri, Abdullah, et al. 2013. "Hospital Outbreak of Middle East Respiratory Syndrome Coronavirus." *The New England Journal of Medicine* 369 (5).

Avanzato, Victoria A., et al. 2020. "Case Study: Prolonged Infectious SARS-CoV-2 Shedding from an Asymptomatic Immunocompromised Individual with Cancer." *Cell* 183 (7).

Barberia, Lorena G., and Eduardo J. Gómez. 2020. "Political and Institutional Perils of Brazil's COVID-19 Crisis." *The Lancet* 396 (10248).

Baric, Ralph S. 2020. "Emergence of a Highly Fit SARS-CoV-2 Variant." *The New England Journal of Medicine* 383 (27).

Baric, Ralph S., et al. 1985. "Characterization of Leader-Related Small RNAs in Coronavirus-Infected Cells: Further Evidence for Leader-Primed Mechanism of Transcription." *Virus Research* 3 (1).

Barry, John M. 2004, 2005. *The Great Influenza: The Epic Story of the Deadliest Plague in History.* London: Penguin.

Bartsch, Yannic C., et al. 2021. "Discrete SARS-CoV-2 Antibody Titers Track with Functional Humoral Stability." *Nature Communications* 12 (1018).

Bedford, Trevor, et al. 2020. "Cryptic Transmission of SARS-CoV-2 in Washington State." *Science* 370 (6516).

Bermingham, A., et al. 2012. "Severe Respiratory Illness Caused by a Novel Coronavirus, in a Patient Transferred to the United Kingdom from the Middle East, September 2012." *Euro Surveillance* 17 (40).

Bertuzzo, Enrico, et al. 2020. "The Geography of COVID-19 Spread in Italy and Implications for the Relaxation of Confinement Measures." *Nature Communications* 11 (4264).

Biek, Roman, et al. 2015. "Measurably Evolving Pathogens in the Genomic Era." *Trends in Ecology & Evolution* 30 (6).

Blanco-Melo, Daniel, et al. 2020. "Imbalanced Host Response to SARS-CoV-2 Drives Development of COVID-19." *Cell* 181 (5).

Bloom, Jesse D. 2021. "Recovery of Deleted Deep Sequencing Data Sheds More Light on the Early Wuhan SARS-CoV-2 Epidemic." *Molecular Biology and Evolution* 38 (12).

Bloom, Jesse D., et al. 2021. "Investigate the Origins of COVID-19." *Science* 372 (6543).

Böhmer, Merle M., et al. 2020. "Investigation of a COVID-19 Outbreak in Germany Resulting from a Single Travel-Associated Primary Case: A Case Series." *The Lancet Infectious Diseases* 20 (8).

Boni, Maciej F., et al. 2020. "Evolutionary Origins of the SARS-CoV-2 Sarbecovirus Lineage Responsible for the COVID-19 Pandemic." *Nature Microbiology* 5 (11).

Borrell, Brendan. 2021. *The First Shots: The Epic Rivalries and Heroic Science Behind the Race to the Coronavirus Vaccine.* New York: Mariner, HarperCollins.

Bowes, James E. 1967. "Rhode Island's End Measles Campaign." *Public Health Report* 82 (5).

Brian, D. A., and R. S. Baric. 2005. "Coronavirus Genome Structure and Replication." *Current Topics in Microbiology and Immunology* 287.

Brilliant, Larry, et al. 2021. "The Forever Virus: A Strategy for the Long Fight Against COVID-19." *Foreign Affairs* 100 (4).

Brook, Cara E., and Andrew P. Dobson. 2015. "Bats as 'Special' Reservoirs for Emerging Zoonotic Pathogens." *Trends in Microbiology* 23 (3).

Brook, Cara E., et al. 2020. "Accelerated Viral Dynamics in Bat Cell Lines, with Implications for Zoonotic Emergence." *eLife* 9 (e48401).

Bryant, Andrew, et al. 2021. "Ivermectin for Prevention and Treatment of COVID-19 Infection: A Systematic Review, Meta-analysis, and Trial Sequential Analysis to Inform Clinical Guidelines." *American Journal of Therapeutics* 28 (4).

Burke, Donald S. 1998. "Evolvability of Emerging Viruses." In *Pathology of Emerging Infections 2.* Edited by Ann Marie Nelson and C. Robert Horsburgh Jr. Washington, D.C.: American Society for Microbiology.

Buss, Lewis F., et al. 2021. "Three-quarters Attack Rate of SARS-CoV-2 in the Brazilian Amazon During a Largely Unmitigated Epidemic." *Science* 371 (6526).

Butler, Colin D., et al. 2021. "Call for a Full and Unrestricted International Forensic Investigation into the Origins of COVID-19." Open letter, posted online March 4, 2021.

Butler, Delcan. 2015. "Engineered Bat Virus Stirs Debate Over Risky Research." *Nature News & Comment,* November 12.

Calisher, Charles H. 2013. *Lifting the Impenetrable Veil: From Yellow Fever to Ebola Hemorrhagic Fever and SARS.* Red Feather Lakes, Colorado: Rockpile Press.

Calisher, Charles H., et al. 2006. "Bats: Important Reservoir Hosts of Emerging Viruses." *Clinical Microbiology Reviews* 19 (3).

Calisher, Charles, et al. 2020. "Statement in Support of the Scientists, Public Health Professionals, and Medical Professionals of China Combatting COVID-19." *The Lancet* 395 (10226).

Callaway, Ewen. 2021. "Rare Reactions Might Hold Key to Variant-Proof COVID Vaccines." *Nature* 592 (7852).

Candido, Darlan S., et al. 2020. "Evolution and Epidemic Spread of SARS-CoV-2 in Brazil." *Science* 369 (6508).

Capua, Ilaria. 2018. "Discovering Invisible Truths." *Journal of Virology* 92 (20).

Capua, Ilaria, and Mario Rasetti. 2020. "Here, the Huge Rainbow Within the COVID-19 Storm." *EClinicalMedicine* 29–30.

Capua, Ilaria, and Carlo Giaquinto. 2021. "The Unsung Virtue of Thermostability." *The Lancet* 397 (10282).

Carlson, Colin J., et al. 2019. "Global Estimates of Mammalian Viral Diversity Accounting for Host Sharing." *Nature Ecology & Evolution* 3 (7).

Carrion, Malwina, and Lawrence C. Madoff. 2017. "Pro-MED-mail: 22 Years of Digital Surveillance of Emerging Infectious Diseases." *International Health* 9 (3).

Carroll, Dennis, et al. 2018. "Building a Global Atlas of Zoonotic Diseases." *Bulletin of the World Health Organization* 96 (4).

Carroll, Dennis, et al. 2019. "The Global Virome Project: Expanded Viral Discovery Can Improve Mitigation." *Science* 359 (6378).

Castro, Marcia C., et al. 2021. "Spatiotemporal Pattern of COVID-19 Spread in Brazil." *Science* 372 (6544).

Celum, Connie, et al. 2020. "COVID-19, Ebola, and HIV—Leveraging Lessons to Maximize Impact." *The New England Journal of Medicine* 383 (19).

Challender, Daniel W. S., Stuart R. Harrop, and Douglas C. MacMillan. 2015. "Understanding Markets to Conserve Trade-Threatened Species in CITES." *Biological Conservation* 187.

Challender, Daniel W. S., Helen C. Nash, and Carly Waterman, volume editors. Philip J. Nyhus, series editor. 2020. *Pangolins: Science, Society and Conservation.* London: Academic Press, Elsevier.

Chan, Alina, and Matt Ridley. 2021. *VIRAL: The Search for the Origin of COVID-19.* New York: HarperCollins.

Chan, Jasper F. W., et al. 2012. "Is the Discovery of the Novel Human Betacoronavirus 2c EMC/2012 (HCoV-EMC) the Beginning of Another SARS-like Pandemic?" *Journal of Infection* 65 (6).

Chan, Jasper Fuk-Woo, et al. 2020. "A Familial Cluster of Pneumonia Associate with the 2019 Novel Coronavirus Indicating Person-to-Person Transmission: A Study of a Family Cluster." *The Lancet* 395 (10223).

Chan, Yujia Alina, and Shing Hei Zhan. 2020. "Single Source of Pangolin CoVs with a Near Identical Spike RBD to SARS-CoV-2." Preprint, bioRxiv, posted October 23, 2020.

Chavarria-Miró, Gemma, et al. 2021. "Time Evolution of Severe Acute Respiratory Syndrome Coronavirus (SARS-CoV-2) in Wastewater During the First Pandemic Wave of COVID-19 in the Metropolitan Area of Barcelona, Spain." *Applied and Environmental Microbiology* 87 (7).

Chen, Albert Tian, et al. 2021. "COVID-19 CG Enables SARS-CoV-2 Mutation and Lineage Tracking by Locations and Dates of Interest." *eLife* 10 (e63409).

Chen, Chi-Mai, et al. 2020. "Containing COVID-19 Among 627,386 Persons in Contact with the Diamond Princess Cruise Ship Passengers Who Disembarked in Taiwan: Big Data Analytics." *Journal of Medical Internet Research* 22 (5).

Chen, Dongsheng, et al. 2020. "Single-Cell Screening of SARS-CoV-2 Target Cells in Pets, Livestock, Poultry and Wildlife." Preprint, bioRxiv, posted June 14, 2020.

Chen, Yu, et al. 2013. "Human Infections with the Emerging Avian Influenza A H7N9 Virus from Wet Market Poultry: Clinical Analysis and Characterisation of Viral Genome." *The Lancet* 381 (9881).

Cheng, Vincent C. C., et al. 2007. "Severe Acute Respiratory Syndrome Coronavirus as an Agent of Emerging and Reemerging Infection." *Clinical Microbiology Reviews* 20 (4).

Cheng, Wenda, Shuang Xing, and Timothy C. Bonebrake. 2017. "Recent Pangolin Seizures in China Reveal Priority Areas for Intervention." *Conservation Letters* 10 (6).

Cherian, Sarah, et al. 2021. "SARS-CoV-2 Spike Mutations, L452R, E478K, E484Q and P681R, in the Second Wave of COVID-19 in Maharashtra, India." *Microorganisms* 9 (7).

Chik, Holly. 2020. "China CDC Chief Defends Early Outbreak Action: 'I Never Said There Was No Human-to-Human Transmission.'" *South China Morning Post*, April 20.

Chinazzi, Matteo, et al. 2020. "The Effect of Travel Restrictions on the Spread of the 2019 Novel Coronavirus (COVID-19) Outbreak." *Science* 368 (6489).

Christakis, Nicholas A. 2020. *Apollo's Arrow: The Profound and Enduring Impact of Coronavirus on the Way We Live.* New York: Little, Brown Spark.

Chu, Daniel K. W., et al. 2014. "MERS Coronaviruses in Dromedary Camels, Egypt." *Emerging Infectious Diseases* 20 (6).

Chu, Hin, et al. 2020. "Comparative Replication and Immune Activation Profiles of SARS-CoV-2 and SARS-CoV in Human Lungs: An *ex vivo* Study with Implications for the Pathogenesis of COVID-19." *Clinical Infectious Diseases* 71 (6).

Clausen, Thomas Mandel, et al. 2020. "SARS-CoV-2 Infection Depends on Cellular Heparan Sulfate and ACE2." *Cell* 183 (4).

Claverie, Jean-Michel. 2006. "Viruses Take Center Stage in Cellular Evolution." *Genome Biology* 7 (6).

Claverie, Jean-Michel. 2020. "All Viruses Are Unconventional." Preprint, posted on Preprints, September 19, 2020.

Claverie, Jean-Michel, et al. 2006. "Mimivirus and the Emerging Concept of 'Giant' Viruses." *Virus Research* 117 (1).

Claverie, Jean-Michel, and Hiroyuki Ogata. 2009. "Ten Good Reasons Not to Exclude Viruses from the Evolutionary Picture." *Nature Reviews Microbiology* 7 (8).

Claverie, Jean-Michel, and Chantal Abergel. 2016. "Giant Viruses: The Difficult Breaking of Multiple Epistemological Barriers." *Studies in History and Philosophy of Biological and Biomedical Sciences* 59.

Cohen, Jon. 2013. "Structural Biology Triumph Offers Hope Against a Childhood Killer." *Science* 342 (6158).

——. 2020a. "Mining Coronavirus Genomes for Clues to the Outbreak's Origins." *Science* 367 (6477).

——. 2020b. "Wuhan Coronavirus Hunter Shi Zhengli Speaks Out. China's 'Bat Woman' Denies Responsibility for the Pandemic, Demands Apology from Trump." *Science* 369 (6503).

——. 2020c. "Wuhan Seafood Market May Not Be Source of Novel Virus Spreading Globally." www.sciencemag.org., January 26.

——. 2021. "Vaccines That Can Protect Against Many Coronaviruses Could Prevent Another Pandemic." www.sciencemag.org., April 15.

Conceicao, Carina, et al. 2020. "The SARS-CoV-2 Spike Protein Has a Broad Tropism for Mammalian ACE2 Proteins." *PLOS Biology* 18 (12).

Corbett, K. S., et al. 2020a. "Evaluation of the mRNA-1273 Vaccine Against SARS-CoV-2 in Nonhuman Primates." *The New England Journal of Medicine* 383 (16).

Corbett, Kizzmekia S., et al. 2020b. "SARS-CoV-2 mRNA Vaccine Design Enabled by Prototype Pathogen Preparedness." *Nature* 586 (7830).

Cottle, Lucy E., et al. 2013. "A Multinational Outbreak of Histoplasmosis Following a Biology Field Trip in the Ugandan Rainforest." *Journal of Travel Medicine* 20 (2).

Cowling, Benjamin J., et al. 2015. "Preliminary Epidemiological Assessment of MERS-CoV Outbreak in South Korea, May to June 2015." *Euro Surveillance* 20 (25).

Cui, Jie, Fang Li, and Zheng-Li Shi. 2019. "Origin and Evolution of Pathogenic Coronaviruses." *Nature Reviews Microbiology* 17 (3).

Cyranoski, David. 2017. "Bat Cave Solves Mystery of Deadly SARS Virus—And Suggests New Outbreak Could Occur." *Nature* 552 (7683).

Dai, Wenhao, et al. 2020. "Design, Synthesis, and Biological Evaluation of Peptidomimetic Aldehydes as Broad-Spectrum Inhibitors Against Enterovirus and SARS-CoV-2." *Journal of Medicinal Chemistry* (0c02258).

Dallmeier, Kai, Geert Meyfroidt, and Johan Neyts. 2021. "COVID-19 and the Intensive Care Unit: Vaccines to the Rescue." *Intensive Care Medicine* 47 (7).

Dance, Amber. 2021. "The Shifting Sands of Gain-of-Function Research." *Nature* 598 (7882).

Daszak, Peter, et al. 2021. "Infectious Disease Threats: A Rebound to Resilience." *Health Affairs* 40 (2).

Davies, Nicholas G., et al. 2021a. "Estimated Transmissibility and Impact of SARS-CoV-2 Lineage B.1.1.7 in England." *Science* 372 (6538).

Davies, Nicholas G., et al. 2021b. "Increased Mortality in Community-Tested Cases of SARS-CoV-2 Lineage B.1.1.7." *Nature* 593 (7858).

Davis, Jessica T., et al. 2021. "Cryptic Transmission of SARS-CoV-2 and the First COVID-19 Wave." *Nature* 600 (7887).

Dearlove, Bethany, et al. 2020. "A SARS-CoV-2 Vaccine Candidate Would Likely Match All Currently Circulating Variants." *Proceedings of the National Academy of Sciences* 117 (38).

Decaro, Nicola, Alessio Lorusso, and Ilaria Capua. 2021. "Erasing the Invisible Line to Empower the Pandemic Response." *Viruses* 13 (348).

Delaune, Deborah, et al. 2021. "A Novel SARS-CoV-2 Related Coronavirus in Bats from Cambodia." *Nature Communications* 12 (1).

Deslandes, A., et al. 2020. "SARS-CoV-2 Was Already Spreading in France in Late December 2019." *International Journal of Antimicrobial Agents* 55 (6).

Devaux, Christian A., et al. 2021. "Spread of Mink SARS-CoV-2 Variants in Humans: A Model of Sarbecovirus Interspecies Evolution." *Frontiers in Microbiology* 12.

Di Giallonardo, Francesca, et al. 2021. "Emergence and Spread of SARS-CoV-2 Lineages B.1.1.7 and P.1 in Italy." *Viruses* 13 (5).

Dinnon III, Kenneth H., et al. 2020. "A Mouse-Adapted Model of SARS-CoV-2 to Test Covid-19 Countermeasures." *Nature* 586 (7830).

Dobson, Andrew P., et al. 2020. "Ecology and Economics for Pandemic Prevention." *Science* 369 (6502).

Duchene, Sebastian, et al. 2020. "Temporal Signal and the Phylodynamic Threshold of SARS-CoV-2." *Virus Evolution* 6 (2).

Duffy, Siobain, Laura A. Shackelton, and Edward C. Holmes. 2008. "Rates of Evolutionary Change in Viruses: Patterns and Determinants." *Nature Reviews Genetics* 9 (4).

du Plessis, Louis, et al. 2021. "Establishment and Lineage Dynamics of the SARS-CoV-2 Epidemic in the UK." *Science* 371 (6530).

Düx, Ariane, et al. 2020. "Measles Virus and Rinderpest Virus Divergence Dated to the Sixth Century BCE." *Science* 368 (6497).

Eban, Katherine. 2021. "The Lab-Leak Theory: Inside the Fight to Uncover COVID-19's Origins." *Vanity Fair*, June 3.

Eckerle, Isabella, and Benjamin Meyer. 2020. "SARS-CoV-2 Seroprevalence in COVID-19 Hotspots." *The Lancet* 396.

Eckerle, Lance D., et al. 2010. "Infidelity of SARS-CoV Nsp 14-Exonuclease Mutant Virus Replication Is Revealed by Complete Genome Sequencing." *PLOS Pathogens* 6 (5).

Eichhorn, Adolph, and George M. Potter. 1917. "Contagious Abortion of Cattle." United States Department of Agriculture *Farmers' Bulletin* 790, January.

Epstein, Jonathan H., et al. 2020. "Nipah Virus Dynamics in Bats and Implications for Spillover to Humans." *Proceedings of the National Academy of Sciences* 117 (46).

Erkens, K., et al. 2002. "Histoplasmosis Group Disease in Bat Researchers Returning from Cuba." *Deutsche Medizinische Wochenschrift* 127 (1–2).

Fang, Fang, translated by Michael Berry. 2020. *Wuhan Diary: Dispatches from a Quarantined City.* New York: HarperVIA, HarperCollins.

Faria, Nuno R., et al. 2021. "Genomics and Epidemiology of the P.1 SARS-CoV-2 Lineage in Manaus, Brazil." *Science* 372 (6544).

Farrar, Jeremy, with Anjana Ahuja. 2021. *Spike: The Virus vs the People, The Inside Story.* London: Profile.

Ferreira, Isabella A. T. M., et al. 2021. "SARS-CoV-2 B.1.617 Mutations L452R and E484Q Are Not Synergistic for Antibody Evasion." *Journal of Infectious Diseases* 224 (6).

Finch, Courtney L., et al. 2020. "Characteristic and Quantifiable COVID-19-like Abnormalities in CT-and PET/CT-imaged Lungs of SARS-CoV-2-infected Crab-Eating Macaques." Preprint, bioRxiv, May 14, 2020.

Fine, Paul, E. M. 1993. "Herd Immunity: History, Theory, Practice. *Epidemiologic Reviews* 15 (2).

Fiorentini, Simona, et al. "First Detection of SARS-CoV-2 Spike Protein N501 Mutation in Italy in August, 2020." *The Lancet Infectious Diseases* 21 (6).

Fongaro, Gislaine, et al. 2021. "The Presence of SARS-CoV-2 RNA in Human Sewage in Santa Catarina, Brazil, November 2019." *Science of the Total Environment* 778.

Forterre, Patrick. 2006. "The Origin of Viruses and Their Possible Roles in Major Evolutionary Transitions." *Virus Research* 117 (1).

———. 2011. "Manipulation of Cellular Synthesis and the Nature of Viruses: The Virocell Concept." *Comptes Rendus Chimie* 14 (4)

———. 2013. "The Virocell Concept and Environmental Microbiology." *International Society for Microbial Ecology Journal* 7 (2).

———. 2016. "To Be or Not to Be Alive: How Recent Discoveries Challenge the Traditional Definitions of Viruses and Life." *Studies in History and Philosophy of Biological and Biomedical Sciences* 59 (100–108).

Forterre, Patrick, and David Prangishvili. 2013. "The Major Role of Viruses in Cellular Evolution: Facts and Hypotheses." *Current Opinion in Virology* 3 (5).

Fox, John P. 1983. "Herd Immunity and Measles." *Reviews of Infectious Diseases* 5 (3).

Fox, John P., et al. 1971. "Herd Immunity: Basic Concept and Relevance to Public Health Immunization Practices." *American Journal of Epidemiology* 94 (3).

French, Rebecca K., and Edward C. Holmes. 2020. "An Ecosystems Perspective on Virus Evolution and Emergence." *Trends in Microbiology* 28 (3).

Frutos, Roger, et al. 2020a. "COVID-19: The Conjunction of Events Leading to the Coronavirus Pandemic and Lessons to Learn for Future Threats." *Frontiers in Medicine* 7 (223).

Frutos, Roger, et al. 2020b. "COVID-19: Time to Exonerate the Pangolin from the Transmission of SARS-CoV-2 to Humans." *Infection, Genetics, and Evolution* 84.

Frutos, Roger, et al. 2021. "There Is No 'Origin' to SARS-CoV-2." *Environmental Research* 30 (40).

Frutos, Roger, Laurent Gavotte, and Christian A. Devaux. 2021. "Understanding the Origin of COVID-19 Requires to Change the Paradigm on Zoonotic Emergence from the Spillover to the Circulation Model." *Infection, Genetics, and Evolution* 95.

Fuller, Thomas, et al. 2020. "A Coronavirus Death in Early February Was 'Probably the Tip of an Iceberg.'" *The New York Times*, April 22.

Gallaher, William R., 2020. "A Palindromic RNA Sequence as a Common Breakpoint Contributor to Copy-Choice Recombination in SARS-CoV-2." *Archives of Virology* 165 (10).

Garde, Damian, and Jonathan Saltzman. 2020. "The Story of mRNA: How a Once-Dismissed Idea Became a Leading Technology in the Covid Vaccine Race." *Boston Globe*, November 10.

Garry, Robert F. 2019. "Ebola Mysteries and Conundrums." *The Journal of Infectious Diseases* 219 (4).

Gautret, Philippe, et al. 2020. "Hydroxychloroquine and Azithromycin as a Treatment of COVID-19: Results of an Open-Label Non-Randomized Clinical Trial." *International Journal of Antimicrobial Agents* 56 (1).

Ge, Xing-Yi, et al. 2013. "Isolation and Characterization of a Bat SARS-like Coronavirus That Uses the ACE2 Receptor." *Nature* 503 (7477).

Ge, Xing-Yi, et al. 2016. "Coexistence of Multiple Coronaviruses in Several Bat Colonies in an Abandoned Mineshaft." *Virologica Sinica* 31 (1).

Ghebreyesus, Tedros Adhanom. 2021. "Five Steps to Solving the Vaccine Inequity Crisis." *PLOS Global Public Health*, October 13.

Giovanetti, Marta, et al. 2020. "A Doubt of Multiple Introduction of SARS-CoV-2 in Italy: A Preliminary Overview." *Journal of Medical Virology* 92 (9).

Gire, Stephen K., et al. 2014. "Genomic Surveillance Elucidates Ebola Virus Origin and Transmission During the 2014 Outbreak." *Science* 345 (6202).

Goes de Jesus, Jaqueline, et al. "Importation and Early Local Transmission of COVID-19 in Brazil, 2020." *Revista do Instituto de Medicina Tropical de São Paulo* 62 (e30).

Goldstein, Tracey, et al. 2018. "The Discovery of Bombali Virus Adds Further Support for Bats as Hosts of Ebolaviruses." *Nature Microbiology* 3 (10).

Gollakner, Rania, and Ilaria Capua. 2020. "Is COVID-19 the First Pandemic That Evolves into a Panzootic?" *Veterinaria Italiana* 56 (1).

Gonzalez-Reiche, Ana S., et al. 2020. "Introductions and Early Spread of SARS-CoV-2 in the New York City Area." *Science* 369 (6501).

Gozzi, Nicolò, et al. 2021. "Estimating the Spreading and Dominance of SARS-CoV-2 VOC 202012/01 (Lineage B.1.1.7) Across Europe." Preprint, medRxiv, posted February 23, 2021.

Graham, Barney S. 2011. "Biological Challenges and Technological Opportunities for Respiratory Syncytial Virus Vaccine Development." *Immunological Review* 239 (1).

———. 2020. "Rapid COVID-19 Vaccine Development." *Science* 368 (6494).

Graham, Barney S., and Nancy J. Sullivan. 2017. "Emerging Viral Diseases from a Vaccinology Perspective: Preparing for the Next Pandemic." *Nature Immunology* 19 (1).

Graham, Barney S., Morgan S. A. Gilman, and Jason S. McLellan. 2019. "Structure-Based Vaccine Antigen Design." *Annual Review of Medicine* 70.

Graham, Rachel L., and Ralph S. Baric. 2010. "Recombination, Reservoirs, and the Modular Spike: Mechanisms of Coronavirus Cross-Species Transmission." *Journal of Virology* 84 (7).

———. 2020. "SARS-CoV-2: Combating Coronavirus Emergence." *Immunity* 52 (5).

Graham, Rachel L., Eric F. Donaldson, and Ralph S. Baric. 2013. "A Decade After SARS: Strategies for Controlling Emerging Coronaviruses." *Nature Reviews Microbiology* 11 (12).

Graham, Rachel L., et al. 2018. "Evaluation of a Recombination-Resistant Coronavirus as a Broadly Applicable, Rapidly Implementable Vaccine Platform." *Communications Biology* 1 (179).

Grange, Zoë I., et al. 2021. "Ranking the Risk of Animal-to-Human Spillover for Newly Discovered Viruses." *Proceedings of the National Academy of Sciences* 118 (15).

Greaney, Allison J., et al. 2021. "Comprehensive Mapping of Mutations in the SARS-CoV-2 Receptor-Binding Domain that Affect Recognition by Polyclonal Human Plasma Antibodies." *Cell Host & Microbe* 29 (3).

Gronvall, Gigi Kwik. 2021. "The Contested Origin of SARS-CoV-2." *Survival* 63 (6).

Grubaugh, Nathan D., et al. 2019. "Tracking Virus Outbreaks in the Twenty-First Century." *Nature Microbiology* 4 (1).

Grubaugh, Nathan D., Mary E. Petrone, and Edward C. Holmes. 2020. "We Shouldn't Worry When a Virus Mutates During Disease Outbreaks." *Nature Microbiology* 5 (4).

Grützmacher, Kim S., et al. 2018. "Human Respiratory Syncytial Virus and *Streptococcus pneumoniae* Infection in Wild Bonobos." *EcoHealth* 15 (2)

Gryseels, Sophie, et al. 2020. "Risk of Human-to-Wildlife Transmission of SARS-CoV-2." *Mammal Review*, October 6, 2020.

Guan, Y., et al. 2003. "Isolation and Characterization of Viruses Related to the SARS Coronavirus from Animals in Southern China." *Science* 302 (5643).

Guo, Hua, et al. 2020. "Evolutionary Arms Race Between Virus and Host Drives Genetic Diversity in Bat Severe Acute Respiratory Syndrome-Related Coronavirus Spike Genes." *Journal of Virology* 94 (20).

Han, Guan-Zhu. 2020. "Pangolins Harbor SARS-CoV-2 Related Coronaviruses." *Trends in Microbiology* 28 (7).

Harcourt, Jennifer, et al. 2020. "Severe Acute Respiratory Syndrome Coronavirus 2 from Patient with Coronavirus Disease, United States." *Emerging Infectious Diseases* 26 (6).

Hasan, Anwarul, et al. 2020. "A Review on the Cleavage Priming of the Spike Protein on Coronavirus by Angiotensin-Converting Enzyme-2 and Furin." *Journal of Biomolecular Structure and Dynamics* 39 (8).

He, Wan-Ting, et al. 2021. "Total Virome Characterization of Game Animals in China Reveals a Spectrum of Emerging Viral Pathogens." Preprint, bioRxiv, posted November 12, 2021.

Heinrich, Sarah, et al. 2016. "Where Did All the Pangolins Go? International CITES Trade in Pangolin Species." *Global Ecology and Conservation* 8.

Hensley, Matthew K., et al. 2021. "Intractable Coronavirus Disease 2019 (COVID-19) and Prolonged Severe Acute Respiratory Syndrome Coronavirus 2 (SARS-CoV-2) Replication in a Chimeric Antigen Receptor-Modified T-Cell Therapy Recipient: A Case Study." *Clinical Infectious Diseases* 73 (3).

Hessler, Peter. 2020. "Nine Days in Wuhan, The Ground Zero of The Coronavirus Pandemic." *The New Yorker*, October 5.

Hodcroft, Emma B., et al. 2021. "Spread of a SARS-CoV-2 Variant Through Europe in the Summer of 2020." *Nature* 595 (7869).

Hodcroft, Emma B., et al. 2021. "Emergence in Late 2020 of Multiple Lineages of SARS-CoV-2 Spike Protein Variants Affecting Amino Acid Position 677." Preprint, medRxiv, posted February 14, 2021.

Holmes, Edward C. 2008. "Evolutionary History and Phylogeography of Human Viruses." *The Annual Review of Microbiology* 62 (307).

———. 2009. *The Evolution and Emergence of RNA Viruses.* Oxford: Oxford University Press.

Holmes, Edward C., and Andrew Rambaut. 2004. "Viral Evolution and the Emergence of SARS Coronavirus." *Philosophical Transactions of the Royal Society of London* 359 (1447).

Holmes, Edward C., et al. 2016. "The Evolution of Ebola Virus: Insights from the 2013–2016 Epidemic." *Nature* 538 (7624).

Holmes, Edward C., Andrew Rambaut, and Kristian G. Andersen. 2018. "Pandemics: Spend on Surveillance, Not Prediction." *Nature* 558 (7709).

Holmes, Edward C., et al. 2021. "The Origins of SARS-CoV-2: A Critical Review." *Cell* 184 (19).

Holmes, Kathryn V. 2005. "Adaptation of SARS Coronavirus to Humans." *Science* 309 (5742).

Holshue, Michelle L., et al. 2020. "First Case of 2019 Novel Coronavirus in the United States." *The New England Journal of Medicine* 382 (10).

Honigsbaum, Mark. 2019. *The Pandemic Century: One Hundred Years of Panic, Hysteria, and Hubris.* New York: W. W. Norton.

Hoong, Chua Mui. 2004. *A Defining Moment: How Singapore Beat SARS.* Singapore: Institute of Policy Studies.

Hotez, Peter J., foreword by Arthur L. Caplan. 2018. *Vaccines Did Not Cause Rachel's Autism: My Journey as a Vaccine Scientist, Pediatrician, and Autism Dad.* Baltimore: Johns Hopkins University Press.

Hotez, Peter. 2020. "COVID-19 in America: An October Plan." *Microbes and Infection* 22 (9).

Hotez, Peter J. 2021. "Anti-Science Kills: From Soviet Embrace of Pseudoscience to Accelerated Attacks on US Biomedicine." *PLOS Biology* 19 (1).

Hotez, Peter J. 2021. *Preventing the Next Pandemic: Vaccine Diplomacy in a Time of Anti-Science.* Baltimore: Johns Hopkins University Press.

Hotez, Peter J., Jorge A. Huete-Perez, and Maria Elena Bottazzi. 2020. "COVID-19 in the Americas and the Erosion of Human Rights for the Poor." *PLOS Neglected Tropical Diseases* 14 (12).

Hotez, Peter J., Alan Fenwick, and David Molyneux. 2021. "The New COVID-19 Poor and the Neglected Tropical Diseases Resurgence." *Infectious Diseases of Poverty* 10 (1).

Hou, Yixuan J., et al. 2020. "SARS-CoV-2 D614G Variant Exhibits Efficient Replication *ex vivo* and Transmission *in vivo*." *Science* 370 (6523).

Hsieh, Ching-Lin, et al. 2020. "Structure-Based Design of Prefusion-Stabilized SARS-CoV-2 Spikes." *Science* 369 (6510).

Hu, Ben, et al. 2017. "Discovery of a Rich Gene Pool of Bat SARS-Related Coronaviruses Provides New Insights into the Origin of SARS Coronavirus." *PLOS Pathogens* 13 (11).

Huang, Canping. 2016. "Novel Virus Discovery in Bat and the Exploration of Receptor of Bat Coronavirus HKU9." PhD dissertation submitted to Chinese Center for Disease Control and Prevention. June.

Huang, Chaolin, et al. 2020. "Clinical Features of Patients Infected with 2019 Novel Coronavirus in Wuhan, China." *The Lancet* 395 (10223).

Huong, Nguyen Quynh, et al. 2020. "Coronavirus Testing Indicates Transmission Risk Increases Along Wildlife Supply Chains for Human Consumption in Viet Nam, 2013–2014." *PLOS ONE* 15 (8).

Hung, Ivan Fan-Ngai, et al. 2020. "SARS-CoV-2 Shedding and Seroconversion Among Passengers Quarantined After Disembarking a Cruise Ship: A Case Series." *The Lancet Infectious Diseases* 20 (9).

Ingram, Daniel J., et al. 2018. "Assessing Africa-Wide Pangolin Exploitation by Scaling Local Data." *Conservation Letters*, 11 (2).

Ingram, Daniel J., et al. 2019. "Characterising Trafficking and Trade of Pangolins in the Gulf of Guinea." *Global Ecology and Conservation* 17.

Irving, Aaron T., et al. 2021. "Lessons from the Host Defences of Bats, a Unique Viral Reservoir." *Nature* 589 (7842).

Jackson, L. A., et al. 2020. "An mRNA Vaccine Against SARS-CoV-2—Preliminary Report." *The New England Journal of Medicine* 383 (20).

Jacobsen, Rowan. 2020. "Could COVID-19 Have Escaped from a Lab?" *Boston Magazine*, September 9.

Jalal, Hawre, Kyueun Lee, and Donald S. Burke. 2021. "Prominent Spatiotemporal Waves of COVID-19 Incidence in the United States: Implications for Causality, Forecasting, and Control." Preprint, medRxiv, posted July 3, 2021.

Ji, Wei, et al. 2020. "Cross-Species Transmission of the Newly Identified Coronavirus 2019-nCoV." *Journal of Medical Virology* 92 (4).

Jimi, Hanako, and Gaku Hashimoto. 2021. "Challenges of COVID-19 Outbreak on the Cruise Ship *Diamond Princess* Docked at Yokohama, Japan: A Real-World Story." *Global Health & Medicine* 2 (2).

Johnson, Bryan A., Rachel L. Graham, and Vineet D. Menachery. 2018. "Viral Metagenomics, Protein Structure, and Reverse Genetics: Key Strategies for Investigating Coronaviruses." *Virology* 517 (30–37).

Johnson, Bryan A., et al. 2020. "Furin Cleavage Site Is Key to SARS-CoV-2 Pathogenesis." Preprint, bioRxiv, posted August 26, 2020.

Johnson, Bryan A., et al. 2021. "Loss of Furin Cleavage Site Attenuates SARS-CoV-2 Pathogenesis." *Nature* 591 (7849).

Johnson, Christine Kreuder, et al. 2015. "Spillover and Pandemic Properties of Zoonotic Viruses with High Host Plasticity." *Scientific Reports* 5.

Johnson, K. M., et al. 1965. "Chronic Infection of Rodents by Machupo Virus." *Science* 150 (3703).

Jonas, Olga, and Richard Seifman. 2019. "Do We Need a Global Virome Project?" *The Lancet Global Health* 7 (10).

Juma, Carl Agisha, et al. 2019. "COVID-19: The Current Situation in the Democratic Republic of Congo." *American Journal of Tropical Medicine and Hygiene* 103 (6).

Karikó, Katalin. 2019. "*In vitro*-transcribed mRNA Therapeutics: Out of the Shadows and into the Spotlight." *Molecular Therapy* 27 (4).

Karikó, Katalin, et al. 2005. "Suppression of RNA Recognition by Toll-Like Receptors: The Impact of Nucleoside Modification and the Evolutionary Origin of RNA." *Immunity* 23 (2).

Karikó, Katalin, et al. 2008. "Incorporation of Pseudouridine into mRNA Yields Superior Nonimmunogenic Vector With Increased Translational Capacity and Biological Stability." *Molecular Therapy* 16 (11).

Kaur, Taranjit, et al. 2008. "Descriptive Epidemiology of Fatal Respiratory Outbreaks and Detection of a Human-Related Metapneumovirus in Wild Chimpanzees (*Pan troglodytes*) at Mahale Mountains National Park, Western Tanzania." *American Journal of Primatology* 70 (8).

Kemp, Steven A., et al. 2021. "SARS-CoV-2 Evolution During Treatment of Chronic Infection." *Nature* 592 (7853).

Kennedy, David A., and Andrew F. Read. 2020. "Monitor for COVID-19 Vaccine Resistance Evolution During Clinical Trials." *PLOS Biology* 18 (11).

Kermack, W. O., and A. G. McKendrick. 1927. "A Contribution to the Mathematical Theory of Epidemics. *Proceedings of the Royal Society*, A, 115.

Khan, Ali S., et al. 1999. "The Reemergence of Ebola Hemorrhagic Fever, Democratic Republic of the Congo, 1995." *The Journal of Infectious Diseases* 179 (Supplement 1).

Khan, Ali S., with William Patrick. 2016. *The Next Pandemic: On the Front Lines Against Humankind's Gravest Dangers.* New York: PublicAffairs.

Ki, Moran. 2015. "2015 MERS Outbreak in Korea: Hospital-to-Hospital Transmission." *Epidemiology and Health* 37.

Kim, K. H., et al. 2017. "Middle East Respiratory Syndrome Coronavirus (MERS-CoV) Outbreak in South Korea, 2015: Epidemiology, Characteristics and Public Health Implications." *Journal of Hospital Infection* 95 (2).

Kirchdoerfer, Robert N., et al. 2016. "Pre-Fusion Structure of a Human Coronavirus Spike Protein." *Nature* 531 (7592).

Kissler, Stephen M., et al. 2020. "Projecting the Transmission Dynamics of SARS-CoV-2 Through the Postpandemic Period." *Science* 368 (6493).

Kogan, Nicole E., et al. 2021. "An Early Warning Approach to Monitor COVID-19 Activity with Multiple Digital Traces in Near Real Time." *Science Advances* 7 (10).

Köndgen, Sophie, et al. 2008. "Pandemic Human Viruses Cause Decline of Endangered Great Apes." *Current Biology* 18 (4).

Koonin, Eugene V., et al. 2020. "Global Organization and Proposed Megataxonomy of the Virus World." *Microbiology and Molecular Biology Reviews* 84 (2).

Koopmans, Marion. 2021. "SARS-CoV-2 and the Human-Animal Interface: Outbreaks on Mink Farms." *The Lancet Infectious Diseases* 21 (1).

Korber, Bette, et al. 2020. "Tracking Changes in SARS-CoV-2 Spike: Evidence That D614G Increases Infectivity of the COVID-19 Virus." *Cell* 182 (4).

Kress, W. John, Jonna A. K. Mazet, and Paul D. N. Hebert. 2020. "Intercepting Pandemics Through Genomics." *Proceedings of the National Academy of Sciences* 17 (25).

Kucharski, Adam. 2020. *The Rules of Contagion: Why Things Spread—And Why They Stop.* New York: Hachette.

Kuchipudi, Suresh V., et al. 2021. "Multiple Spillovers and Onward Transmission of SARS-CoV-2 in Free-Living and Captive White-Tailed Deer (*Odocoileus virginianus*). Preprint, bioRxiv, posted November 1, 2021.

Kuhn, Jens H., edited by Charles H. Calisher. 2008. *Filoviruses: A Compendium of Forty Years of Epidemiological, Clinical, and Laboratory Studies.* New York: Springer.

Kuhn, Jens H., et al. 2019. "Classify Viruses—The Gain Is Worth the Pain." *Nature* 566 (7744).

Kupferschmidt, Kai. 2021a. "Viral Evolution May Herald New Pandemic Phase." *Science* 371 (6525).

———. 2021b. "Patience Is Crucial: Why We Won't Know for Weeks How Dangerous Omicron Is." *Science* 374 (6572).

Kuppalli, Krutika, and Angela L. Rasmussen. 2020. "A Glimpse into the Eye of the COVID-19 Cytokine Storm." *EBioMedicine* 55 (102789).

Lai, Alessia, et al. 2020. "Early Phylogenetic Estimate of the Effective Reproduction Number of SARS-CoV-2." *Journal of Medical Virology* 92 (6).

Lam, Tommy Tsan-Yuk, et al. 2020. "Identifying SARS-CoV-2 Related Coronaviruses in Malayan Pangolins." *Nature* 583 (7815).

La Rosa, Giuseppina, et al. 2020. "SARS-CoV-2 Has Been Circulating in Northern Italy Since December 2019: Evidence from Environmental Monitoring." *Science of the Total Environment* 750.

Larson, Heidi J. 2020. *Stuck: How Vaccine Rumors Start—And Why They Don't Go Away.* New York: Oxford University Press.

Larson, Heidi, and David A. Broniatowski. 2021. "Why Debunking Misinformation Is Not Enough to Change People's Minds About Vaccines." *American Journal of Public Health* 111 (6).

La Scola, Bernard, et al. 2003. "A Giant Virus in Amoebae." *Science* 299 (5615).

Latinne, Alice, et al. 2020. "Origin and Cross-Species Transmission of Bat Coronaviruses in China." *Nature Communications* 11 (4235).

Lau, Susanna K. P., et al. 2005. "Severe Acute Respiratory Syndrome Coronavirus-like Virus in Chinese Horseshoe Bats." *Proceedings of the National Academy of Sciences* 102 (39).

Laxminarayan, Ramanan, and T. Jacob John. 2020. "Is Gradual and Controlled Approach to Herd Protection a Valid Strategy to Curb the COVID-19 Pandemic?" *Indian Pediatrics* 57 (6).

Laxminarayan, Ramanan, Shahid Jameel, and Swarup Sarkar. 2020. "India's Battle Against COVID-19: Progress and Challenges." *American Journal of Tropical Medicine and Hygiene* 103 (4).

Laxminarayan, Ramanan, et al. 2020. "Epidemiology and Transmission Dynamics of COVID-19 in Two Indian States." *Science* 370 (6517).

Lee, Jimmy, et al. 2020. "No Evidence of Coronaviruses or Other Potentially Zoonotic Viruses in Sunda Pangolins (*Manis javanica*) Entering the Wildlife Trade via Malaysia." *EcoHealth* 17 (3).

Lee, Juhye M., et al. 2021. "Mapping Person-to-Person Variation in Viral Mutations That Escape Polyclonal Serum Targeting Influenza Hemagglutinin." *eLife* 8.

Lehman, David, et al. 2020. "Pangolins and Bats Living Together in Underground Burrows in Lopé National Park, Gabon." *African Journal of Ecology* 10 (1111).

Lemieux, Jacob E., et al. 2020. "Phylogenetic Analysis of SARS-CoV-2 in Boston Highlights the Impact of Superspreading Events." *Science* 371 (6529).

Lentzos, Filippa. 2020. "Natural Spillover or Research Lab Leak? Why a Credible Investigation Is Needed to Determine the Origin of the Coronavirus Pandemic." *Bulletin of the Atomic Scientists*. May 1.

Lescure, Francois-Xavier, et al. 2020. "Clinical and Virological Data of the First Cases of COVID-19 in Europe: A Case Series." *The Lancet Infectious Disease* 20 (6).

Letko, Michael, Andrea Marzi, and Vincent Munster. 2020. "Functional Assessment of Cell Entry and Receptor Usage for SARS-CoV-2 and Other Lineage B Betacoronaviruses." *Nature Microbiology* 5 (4).

Lewis, Gregory, et al. 2020. "The Biosecurity Benefits of Genetic Engineering Attribution." *Nature Communications* 11 (1).

Lewis, Michael. 2021. *The Premonition: A Pandemic Story*. New York: W. W. Norton.

Li, Fang, et al. 2005. "Structure of SARS Coronavirus Spike Receptor-Binding Domain Complexed with Receptor." *Science* 309 (5742).

Li, Qun. 2020. "An Outbreak of NCIP (2019-nCoV) Infection in China—Wuhan, Hubei Province, 2019–2020." *Chinese Center for Disease Control and Prevention Weekly* 2 (5).

Li, Qun, et al. 2020. "Early Transmission Dynamics in Wuhan, China, of Novel Coronavirus-Infected Pneumonia." *The New England Journal of Medicine* 382 (13).

Li, Wendong, et al. 2005. "Bats Are Natural Reservoirs of SARS-like Coronaviruses." *Science* 310 (5748).

Li, Yize, et al. 2021. "SARS-CoV-2 Induces Double-Stranded RNA-Mediated Innate Immune Responses in Respiratory Epithelial-Derived Cells and Cardiomyocytes." *Proceedings of the National Academy of Sciences* 118 (16).

Li, Zhencui, et al. "Notes from the Field: Genome Characterization of the First Outbreak of COVID-19 Delta Variant B.1.617.2—Guangzhou City, Guangdong Province, China, May 2021." *China Center for Disease Control and Prevention Weekly* 3 (27).

Lim, Poh Lian. 2015. "Middle East Respiratory Syndrome (MERS) in Asia: Lessons Gleaned from the South Korean Outbreak." *Transactions of the Royal Society of Tropical Medicine and Hygiene* 109 (9).

Lim, Poh Lian, et al. 2004. "Laboratory-Acquired Severe Acute Respiratory Syndrome." *The New England Journal of Medicine* 350 (17).

Lin, Xian-Dan, et al. 2017. "Extensive Diversity of Coronaviruses in Bats from China." *Virology* 507 (1–10).

Lipkin, W. Ian. 2013. "The Changing Face of Pathogen Discovery and Surveillance." *Nature Reviews Microbiology* 11 (2).

———. 2021. "The Known Knowns, Known Unknowns, and Unknown Unknowns of COVID-19." *Bulletin of the Atomic Scientists* July 21.

Lipsitch, Marc. 2018. "Why Do Exceptionally Dangerous Gain-of-Function Experiments in Influenza?" In *Influenza Virus: Methods and Protocols.* Methods in Molecular Biology. Yohei Yamauchi, editor. Berlin: Springer Science+Business Media.

Lipsitch, Marc, and Alison P. Galvani. 2014. "Ethical Alternatives to Experiments with Novel Potential Pandemic Pathogens." *PLOS Medicine* 11 (5).

Liu, Ping, Wu Chen, and Jin-Ping Chen. 2019. "Viral Metagenomics Revealed Sendai Virus and Coronavirus Infection of Malayan Pangolins (*Manis javanica*)." *Viruses* 11 (11).

Liu, Ping, et al. 2020. "Are Pangolins the Intermediate Host of the 2019 Novel Coronavirus (SARS-CoV-2)?" *PLOS Pathogens* 16 (5).

Liu, Shan-Lu, et al. 2020. "No Credible Evidence Supporting Claims of the Laboratory Engineering of SARS-CoV-2." *Emerging Microbes & Infections* 9 (1).

Liu, Yinghui, et al. 2021. "Functional and Genetic Analysis of Viral Receptor ACE2 Orthologs Reveals a Broad Potential Host Range of SARS-CoV-2." *Proceedings of the National Academy of Sciences* 118 (12).

Lloyd-Smith, J. O., et al. 2005. "Superspreading and the Effect of Individual Variation on Disease Emergence." *Nature* 438 (7066).

Loomba, Sahil, et al. 2021. "Measuring the Impact of COVID-19 Vaccine Misinformation on Vaccination Intent in the UK and USA." *Nature Human Behaviour* 5 (3).

Lu, Guangwen, Qihui Wang, and George F. Gao. 2015. "Bat-to-Human: Spike Features Determining 'Host Jump' of Coronaviruses SARS-CoV, MERS-CoV, and Beyond." *Trends in Microbiology* 23 (8).

Lu, Hongzhou, Charles W. Stratton, and Yi-Wei Tang. 2020. "Outbreak of Pneumonia of Unknown Etiology in Wuhan, China: The Mystery and the Miracle." *Journal of Medical Virology* 92 (4).

Lucey, Daniel, and Annie Sparrow. 2020. "China Deserves Some Credit for Its Handling of the Wuhan Pneumonia." *Foreign Policy*, January 14.

Lucey, Daniel, and Kristen Kent. 2020. "Coronavirus—Unknown Source, Unrecognized Spread, and Pandemic Potential." *Think Global Health*, February 6.

Lurie, Nicole, and Gerald T. Keusch. 2020. "The R&D Preparedness Ecosystem: Preparedness for Health Emergencies." Report to the US National Academy of Medicine. August 9.

Lurie, Nicole, Gerald T. Keusch, and Victor J. Dzau. 2021. "Urgent Lessons from COVID 19: Why the World Needs a Standing, Coordinated System and Sustainable Financing for Global Research and Development." *The Lancet* 397 (10280).

Lwoff, A. 1957. "The Concept of Virus." *Journal of General Microbiology* 17 (1).

Lytras, Spyros, et al. 2021. "Exploring the Natural Origins of SARS-CoV-2 in the Light of Recombination." Preprint, bioRxiv, posted May 27, 2021.

MacLean, Oscar A., et al. 2021. "Natural Selection in the Evolution of SARS-CoV-2 in Bats Created a Generalist Virus and Highly Capable Human Pathogen." *PLOS Biology* 19 (3).

Maganga, Gael Darren, et al. 2020. "Genetic Diversity and Ecology of Coronaviruses Hosted by Cave-Dwelling Bats in Gabon." *Scientific Reports* 10 (1).

Mai, Jun. 2020. "Paper on Human Transmission of Coronavirus Sets Off Social Media Storm in China." *South China Morning Post*, January 31.

Manes, Costanza, Rania Gollakner, and Ilaria Capua. 2020. "Could Mustelids Spur COVID-19 into a Panzootic?" *Veterinaria Italiana* 56 (2–3).

Mari, Lorenzo, et al. 2021. "The Epidemicity Index of Recurrent SARS-CoV-2 Infections." *Nature Communications* 12 (1).

Mbala-Kingebeni, P., et al. 2021. "Ebola Virus Transmission Initiated by Relapse of Systemic Ebola Virus Disease." *The New England Journal of Medicine* 384 (13).

McCarthy, Kevin R., et al. 2021. "Recurrent Deletions in the SARS-CoV-2 Spike Glycoprotein Drive Antibody Escape." *Science* 371 (6534).

McKee, Clifton D., et al. 2021. "The Ecology of Nipah Virus in Bangladesh: A Nexus of Land-Use Change and Opportunistic Feeding Behavior in Bats." *Viruses* 13 (2).

McLellan, Jason S., et al. 2013a. "Structure of RSV Fusion Glycoprotein Trimer Bound to a Prefusion-Specific Neutralizing Antibody." *Science* 340 (6136).

McLellan, Jason S., et al. 2013b. "Structure-Based Design of a Fusion Glycoprotein Vaccine for Respiratory Syncytial Virus." *Science* 342 (6158).

McNeil, Donald G. Jr. 2021. "How I Learned to Stop Worrying and Love the Lab-Leak Theory." Medium, May 17.

Medawar, P. B., and J. S. Medawar. 1983. *Aristotle to Zoos: A Philosophical Dictionary of Biology*. Cambridge: Harvard University Press.

Memish, Ziad A., et al. "Middle East Respiratory Syndrome Coronavirus in Bats, Saudi Arabia." *Emerging Infectious Diseases* 19 (11).

Menachery, Vineet D., et al. 2015. "A SARS-like Cluster of Circulating Bat Coronaviruses Shows Potential for Human Emergence." *Nature Medicine* 21 (12).

Menachery, Vineet D., et al. 2016. "SARS-like WIV1-CoV Poised for Human Emergence." *Proceedings of the National Academy of Sciences Early Edition* 113 (11).

Menachery, Vineet D., Rachel L. Graham, and Ralph S. Baric. 2017. "Jumping Species—A Mechanism for Coronavirus Persistence and Survival." *ScienceDirect* 23 (1–7).

Menachery, Vineet D., et al. 2020. "Trypsin Treatment Unlocks Barrier for Zoonotic Bat Coronavirus Infection." *Journal of Virology* 94 (5).

Meredith, Luke W., et al. 2020. "Rapid Implementation of SARS-CoV-2 Sequencing to Investigate Cases of Health-Care Associated COVID-19: A Prospective Genomic Surveillance Study." *The Lancet Infectious Diseases* 20 (11).

Mistry, Dina, et al. 2021. "Inferring High-Resolution Human Mixing Patterns for Disease Modeling." *Nature Communications* 12 (1).

Mitjà, O., et al. 2021. "A Cluster-Randomized Trial of Hydroxychloroquine for Prevention of Covid-19." *The New England Journal of Medicine* 384 (5).

Moore, Kristine A. 2020. "COVID-19: The CIDRAP Viewpoint, Part 1: The Future of the COVID-19 Pandemic: Lessons Learned from Pandemic Influenza." Center for Infectious Disease Research and Policy, University of Minnesota. April 30.

Morens, David M., and Anthony S. Fauci. 2020. "Emerging Pandemic Diseases: How We Got to COVID-19." *Cell* 182 (5).

Morens, David M., et al. 2020. "The Origin of COVID-19 and Why It Matters." *American Journal of Tropical Medicine and Hygiene* 103 (3).

Morens, David M., Peter Daszak, and Jeffery K. Taubenberger. 2020. "Escaping Pandora's Box—Another Novel Coronavirus." *The New England Journal of Medicine* 382 (14).

Morse, Stephen S., editor. 1993. *Emerging Viruses*. New York: Oxford University Press.

Mughal, Fizza, Arshan Nasir, and Gustavo Caetano-Anollés. 2020. "The Origin and Evolution of Viruses Inferred from Fold Family Structure." *Archives of Virology* 165 (10).

Munnink, Bas B. Oude, et al. 2021. "Transmission of SARS-CoV-2 on Mink Farms Between Humans and Mink and Back to Humans." *Science* 371 (6525).

Murray, Christopher J. L., and Peter Piot. 2021. "The Potential Future of the COVID-19 Pandemic: Will SARS-CoV-2 Become a Recurrent Seasonal Infection?" *Journal of the American Medical Association* 325 (13).

Nachega, Jean B., et al. 2020. "Responding to the Challenge of the Dual COVID-19 and Ebola Epidemics in the Democratic Republic of Congo—Priorities for Achieving Control." *American Journal of Tropical Medicine and Hygiene.* 103 (2).

Nakazawa, Eisuke, Hiroyasu Ino, and Akira Akabayashi. 2020. "Chronology of COVID-19 Cases on the Diamond Princess Cruise Ship and Ethical Considerations: A Report from Japan." *Disaster Medicine and Public Health Preparedness* 14 (4).

Nasir, Arshan, and Gustavo Caetano-Anollés. 2015. "A Phylogenetic Data-Driven Exploration of Viral Origins and Evolution." *Science Advances* 1 (8).

Nasir, Arshan, Kyung Mo Kim, and Gustavo Caetano-Anollés. 2017. "Long-Term Evolution of Viruses: A Janus-Faced Balance." *BioEssays* 39 (8).

Nasir, Arshan, Ethan Romero-Severson, and Jean-Michel Claverie. 2020. "Investigating the Concept and Origin of Viruses." *Trends in Microbiology* 28 (12).

National Intelligence Council. 2021. "Updated Assessment on COVID-19 Origins." October 29.

Neher, Richard A., et al. 2020. "Potential Impact of Seasonal Forcing on a SARS-CoV-2 Pandemic." *Swiss Medical Weekly* 150.

Norton, Alice, et al. 2020. "The Remaining Unknowns: A Mixed Methods Study of the Current and Global Health Research Priorities for COVID-19." *BMJ Global Health* 5 (7).

Nsoesie, Elaine Okanyene, et al. 2020. "Analysis of Hospital Traffic and Search Engine Data in Wuhan China Indicates Early Disease Activity in the Fall of 2019." Digital Access to Scholarship at Harvard (DASH), June 8.

Offit, M.D., Paul A. 2007. *Vaccinated: One Man's Quest to Defeat the World's Deadliest Diseases.* New York: HarperCollins.

Okada, Pilailuk, et al. 2020. "Early Transmission Patterns of Coronavirus Disease 2019 (COVID-19) in Travellers from Wuhan to Thailand, January 2020." *Euro Surveillance* 25 (8).

Okba, Nisreen M. A., et al. 2020. "Severe Acute Respiratory Syndrome Coronavirus 2—Specific Antibody Responses in Coronavirus Disease Patients." *Emerging Infectious Diseases* 26 (7).

Olival, Kevin J., et al. 2020. "Possibility for Reverse Zoonotic Transmission of SARS-CoV-2 to Free-Ranging Wildlife: A Case Study of Bats." *PLOS Pathogens* 16 (9).

Omrani, Ali S., Jaffar A. Al-Tawfiq, and Ziad A. Memish. 2015. "Middle East Respiratory Syndrome Coronavirus (MERS-CoV): Animal to Human Interaction." *Pathogens and Global Health* 109 (8).

Ortiz, Nancy, et al. 2021. "Epidemiologic Findings from Case Investigations and Contact Tracing for First 200 Cases of Coronavirus Disease, Santa Clara County, California, USA." *Emerging Infectious Diseases* 27 (5).

Osnas, Erik E., Paul J. Hurtado, and Andrew P. Dobson. 2015. "Evolution of Pathogen Virulence Across Space During an Epidemic." *The American Naturalist* 185 (3).

Osterholm, Michael T., and Mark Olshaker. 2017. *Deadliest Enemy: Our War Against Killer Germs.* New York: Little, Brown Spark.

———. 2020. "Chronicle of a Pandemic Foretold. Learning from the COVID-19 Failure—Before the Next Outbreak Arrives." *Foreign Affairs*, July/August 2020.

———. 2021. "The Pandemic That Won't End. COVID-19 Variants and the Peril of Vaccine Inequity." *Foreign Affairs*, March 8.

Pagani, Gabriele, et al. 2020. "Seroprevalence of SARS-CoV-2 Significantly Varies with Age: Preliminary Results from a Mass Population Screening." *Journal of Infection* 81 (6).

Pagani, Gabriele, et al. 2021a. "Human-to-Cat SARS-CoV-2 Transmission: Case Report and Full-Genome Sequencing from an Infected Pet and Its Owner in Northern Italy." *Pathogens* 10 (2).

Pagani, Gabriele, et al. 2021b. "Prevalence of SARS-CoV-2 in an Area of Unrestricted Viral Circulation: Mass Seroepidemiological Screening in Castiglione d'Adda, Italy." *PLOS ONE* 16 (2).

Palacios, Gustavo, et al. 2011. "Human Metapneumovirus Infection in Wild Mountain Gorillas, Rwanda." *Emerging Infectious Diseases* 17 (4).

Pallesen, Jesper, et al. 2017. "Immunogenicity and Structures of a Rationally Designed Prefusion MERS-CoV Spike Antigen." *Proceedings of the National Academy of Sciences* 114 (35).

Pardi, Norbert, et al. 2018. "mRNA Vaccines—A New Era in Vaccinology." *Nature Reviews Drug Discovery* 17 (4).

Patrono, Livia V., et al. 2018. "Human Coronavirus OC43 Outbreak in Wild Chimpanzees, Côte d'Ivoire, 2016." *Emerging Microbes & Infections* 7 (1).

Patrono, Livia Victoria, et al. 2021. "Archival Influenza Virus Genomes from Europe Reveal Genomic and Phenotypic Variability During the 1918 Pandemic." Preprint, bioRxiv, posted May 14, 2021.

Peacock, Sharon. 2020. "History of COG-UK: A Short History of the COG-UK Consortium." COVID-19 Genomics UK Consortium website, December 17.

Peacock, Thomas P., et al. 2021. "The SARS-CoV-2 Variants Associated with Infections in India, B.1.617, Show Enhanced Spike Cleavage by Furin." Preprint, bioRxiv, posted May 28, 2021.

Pekar, Jonathan, et al. 2021. "Timing the SARS-CoV-2 Index Case in Hubei Province." *Science* 372 (6540).

Pereira, H. G., Bela Tumova, and R. G. Webster. 1967. "Antigenic Relationship Between Influenza A Viruses of Human and Avian Origins." *Nature* 215 (5104).

Peters, C. J., and Mark Olshaker. 1997. *Virus Hunter: Thirty Years of Battling Hot Viruses Around the World.* New York: Doubleday.

Philippe, Nadège, et al. 2013. "Pandoraviruses: Amoeba Viruses with Genomes Up to 2.5 Mb Reaching That of Parasitic Eukaryotes." *Science* 341 (6143).

Piot, Peter, with Ruth Marshall. 2012. *No Time to Lose: A Life in Pursuit of Deadly Viruses.* New York: W. W. Norton.

Piot, Peter, Moses J. Soka, and Julia Spencer. 2019. "Emergent Threats: Lessons Learnt from Ebola." *International Health* 11 (5).

Plante, Jessica A., et al. 2020. "Spike Mutation D614G Alters SARS-CoV-2 Fitness." *Nature* 592 (7852).

Plowright, Raina K., et al. 2008. "Reproduction and Nutritional Stress Are Risk Factors for Hendra Virus Infection in Little Red Flying Foxes (*Pteropus scapulatus*)." *Proceedings of the Royal B Society Biological Sciences* 275 (1636).

Plowright, Raina K., et al. 2021. "Land Use-Induced Spillover: A Call to Action to Safeguard Environmental, Animal, and Human Health." *The Lancet Planet Health* 5 (4).

Pollack, Marjorie P., et al. 2012. "Latest Outbreak News from ProMED-Mail Novel Coronavirus—Middle East." *International Journal of Infectious Diseases* 17 (2).

Pradhan, Prashant, et al. 2020. "Uncanny Similarity of Unique Inserts in the 2019-nCoV Spike Protein to HIV-1 gp120 and Gag." Preprint, bioRxiv, posted January 31, 2020. Later withdrawn.

Putcharoen, Opass, et al. 2021. "Early Detection of Neutralizing Antibodies Against SARS-CoV-2 in COVID-19 Patients in Thailand." *PLOS ONE* 16 (2).

Qiu, Jane. 2020. "How China's 'Bat Woman' Hunted Down Viruses from SARS to the New Coronavirus." *Scientific American*, April 27.

————. 2021. "This Scientist Now Believes COVID Started in Wuhan's Wet Market. Here's Why." *MIT Technology Review*, November 19.

Rahalkar, Monali C., and Rahul A. Bahulikar. 2020. "Lethal Pneumonia Cases in Mojiang Miners (2012) and the Mineshaft Could Provide Important Clues to the Origin of SARS-CoV-2." *Frontiers in Public Health* 8.

Rambaut, Andrew, et al. 2020. "A Dynamic Nomenclature Proposal for SARS-CoV-2 Lineages to Assist Genomic Epidemiology." *Nature Microbiology* 5 (11).

Raoult, Didier, et al. 2004. "The 1.2-Megabase Genome Sequence of Mimivirus." *Science* 306 (5700).

Raoult, Didier, and Patrick Forterre. 2008. "Redefining Viruses: Lessons from Mimivirus." *Nature Reviews Microbiology* 6 (4).

Rasmussen, Angela L. 2020. "Vaccination Is the Only Acceptable Path to Herd Immunity." *Med* 1 (1).

————. 2021. "On the Origins of SARS-CoV-2." *Nature Medicine* 27 (1).

Rausch, Jason W., et al. 2020. "Low Genetic Diversity May Be an Achilles Heel of SARS-CoV-2." *Proceedings of the National Academy of Sciences* 117 (40).

Relman, David A. 2020. "To Stop the Next Pandemic, We Need to Unravel the Origins of COVID-19." *Proceedings of the National Academy of Sciences* 117 (47).

Ren, Wuze, et al. 2006. "Full-Length Genome Sequences of Two SARS-like Coronaviruses in Horseshoe Bats and Genetic Variation Analysis." *Journal of General Virology* 87 (Pt 11).

Ren, Wuze, et al. 2008. "Difference in Receptor Usage Between Severe Acute Respiratory Syndrome (SARS) Coronavirus and SARS-like Coronavirus of Bat Origin." *Journal of Virology* 82 (4).

Reusken, Chantal B. E. M., et al. 2013. "Middle East Respiratory Syndrome Coronavirus Neutralising Serum Antibodies in Dromedary Camels: A Comparative Serological Study." *The Lancet Infectious Diseases* 13 (10).

Richard, Mathilde, et al. 2020 "SARS-CoV-2 Is Transmitted Via Contact and Via the Air Between Ferrets." *Nature Communications* 11 (1).

Rimoin, Anne W., et al. 2018. "Ebola Virus Neutralizing Antibodies Detectable in Survivors of the Yambuku, Zaire Outbreak 40 Years After Infection." *The Journal of Infectious Diseases* 217 (2).

Robertson, David. 2021. "Of Mice and Schoolchildren: A Conceptual History of Herd Immunity." *American Journal of Public Health* 111 (8).

Rocha, Rudi, et al. 2021. "Effect of Socioeconomic Inequalities and Vulnerabilities on Health-System Preparedness and Response to COVID-19 in Brazil: A Comprehensive Analysis." *The Lancet Global Health* 9 (6).

Rocklöv, J., H. Sjödin, and A. Wilder-Smith. 2020. "COVID-19 Outbreak on the Diamond Princess Cruise Ship: Estimating the Epidemic Potential and Effectiveness of Public Health Countermeasures." *Journal of Travel Medicine* 27 (3).

Rohwer, Forest, and Katie Barott. 2013. "Viral Information." *Biology and Philosophy* 28 (2).

Rojas, Maria, et al. 2021. "Swabbing the Urban Environment—A Pipeline for Sampling and Detection of SARS-CoV-2 from Environmental Reservoirs." *Journal of Visualized Experiments* 170.

Rothe, Camilla, et al. 2020. "Transmission of 2019-nCoV Infection from an Asymptomatic Contact in Germany." *The New England Journal of Medicine* 382 (10).

Sabeti, Pardis, and Lara Salahi. 2021. *Outbreak Culture: The Ebola Crisis and the Next Epidemic.* Cambridge: Harvard University Press.

Sabino, Ester C., et al. 2021. "Resurgence of COVID-19 in Manaus, Brazil, Despite High Seroprevalence." *The Lancet* 397 (10273).

Sahin, Ugur, Katalin Karikó, and Özlem Türeci. 2014. "mRNA-Based Therapeutics—Developing a New Class of Drugs." *Nature Reviews Drug Discovery* 13 (10).

Salvatore, Maxwell, et al. 2021. "Resurgence of SARS-CoV-2 in India: Potential Role of the B.1.617.2 (Delta) Variant and Delayed Interventions." Preprint, medRxiv, posted June 30, 2021.

Santini, Joanne M., and Sarah J. L. Edwards. 2020. "Host Range of SARS-CoV-2 and Implications for Public Health." *The Lancet Microbe* 1 (4).

Scott, H. Denman, MD. 1971. "The Elusiveness of Measles Eradication: Insights Gained from Three Years of Intensive Surveillance in Rhode Island." *American Journal of Epidemiology* 94 (1).

Shairp, Rachel, et al. 2016. "Understanding Urban Demand for Wild Meat in Vietnam: Implications for Conservation Actions." *PLOS ONE* 11 (1).

Sheahan, Timothy P., et al. 2017. "Broad-Spectrum Antiviral GS-5734 Inhibits Both Epidemic and Zoonotic Coronavirus." *Science Translational Medicine* 9 (396).

Sheahan, Timothy P., et al. 2020. "An Orally Bioavailable Broad-Spectrum Antiviral Inhibits SARS-CoV-2 in Human Airway Epithelial Cell Cultures and Multiple Coronaviruses in Mice." *Science Translational Medicine* 12 (541).

Shi, Jianzhong, et al. 2020. "Susceptibility of Ferrets, Cats, Dogs, and Other Domesticated Animals to SARS-coronavirus 2." *Science* 368 (6494).

Shi, Zhengli. 2021. "Origins of SARS-CoV-2: Focusing on Science." *Infectious Diseases & Immunity* 1 (1).

Shi, Zhengli, and Zhihong Hu. 2008. "A Review of Studies on Animal Reservoirs of the SARS Coronavirus." *Virus Research* 133 (1).

Siciliano, Bruno, et al. 2020. "The Impact of COVID-19 Partial Lockdown on Primary

Pollutant Concentrations in the Atmosphere of Rio de Janeiro and São Paulo Megacities (Brazil)." *Bulletin of Environmental Contamination and Toxicology* 105 (1).

Siegel, Dustin, et al. 2017. "Discovery and Synthesis of a Phosphoramidate Prodrug of a Pyrrolo[2, 1-*f*][triazin-4-amino] Adenine C-Nucleoside (GS-5734) for the Treatment of Ebola and Emerging Viruses." *Journal of Medicinal Chemistry* 60 (5).

Sirotkin, Karl, and Dan Sirotkin. 2020. "Might SARS-CoV-2 Have Arisen Via Serial Passage Through an Animal Host or Cell Culture?" *BioEssays* 42 (10).

Sit, Thomas H. C., et al. 2020. "Infection of Dogs with SARS-CoV-2." *Nature* 586 (7831).

Slavitt, Andy. 2021. *Preventable: The Inside Story of How Leadership Failures, Politics, and Selfishness Doomed the U.S. Coronavirus Response.* New York: St. Martin's Press.

Souza, Thiago Moreno L., and Carlos Medicis Morel. 2021. "The COVID-19 Pandemics and the Relevance of Biosafety Facilities For Metagenomics Surveillance, Structured Disease Prevention and Control." *Biosafety and Health* 3 (1).

Specter, Michael. 2020. "The Good Doctor: How Anthony Fauci Became the Face of a Nation's Crisis Response." *The New Yorker*, April 20.

Starr, Tyler N., et al. 2020. "Deep Mutational Scanning of SARS-CoV-2 Receptor Binding Domain Reveals Constraints on Folding and ACE2 Binding." *Cell* 182 (5).

Stein, Richard A. 2011. "Super-Spreaders in Infectious Diseases." *International Journal of Infectious Diseases* 15 (8).

Sugerman, David E., et al. 2010. "Measles Outbreak in a Highly Vaccinated Population, San Diego, 2008: Role of the Intentionally Undervaccinated." *Pediatrics* 125 (4).

Swanepoel, Robert, et al. 2007. "Studies of Reservoir Hosts for Marburg Virus." *Emerging Infectious Diseases* 13 (12).

Tan, Chee Wah, et al. 2021. "A SARS-CoV-2 Surrogate Virus Neutralization Test Based on Antibody-Mediated Blockage of ACE2-Spike Protein-Protein Interaction." *Nature Biotechnology* 38 (9).

Tang, Xiaolu, et al. 2020. "On the Origin and Continuing Evolution of SARS-CoV-2." *National Science Review* 7 (6).

Taubenberger, Jeffery K., and David M. Morens. 2006. "1918 Influenza: The Mother of All Pandemics." *Emerging Infectious Diseases* 12 (1).

Tchesnokova, Veronika, et al. 2021. "Acquisition of the L452R Mutation in the ACE2-Binding Interface of Spike Protein Triggers Recent Massive Expansion of SARS-CoV-2 Variants." *Journal of Clinical Microbiology* 59 (11).

Tegally, Houriiyah, et al. 2020. "Emergence and Rapid Spread of a New Severe Acute Respiratory Syndrome-Related Coronavirus 2 (SARS-CoV-2) Lineage with Multiple Spike Mutations in South Africa." Preprint, medRxiv, posted December 22, 2020.

Temmam, Sarah, et al. 2021. "Coronaviruses with a SARS-CoV-2-like Receptor-Binding Domain Allowing ACE2-Mediated Entry into Human Cells Isolated from Bats of Indochinese Peninsula." Preprint, Research Square, posted September 17, 2021.

Thomson, Emma C., et al. 2021. "Circulating SARS-CoV-2 Spike N439K Variants Maintain Fitness While Evading Antibody-Mediated Immunity." *Cell* 184 (5).

Topley, W. W. C., and G. S. Wilson. 1923. "The Spread of Bacterial Infection. The Problem of Herd-Immunity." *Journal of Hygiene* 21 (3).

Towner, Jonathan S., et al. 2009. "Isolation of Genetically Diverse Marburg Viruses from Egyptian Fruit Bats." *PLOS Pathogens* 5 (7).

Traynor, Bryan J. 2009. "The Era of Genomic Epidemiology." *Neuroepidemiology* 33 (3).

Tumpey, Terrence M., et al. 2005. "Characterization of the Reconstructed 1918 Spanish Influenza Pandemic Virus." *Science* 310 (5745).

Urakova, Nadya, et al. 2018. "β-D-N^4-Hydroxycytidine Is a Potent Anti-Alphavirus Compound That Induces a High Level of Mutations in the Viral Genome." *Journal of Virology* 92 (3).

van Aken, J. 2007. "Ethics of Reconstructing Spanish Flu: Is it Wise to Resurrect a Deadly Virus?" *Heredity* 98 (1).

van Dorp, Lucy, et al. "Emergence of Genomic Diversity and Recurrent Mutations in SARS-CoV-2." *Infection, Genetics and Evolution* 83.

Vandyck, Koen, et al. 2021. "ALG-09711, A Potent and Selective SARS-CoV-2 3-Chymotrypsin-like Cysteine Protease Inhibitor Exhibits *in vivo* Efficacy in a Syrian Hamster Model." *Biochemical and Biophysical Research Communications* 555.

Vetter, Pauline, et al. 2020. "Daily Viral Kinetics and Innate and Adaptive Immune Response Assessment in COVID-19: A Case Series." *mSphere* 5 (6).

Vijgen, Leen, et al. 2005. "Complete Genomic Sequence of Human Coronavirus OC43: Molecular Clock Analysis Suggests a Relatively Recent Zoonotic Coronavirus Transmission Event." *Journal of Virology* 79 (3).

Vincent, Martin J., et al. 2005. "Chloroquine Is a Potent Inhibitor of SARS Coronavirus Infection and Spread." *Virology Journal* 2 (69).

Vlasova, Anastasia N., et al. 2021. "Novel Canine Coronavirus Isolated from a Hospitalized Pneumonia Patient, East Malaysia." *Clinical Infectious Diseases*, May 20.

Voight, Christian C., and Tigga Kingston, editors. 2016. *Bats in the Anthropocene: Conservation of Bats in a Changing World.* New York: Springer Open.

Volz, Erik, et al. 2021. "Evaluating the Effects of SARS-CoV-2 Spike Mutation D614G on Transmissibility and Pathogenicity." *Cell* 184 (1).

Wacharapluesadee, Supaporn, et al. 2013. "Group C Betacoronavirus in Bat Guano Fertilizer, Thailand." *Emerging Infectious Diseases* 19 (8).

Wacharapluesadee, Supaporn, et al. 2015. "Diversity of Coronavirus in Bats from Eastern Thailand." *Virology Journal* 12 (57).

Wacharapluesadee, Supaporn, et al. 2020. "Identification of a Novel Pathogen Using Family-Wide PCR: Initial Confirmation of COVID-19 in Thailand." *Frontiers in Public Health* 8.

Wacharapluesadee, Supaporn, et al. 2021. "Evidence for SARS-CoV-2 Related Coronaviruses Circulating in Bats and Pangolins in Southeast Asia." *Nature Communications* 12 (1).

Wade, Nicholas. 2021. "The Origin of COVID: Did People or Nature Open Pandora's Box at Wuhan?" *Bulletin of the Atomic Scientists*, May 5.

Wahl, Angela, et al. 2021. "SARS-CoV-2 Infection Is Effectively Treated and Prevented by EIDD-2801." *Nature* 591 (7850).

Wan, Yushun, et al. 2020. "Receptor Recognition by the Novel Coronavirus from Wuhan: An Analysis Based on Decade-Long Structural Studies of SARS Coronavirus." *Journal of Virology* 94 (7).

Wang, Lin-Fa, and Christopher Cowled, editors. 2015. *Bats and Viruses: A New Frontier of Emerging Infectious Diseases.* Hoboken, N.J.: Wiley Blackwell.

Wang, Lin-Fa, et al. 2020. "From Hendra to Wuhan: What Has Been Learned in Responding to Emerging Zoonotic Viruses." *The Lancet* 395 (10224).

Wang, Manli, et al. 2020. "Remdesivir and Chloroquine Effectively Inhibit the Recently Emerged Novel Coronavirus (2019-nCoV) *in vitro.*" *Cell Research* 30 (3).

Wang, Ning, et al. 2018. "Serological Evidence of Bat SARS-Related Coronavirus Infection in Humans, China." *Virologica Sinica* 33 (1).

Wang, Weier, Jianming Tang, and Fangqiang Wei. 2020. "Updated Understanding of the Outbreak of 2019 Novel Coronavirus (2019-nCoV) in Wuhan, China." *Journal of Medical Virology* 92 (4).

Wang, Yeming, et al. 2020. "Remdesivir in Adults with Severe COVID-19: A Randomised, Double-Blind, Placebo-Controlled, Multicentre Trial." *The Lancet* 395 (10236).

Washington, Nicole L., et al. 2021. "Emergence and Rapid Transmission of SARS-CoV-2 B.1.1.7 in the United States. *Cell* 184.

Webb, P. A., et al. 1967. "Some Characteristics of Machupo Virus, Causative Agent of Bolivian Hemorrhagic Fever." *The American Journal of Tropical Medicine and Hygiene* 16 (4).

Webster, Robert G. 2018. *Flu Hunter: Unlocking the Secrets of a Virus.* Dunedin, New Zealand: Otago University Press.

Weisblum, Yiska, et al. 2020. "Escape from Neutralizing Antibodies by SARS-CoV-2 Spike Protein Variants." *eLife* 9.

Weiss, Susan R. 2020. "Forty Years with Coronaviruses." *Journal of Experimental Medicine* 217 (5).

Welkers, Matthijs R. A., et al. 2021. "Possible Host-Adaptation of SARS-CoV-2 Due to Improved ACE2 Receptor Binding in Mink." *Virus Evolution* 7 (1).

Wells, H. L., et al. 2021. "The Evolutionary History of ACE2 Usage Within the Coronavirus Subgenus *Sarbecovirus.*" *Virus Evolution* 7 (1).

Wertheim, Joel O., and Michael Worobey. 2009. "Dating the Age of the SIV Lineages That Gave Rise to HIV-1 and HIV-2." *PLOS Computational Biology* 5 (5)

Wertheim, Joel O., et al. 2013. "A Case for the Ancient Origin of Coronaviruses." *Journal of Virology* 87 (12).

White, Tracie. 2020. "The Virus Hunter Becomes Prey: Renowned Microbiologist's Battle Against the Coronavirus Gets Personal." *Stanford Medicine* 2.

WHO-China Study. 2021. "WHO-convened Global Study of Origins of SARS-CoV-2: China Part." March 30.

Wolfe, Nathan D., et al. 2005. "Bushmeat Hunting, Deforestation, and Prediction of Zoonoses Emergence." *Emerging Infectious Diseases* 11 (12).

Wolff, Jon A., et. al. 1990. "Direct Gene Transfer into Mouse Muscle *in vivo.*" *Science* 247 (4949 Pt 1).

Wong, Gary, et al. 2015. "MERS, SARS, and Ebola: The Role of Super-Spreaders in Infectious Disease." *Cell Host & Microbe* 18 (4).

Wong, Matthew C., et al. 2020. "Evidence of Recombination in Coronaviruses Implicating Pangolin Origins of nCoV-2019." Preprint, bioRxiv, posted February 13, 2020.

Woo, Patrick C. Y., Susanna K. P. Lau, and Kwok-yung Yuen. 2006. "Infectious Diseases Emerging from Chinese Wet-Markets: Zoonotic Origins of Severe Respiratory Viral Infections." *Current Opinion in Infectious Diseases* 19 (5).

Woo, Patrick C. Y., et al. 2006. "Molecular Diversity of Coronaviruses in Bats." *Virology* 351 (1).

Worobey, Michael. 2021. "Dissecting the Early COVID-19 Cases in Wuhan." *Science* 374 (6572).

Worobey, Michael, et al. 2004. "Origin of AIDS: Contaminated Polio Vaccine Theory Refuted." *Nature* 428 (6985).

Worobey, Michael, et al. 2008. "Direct Evidence of Extensive Diversity of HIV-1 in Kinshasa by 1960." *Nature* 455 (7213).

Worobey, Michael, Jim Cox, and Douglas Gill. 2019. "The Origins of the Great Pandemic." *Evolution, Medicine, & Public Health* 2019 (1).

Worobey, Michael, et al. 2020. "The Emergence of SARS-CoV-2 in Europe and North America." *Science* 370 (6516).

Worobey, Michael, et al. 2022. "The Huanan Market Was the Epicenter of SARS-CoV-2 Emergence." Preprint, Zenodo, posted February 26, 2022.

Wrapp, Daniel, et al. 2020. "Cryo-EM Structure of the 2019-nCoV Spike in the Prefusion Conformation." *Science* 367 (6483).

Wright, Lawrence. 2021. *The Plague Year: America in the Time of COVID.* New York: Alfred A. Knopf.

Wrobel, Antoni G., et al. 2020. "SARS-CoV-2 and Bat RaTG13 Spike Glycoprotein Structures Inform on Virus Evolution and Furin-Cleavage Effects." *Nature Structural & Molecular Biology* 27 (8).

Wu, Fan, et al. 2020. "A New Coronavirus Associated with Human Respiratory Disease in China." *Nature* 579 (7798).

Wu, Kai, et al. 2021. "Serum Neutralizing Activity Elicited by mRNA-1273 Vaccine." *The New England Journal of Medicine* 384 (15).

Wu, Zhiqiang, et al. 2014. "Novel Henipa-like Virus, Mojiang Paramyxovirus, in Rats, China, 2012." *Emerging Infectious Diseases* 20 (6).

Xia, Hongjie, et al. 2020. "Evasion of Type 1 Interferon by SARS-CoV-2." *Cell Reports* 33 (1).

Xia, Wei, et al. 2021. "How One Pandemic Led to Another: ASFV, the Disruption Contributing to SARS-CoV-2 Emergence in Wuhan." Preprint, on Preprints, posted February 25, 2021.

Xia, Yuanqing, et al. 2020. "How to Understand 'Herd Immunity' in COVID-19 Pandemic." *Frontiers in Cell and Developmental Biology* 8.

Xiao, Botao, and Lei Xiao. 2020. "The Possible Origins of 2019-nCoV Coronavirus." Preprint, Research Gate, posted February 6, 2020. Later withdrawn.

Xiao, Chuan, et al. 2020. "HIV-1 Did Not Contribute to the 2019-nCoV Genome." *Emerging Microbes & Infections* 9 (1).

Xiao, Kangpeng, et al. 2020. "Isolation of SARS-CoV-2-related Coronavirus from Malayan Pangolins." *Nature* 583 (7815).

Xiao, Xiao, et al. 2021. "Animal Sales from Wuhan Wet Markets Immediately Prior to the COVID-19 Pandemic." *Scientific Reports* 11 (1).

Xie, Xuping, et al. 2021. "Engineering SARS-CoV-2 Using a Reverse Genetic System." *Nature Protocols* 16 (3).

Xu, Li. 2013. "The Analysis of Six Patients with Severe Pneumonia Caused by Unknown Viruses." Master's thesis, Kunming Medical University, Kunming, China.

Yan, Li-Meng, et al. 2020. "Unusual Features of the SARS-CoV-2 Genome Suggesting Sophisticated Laboratory Modification Rather Than Natural Evolution and Delineation of Its Probable Synthetic Route." Preprint, Zenodo, posted September 14, 2020.

Yang, Xing-Lou, et al. 2016. "Isolation and Characterization of a Novel Bat Coronavirus Closely Related to the Direct Progenitor of Severe Acute Respiratory Syndrome Coronavirus." *Journal of Virology* 90 (6).

Yount, Boyd, et al. 2003. "Reverse Genetics with a Full-Length Infectious cDNA of Severe Acute Respiratory Syndrome Coronavirus." *Proceedings of the National Academy of Sciences* 100 (22).

Yu, Wufei. 2020. "Coronavirus: Revenge of the Pangolins?" *The New York Times*, March 5.

Yuen, Kwok-Yung. 2020. "Reflections from a Clinician-Scientist During COVID-19 Pandemic: Facing Unknowns, Breaking Dogmas." *Synapse*, October.

Yurkovetskiy, Leonid, et al. "Structural and Functional Analysis of the D614G SARS-CoV-2 Spike Protein Variant." *Cell* 183 (3).

Zakj, Ali Moh, et al. 2012. "Isolation of Novel Coronavirus from a Man with Pneumonia in Saudi Arabia." *The New England Journal of Medicine* 367 (19).

Zehender, Gianguglielmo, et al. 2020. "Genomic Characterization and Phylogenetic Analysis of SARS-CoV-2 in Italy." *Journal of Medical Virology* 92 (9).

Zeng, Lei-Ping, et al. 2016. "Bat Severe Acute Respiratory Syndrome-like Coronavirus WIV1 Encodes an Extra Accessory Protein, ORFX, Involved in Modulation of the Host Immune Response." *Journal of Virology* 90 (14).

Zhan, Shing Hei, Benjamin E. Deverman, and Yujia Alina Chan. 2020. "SARS-CoV-2 Is Well Adapted for Humans. What Does This Mean for Re-Emergence?" Preprint, bioRxiv, posted May 2, 2020.

Zhang, Meng, et al. 2021. "Transmission Dynamics of an Outbreak of the COVID-19 Delta Variant B.1.617.2—Guangdong Province, China, May–June 2021." *China Center for Disease Control and Prevention Weekly* 3 (27).

Zhang, Qiang, et al. 2020. "A Serological Survey of SARS-CoV-2 in Cat in Wuhan." *Emerging Microbes & Infections* 9 (1).

Zhang, Tao, Qunfu Wu, and Zhigang Zhang. 2020. "Probable Pangolin Origin of SARS-CoV-2 Associated with the COVID-19 Outbreak." *Current Biology* 30 (7).

Zhang, Yong-Zhen, and Edward C. Holmes. 2020. "A Genomic Perspective on the Origin and Emergence of SARS-CoV-2." *Cell* 181 (2).

Zhao, Guo-ping. 2007. "SARS Molecular Epidemiology: A Chinese Fairy Tale of Controlling an Emerging Zoonotic Disease in the Genomics Era." *Philosophical Transactions of the Royal Society* 362 (1482).

Zhou, Hong, et al. 2020. "A Novel Bat Coronavirus Closely Related to SARS-CoV-2 Contains Natural Insertions at the S1/S2 Cleavage Site of the Spike Protein." *Current Biology* 30 (11).

Zhou, Hong, et al. 2021. "Identification of Novel Bat Coronaviruses Sheds Light on the Evolutionary Origins of SARS-CoV-2 and Related Viruses." *Cell* 184 (17).

Zhou, Peng, et al. 2020a. "A Pneumonia Outbreak Associated with a New Coronavirus of Probable Bat Origin." *Nature* 579 (7798).

Zhou, Peng, et al. 2020b. "Addendum: A Pneumonia Outbreak Associated with a New Coronavirus of Probable Bat Origin." *Nature* 588 (7836).

Zhou, Shuntai, et al. 2021. "β-D-N^4-hydroxycytidine Inhibits SARS-CoV-2 Through Lethal Mutagenesis but Is Also Mutagenic to Mammalian Cells." *The Journal of Infectious Diseases* 224 (3).

Zhu, Na, et al. 2020. "A Novel Coronavirus from Patients with Pneumonia in China, 2019." *The New England Journal of Medicine* 382 (8).

Zuckerman, Gregory. 2021. *A Shot to Save the World: The Inside Story of the Life-or-Death Race for a COVID-19 Vaccine.* London: Portfolio Penguin.

INDEX